Reciprocating Compressors

G|P
P|

Gulf Professional Publishing
An Imprint of Elsevier

Reciprocating Compressors

Heinz P. Bloch
and
John J. Hoefner

This book is dedicated to the mechanics, technicians, engineers, and managers whose quest for excellence leads them to be resourceful . . . and who recognize that resourcefulness includes reaching for the written word.

Originally published by Gulf Professional Publishing, Houston, TX.

For information, please contact:
Manager of Special Sales
Butterworth–Heinemann
An Imprint of Elsevier
225 Wildwood Avenue
Woburn, MA 01801–2041
Tel: 781-904-2500
Fax: 781-904-2620
For information on all Butterworth–Heinemann publications available, contact our World Wide Web home page at: http://www.bh.com

Library of Congress Cataloging-in-Publication Data

Bloch, Heinz P., 1933–
Reciprocating compressors : operation and maintenance / Heinz P. Bloch, John J. Hoefner.
p. cm.
Includes bibliographical references and index.
ISBN-13: 978-0-88415-525-6 ISBN-10: 0-88415-525-0 (alk. paper)
1. Compressors. 2. Compressors—Maintenance and repair. I. Hoefner, John J. II. Title.
TJ990.B547 1996
621.5′1—dc20 96-3059
CIP

Transferred to Digital Printing 2009.

Printed on Acid-Free Paper (∞)

Contents

Acknowledgments

The authors are indebted to a number of compressor manufacturers for granting permission to use copyrighted material for this text. First among these is Dresser-Rand's Olean, New York, facility—successor company to John Hoefner's original employer, The Worthington Compressor Company of Buffalo, New York. We also acknowledge Ingersoll-Dresser (Painted Post, New York), Sulzer-Burckhardt (Winterthur, Switzerland) for allowing us to use material on labyrinth piston machines, and Nuovo Pignone (Florence, Italy) whose input on ethylene hyper-compressors originated from one of their technical publications, *Quaderni Pignone.*

The reader should note that this text was originally compiled for a three-day intense course presented by the authors for the Center for Professional Advancement, East Brunswick, New Jersey. Please contact the center for information on in-plant and public presentations.

Disclaimer

Preface

One doesn't have to do too much research to establish how old reciprocating compressor technology really is. If steam turbines ushered in the Industrial Revolution over 200 years ago, reciprocating compressors couldn't have been far behind.

On a visit to an Iowa-based equipment manufacturer in 1989, I was amazed to see how a 1908-vintage reciprocating compressor satisfied their around-the-clock plant air requirements dependably and efficiently. Eighty years with nothing but routine, albeit conscientious, "tender loving care" maintenance! What an endorsement of the skill of the original designers, machine builders, and generations of maintenance craftsmen.

It's only fair to say that the old turn-of-the-century compressor was designed with greater margins of safety, or strength, or capacity to survive abuse than today's higher rotating speed and higher linear piston velocity reciprocating machines. Many of today's compressors are likely to have been designed with the emphasis on reduced weight, less floor space and, let's face it, least cost. The concepts of maintainability, surveillability, and true life cycle cost are too new to be taught in modern universities and engineering colleges. The reward system for project managers, process design contractors, and project engineers is largely based on capital cost savings and rapid schedules. Regrettably, even the commitment to maintenance excellence of many of today's managers and mechanic/technicians is not always as sound, or as rigorous and consistent, as it perhaps was a few decades ago.

Today, everyone speaks of reliability, but many of these well-meaning folks seem to be "forgetful hearers" instead of "doers." There are precious few instances where the maintenance or reliability technician is given either the time or the training to determine the true root causes of equipment failures. Scores of workers are instructed to find the defective part, replace it with a new one, and get the machine back in service. But the part failed for a reason, and if we don't find the reason for its failure, we are certain to set ourselves up for a repeat event.

Whenever we rush a maintenance task, we are likely to omit taking the types of measurements that are critically important to the achievement of run length extensions and increased reliability and safety. What is needed is more attention to detail; the notion that equipment reliability can be upheld by fixing only those components that are visibly defective may not always be correct. There may be compelling reasons to call for a restoration of *all* fits, clearances, and dimensions to as-designed values. This takes time and planning. It requires access to authoritative data and a fundamental shift away from business-as-usual, quick-fix, or big-picture attitudes.

Time and again we have seen reciprocating compressor owners/users engage in the search for the high technology solution. When a succession of broken valves is encountered, the hunt concentrates on better valve materials instead of the elimination of moisture condensation and flow-induced liquid slugging. When piston rods wear unevenly, some users pursue superior metallic coatings, but close their ears to the possibility of tolerance stackup being the real culprit. This progressive move towards out-of-roundness or not-so-perfect perpendicularity of mating parts could well be the root cause of equipment distress and would have to be rectified before it makes economic sense to install components with advanced configurations or metallurgical compositions. And we might add that it wouldn't hurt if someone took the time to carefully read and implement the original equipment manufacturer's maintenance manual.

With the downsizing and re-engineering of organizations in the United States and most other industrialized countries came the attrition of experienced personnel. Less time is spent on rigorous training, and outside contractors are asked to step into the gap. Where do they get their training? How diligently will they perform the tasks at hand if cost and schedule are emphasized to the detriment of long-term reliability goals? Well, that is perhaps the primary reason why we set out to compile this text. There is clearly a need to provide guidance and direction to compressor maintenance and rebuilding efforts. Having comprehensive failure analysis and troubleshooting instructions readily available and widely distributed makes economic sense, and it has certainly been our goal to address these needs by pulling together as much pertinent information as seems relevant in support of these tasks.

Books should be written with audience awareness in mind. Our audience is clearly multifunctional. We deal with compression theory only peripherally, perhaps duplicating some aspects of the high school physics curriculum. We describe compressor operation in terms that the entry-level operator will find of interest, and we go into considerable detail whenever the main topic, reciprocating compressor maintenance, is explained. This would surely be the part of this book that should be read and absorbed by maintenance technicians, mechanic/machinist/fitters, and millwright personnel, regardless of background and experience levels. Reasonable people will agree that we don't know everything, that humans are creatures of habit, that we may not have been taught by a perfect teacher, that we are prone to forget something, that we can always learn. Certainly the co-authors feel that way and will admit at the outset that this book is not perfect. But, we believe it's a worthwhile start.

Most of the credit for assembling and organizing this material must go to my able mentor and co-author John J. ("Jack") Hoefner, of West Seneca, New York. Born in 1919, Jack qualifies as a member of the old guard. He spent a career in the compressor technology field, retiring as Field Service Manager from one of the world's foremost manufacturers of reciprocating compressors. In his days, he has seen and solved more compressor problems than most of us knew existed, and I continue to express gratitude for his agreeing to share his extensive knowledge with me and our readers.

Jack joins me in giving thanks to the various companies and contributors whose names can be found in the source descriptions beneath many of the illustrations in this text. Listed alphabetically, they include Anglo Compression, Mount Vernon, Ohio; Babcock-Borsig, Berlin, Germany; Bently-Nevada Corporation, Minden, Nevada; C. Lee Cook Company, Louisville, Kentucky; Caldwell, James H., as published in Cooper Bessemer Bulletin No. 129, 10M69; Cook-Manley Company, Houston, Texas; Cooper Energy Services, Mount Vernon, Ohio; Exxon Corporation, Marketing Technical Department, Houston, Texas; France Compressor Products, Newtown, Pennsylvania; In-Place Machining Company, Milwaukee, Wisconsin; Indikon Company, Somerville, Massachusetts; Joy Manufacturing Company, Division of Gardner-Denver, Quincy, Illinois; Lubriquip, Inc., Cleveland, Ohio; Nuovo Pignone, Florence, Italy; Pennsylvania Process Compressors, Easton, Pennsylvania; PMC/Beta, Natick, Massachusetts; Sloan Brothers Company, Oakmont, Pennsylvania; Sulzer-Burckhardt and Sulzer Roteq, Winterthur, Switzerland and New York,

New York. Very special thanks are extended to Dresser-Rand's Engine and Process Compressor Division in Painted Post, New York. Their numerous maintenance and service manuals were made available to us; needless to say, they greatly facilitated our task.

Our sincere thanks are reserved for Ms. Joyce Alff, Managing Editor, Book Division, Gulf Publishing Company. We gave her a manuscript which needed far more than the usual attention, but she managed to convert it into a solid, permanent text by being efficient and resourceful.

Heinz P. Bloch, P.E.

CHAPTER 1

Reciprocating Compressors and Their Applications

INTRODUCTION

The purpose of compressors is to move air and other gases from place to place. Gases, unlike liquids, are compressible and require compression devices, which although similar to pumps, operate on somewhat different principles. Compressors, blowers, and fans are such compression devices.

- Compressors. Move air or gas in higher differential pressure ranges from *35 psi* to as high as *65,000 psi* in extreme cases.
- Blowers. Move large volumes of air or gas at pressures up to *50 pounds per square inch.*
- Fans. Move air or gas at a sufficient pressure to overcome static forces. Discharge pressures range from a few *inches of water* to about *1 pound per square inch.*

WHAT IS A COMPRESSOR?

BASIC GAS LAWS

Before discussing the types of compressors and how they work, it will be helpful to consider some of the basic gas laws and the manner in which they affect compressors.

By definition, *a gas is a fluid having neither independent shape nor form, which tends to expand indefinitely.*

Gases may be composed of only one specific gas maintaining its own identity in the gas mixture. Air, for example, is a mixture of several gases, primarily nitrogen (78% by volume), oxygen (21%), argon (about 1%), and some water vapor. Air may also, due to local conditions, contain varying small percentages of industrial gases not normally a part of air.

The First Law of Thermodynamics

This law states that energy cannot be created or destroyed during a process, such as compression and delivery of a gas. In other words, whenever a quantity of one kind of energy disappears, an exactly equivalent total of other kinds of energy must be produced.

The Second Law of Thermodynamics

This law is more abstract, but can be stated in several ways:

1. Heat cannot, of itself, pass from a colder to a hotter body.
2. Heat can be transferred from a body at a lower temperature to one at a higher temperature only if external work is performed.
3. The available energy of the isolated system decreases in all real processes.
4. By itself, heat or energy (like water), will flow only downhill (i.e., from hot to cold).

Basically, then, these statements say that *energy which exists at various levels is available for use only if it can move from a higher to a lower level.*

Ideal or Perfect Gas Laws

An ideal or perfect gas is one to which the laws of *Boyle, Charles,* and *Amonton* apply. Such perfect gases do not really exist, but these three laws of thermodynamics can be used if corrected by compressibility factors based on experimental data.

Boyle's Law states that at a constant temperature, the volume of an ideal gas decreases with an increase in pressure.

For example, if a given amount of gas is compressed at a constant temperature to half its volume, its pressure will be doubled.

$$\frac{V_2}{V_1} = \frac{P_1}{P_2} \text{ or } P_2V_2 = P_1V_1 = \text{constant}$$

Charles' Law states that at constant pressure, the volume of an ideal gas will increase as the temperature increases.

If heat is applied to a gas it will expand, and the pressure will remain the same. This law assumes the absence of friction or the presence of an applied force.

$$\frac{V_2}{V_1} = \frac{T_2}{T_1} \text{ or } \frac{V_2}{T_2} = \frac{V_1}{T_1}$$

Amonton's Law states that at constant volume, the pressure of an ideal gas will increase as the temperature increases.

$$\frac{P_2}{P_1} = \frac{T_2}{T_1} \text{ or } \frac{P_2}{T_2} = \frac{P_1}{T_1}$$

Gas and Vapor

By definition, a *gas* is that fluid form of substance in which the substance can expand indefinitely and completely fill its container. A *vapor* is a gasified liquid or solid—a substance in gaseous form.

The terms gas and vapor are generally used interchangeably.

HOW COMPRESSORS WORK

To understand how gases and gas mixtures behave, it is necessary to recognize that gases consist of individual molecules of the various gas components, widely separated compared to their size. These molecules are always traveling at high speed; they strike against the walls of the enclosing vessel and produce what we know as *pressure.* Refer to Figure 1-1.

Temperature affects average molecule speed. When heat is added to a fixed volume of gas, the molecules travel faster, and hit the containing walls of the vessel more often and with greater force. See Figure 1-2. This then produces a *greater pressure.* This is consistent with *Amonton's Law.*

If the enclosed vessel is fitted with a piston so that the gas can be squeezed into a smaller space, the molecule travel is now restricted. The molecules now hit the walls with a greater frequency, *increasing the pressure,* consistent with *Boyle's Law.* See Figure 1-3.

However, moving the piston also delivers energy to the molecules, causing them to move with increasing velocity. As with heating, this

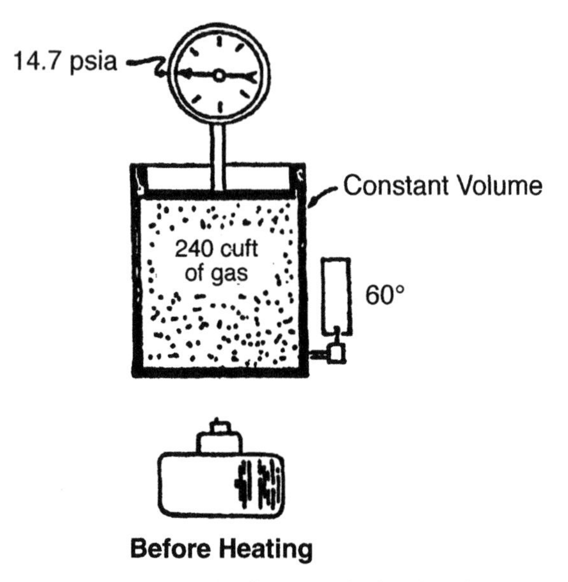

FIGURE 1-1. Confined gas before heating.

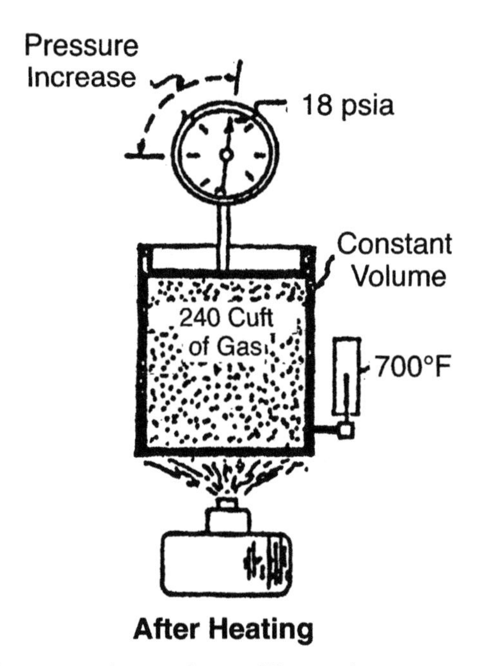

FIGURE 1-2. Constant volume of gas will experience pressure increase when heated.

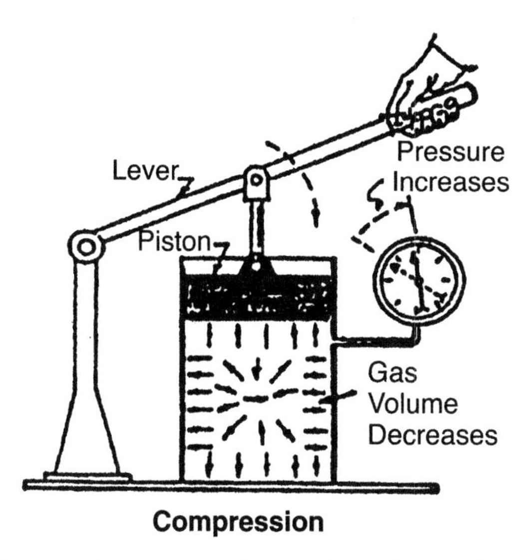

FIGURE 1-3. Compression process reduces volume of gas and increases pressure.

results in a temperature increase. Furthermore, all the molecules have been forced into a smaller space, which results in an increased number of collisions on a unit area of the wall. This, together with the increased velocity, results in increased pressure.

The compression of gases to higher pressures results in higher temperatures, creating problems in compressor design. All basic compressor elements, regardless of type, have certain design-limiting operating conditions. When any limitation is involved, it becomes necessary to perform the work in more than one step of the compression process. This is termed *multistaging* and uses one basic machine element designed to operate in series with other elements of the machine.

This limitation varies with the type of compressor, but the most important limitations include:

1. Discharge pressure—all types.
2. Pressure rise or differential—dynamic units and most displacement types.
3. Compression ratio—dynamic units.
4. Effect of clearance—reciprocating units (this is related to the compression ratio).
5. Desirability of saving power.

METHODS OF COMPRESSION

Four methods are used to compress gas. Two are in the intermittent class, and two are in the continuous flow class. (These are descriptive, not thermodynamic or duty classification terms.)

1. Trap consecutive quantities of gas in some type of enclosure, reduce the volume (thus increasing the pressure), then push the compressed gas out of the enclosure.
2. Trap consecutive quantities of gas in some type of enclosure, carry it without volume change to the discharge opening, compress the gas by backflow from the discharge system, then push the compressed gas out of the enclosure.
3. Compress the gas by the mechanical action of rapidly rotating impellers or bladed rotors that impart velocity and pressure to the flowing gas. (Velocity is further converted into pressure in stationary diffusers or blades.)
4. Entrain the gas in a high velocity jet of the same or another gas (usually, but not necessarily, steam) and convert the high velocity of the mixture into pressure in a diffuser.

Compressors using methods 1 and 2 are in the intermittent class and are known as positive displacement compressors. Those using method 3 are known as dynamic compressors. Compressors using method 4 are known as ejectors and normally operate with an intake below atmospheric pressure.

Compressors change mechanical energy into gas energy. This is in accordance with the First Law of Thermodynamics, which states that energy cannot be created or destroyed during a process (such as compression of a gas), although the process may change mechanical energy into gas energy. Some of the energy is also converted into nonusable forms such as heat losses.

Mechanical energy can be converted into gas energy in one of two ways:

1. By positive displacement of the gas into a smaller volume. Flow is directly proportional to speed of the compressor, but the pressure ratio is determined by pressure in the system into which the compressor is pumping.

2. By dynamic action imparting velocity to the gas. This velocity is then converted into pressure. Flow rate and pressure ratio both vary as a function of speed, but only within a very limited range and then only with properly designed control systems. Figure 1-4 shows the basic idea.

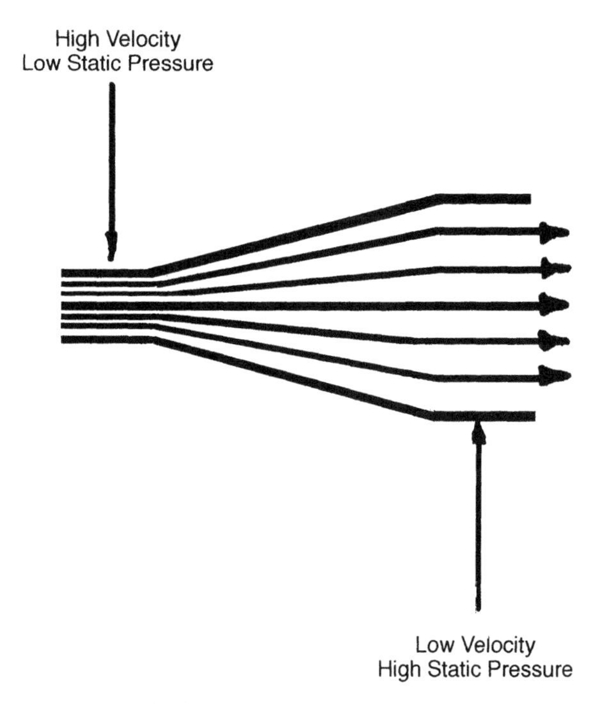

FIGURE 1-4. Velocity energy being converted to pressure energy.

Total energy in a flowing air stream is constant. Entering an enlarged section, flow speed is reduced and some of the velocity energy turns into pressure energy. Thus static pressure is higher in the enlarged section.

TYPES OF COMPRESSORS

The principal types of compressors are shown in Figure 1-5 and are defined below. Cam, diaphragm, and diffusion compressors are not shown because of their specialized applications and relatively small size.

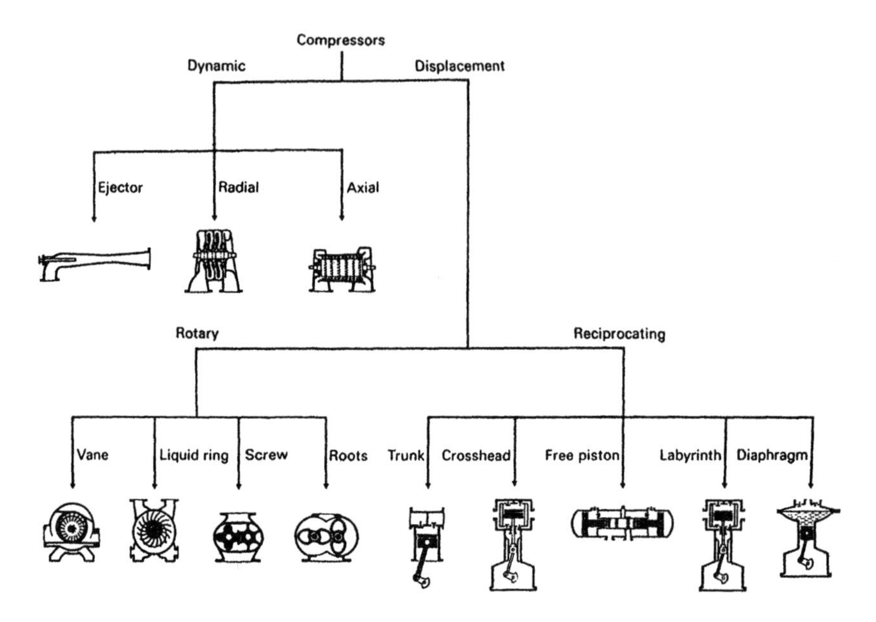

FIGURE 1-5. Principal compressor types used in industry.

- **Positive displacement** units are those in which successive volumes of gas are confined within a closed space and elevated to a higher pressure.
- **Rotary positive displacement** compressors are machines in which compression and displacement result from the positive action of rotating elements.
- **Sliding vane compressors** are rotary positive displacement machines in which axial vanes slide radially in a rotor eccentrically mounted in a cylindrical casing. Gas trapped between vanes is compressed and displaced.
- **Liquid piston compressors** are rotary positive displacement machines in which water or other liquid is used as the piston to compress and displace the gas handled.
- **Two-impeller straight-lobe** compressors are rotary positive displacement machines in which two straight mating lobed impellers trap gas and carry it from intake to discharge. There is no internal compression.
- **Helical or spiral lobe** compressors are rotary positive displacement machines in which two intermeshing rotors, each with a helical form, compress and displace the gas.

- **Dynamic compressors** are rotary continuous-flow machines in which the rapidly rotating element accelerates the gas as it passes through the element, converting the velocity head into pressure. This occurs partially in the rotating element and partially in stationary diffusers or blades.
- **Centrifugal compressors** are dynamic machines in which one or more rotating impellers, usually shrouded on the sides, accelerate the gas. Main gas flow is radial.
- **Axial compressors** are dynamic machines in which gas acceleration is obtained by the action of the bladed rotor. Main gas flow is axial.
- **Mixed flow compressors** are dynamic machines with an impeller form combining some characteristics of both the centrifugal and axial types.

Compressor Definitions

Gauge pressure (psig) is the pressure, in pounds per square inch, above local atmospheric pressure.

Absolute pressure (psia) is the existing gauge pressure plus local atmospheric or barometric pressure. At sea level, absolute pressure is gauge pressure plus 14.7 psi. At elevations above sea level, the atmospheric pressure or barometric pressure becomes less. For example, it is approximately 12.2 psia at 5,000 feet elevation.

Displacement of a compressor is the volume swept through the first-stage cylinder or cylinders and is usually expressed in cubic feet per minute.

Free air is air at normal atmospheric conditions. Because the altitude, barometric pressure, and temperature vary at different localities and at different times, it follows that this term does not mean air under identical conditions.

Standard air unfortunately does not mean the same to everyone.

1. ASME power test code defines air at:
   68°F; 14.7 psia; RH of 36%
2. Compressed Air Institute defines air at:
   60°F; 14.7 psia and dry
3. Natural gas pipeline industry defines air at:
   14.4 psia; @ suction temperature

Unless otherwise specified, the second definition is generally used in reference to reciprocating compressors.

Actual capacity is a term that is sometimes applied to the capacity of a compressor at intake conditions. It is commonly expressed by either the term *ICFM* (intake cubic feet per minute), or *ACFM* (actual cubic feet per minute).

Volumetric efficiency is the ratio of the actual capacity of the compressor to displacement and is expressed as a percentage.

Compression efficiency is the ratio of the theoretical horsepower to the actual indicated horsepower required to compress a definite amount of gas.

Mechanical efficiency is the ratio of the indicated horsepower in the compressing cylinders to the brake horsepower delivered to the shaft. It is expressed as a percentage.

Overall efficiency is the product of the compression efficiency and the mechanical efficiency.

Adiabatic compression occurs when no heat is transferred to or from the gas during compression ($PV^k = C$).

Isothermal compression occurs when the temperature of the gas remains constant during compression ($PV = C$).

Polytropic compression occurs when heat is transferred to or from the gas at a precise rating and where the compression and expansion lines follow the general law $PV^n = C$.

Frame load is the amount of load or force the compressor frame and running gear (i.e., the connecting rod, bolts, crosshead, crosshead pin, piston rod, connecting rod bearings, and crankshaft) can safely carry in tension and compression, expressed in pounds. This is a design factor, and any changes in cylinder bore size, suction or discharge pressure, or the type of gas handled may adversely affect component life. Also, mechanical failure of the parts could result.

Compression ratio is the ratio of the absolute discharge pressure (psia) and the absolute inlet pressure (14.696). Thus, a compressor operating at sea level on plant air service with a 100 psi discharge pressure would have a compression ratio of 7.8:

$$(100 \text{ psig} + 14.7) \div 14.7 = 7.8$$

Piston displacement is the net volume displaced by the piston at rated compressor speed. On double-acting cylinders, it is the total of both head and crank end of the stroke. It is expressed in cubic feet per minute, or CFM.

PRESSURE

Pressure is expressed as a force per unit of area exposed to the pressure. Because weight is really the force of gravity on a mass of material, the weight necessary to balance the pressure force is used as a measure. Hence, as examples:

Pounds/sq in = (psi)
Pounds/sq ft = (lb/sq ft)
Grams/sq cm = (gr/sq cm)
Kilograms/sq cm = (kg/sq cm)

Pressure is usually measured by a gauge that registers the difference between the pressure in a vessel and the current atmospheric pressure. Therefore, a gauge (psig) *does not indicate the true total gas pressure.*

To obtain the true pressure, or pressure above zero, it is necessary to add the current atmospheric or barometric pressure, expressed in proper units. This sum is the absolute pressure (psia). See Figure 1-6. For all compressor calculations the absolute pressure is required.

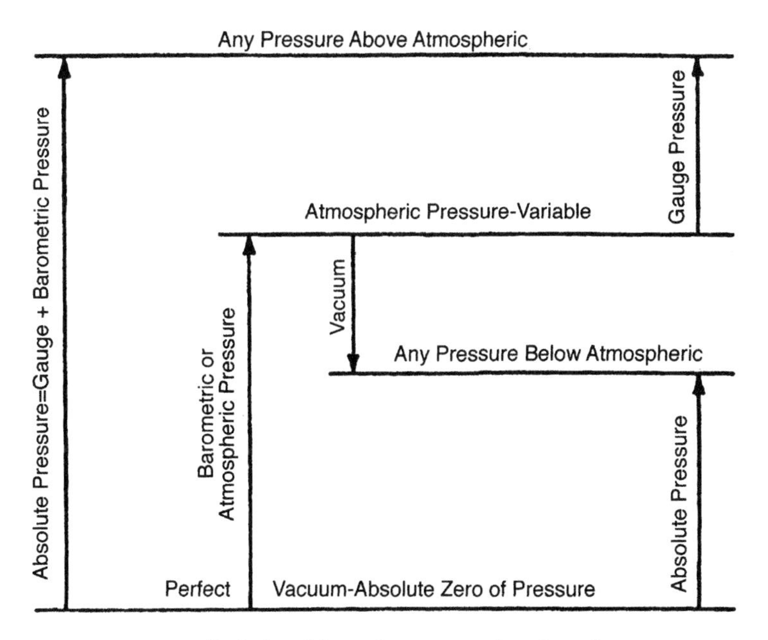

FIGURE 1-6. Relationship and terms used to describe pressure.

Note: There is frequent confusion in transmission of pressure data. It is recommended that specific notation be made after each pressure as to whether it is gauge or absolute. Use the symbol psig or psia. If psig is given, be sure the barometric pressure is also specified.

Also, because a column of a material of a specified height will have a weight proportional to its height, the height can be used as a force measure. It is reduced to a unit area basis automatically, since the total weight and the area are proportional. For example:

Feet of water = (ft H_2O)
Inches of water = (in. H_2O)
Inches of mercury = (in. Hg)
Millimeters of mercury = (mm Hg)

With the exception of barometric pressure, when pressures are expressed in the above terms, they are gauge pressures unless specifically noted as absolute values.

Atmospheric pressure is measured by a barometer. It is designed to read the height of a column of mercury. The upper end of the tube containing the mercury is closed and is at zero absolute pressure. The lower end of the tube is submerged in a pot of mercury, the surface of which is open to the atmosphere. The weight of this column of mercury exactly balances the weight of a similar column of atmospheric air.

Although this gauge really measures a *differential* pressure, by design one of those pressures is *zero,* and the actual reading is true *absolute* or total pressure of the atmosphere. 14.696 psia sea level measure is equal to 29.92 in. Hg.

PRESSURE DEFINITIONS ASSOCIATED WITH COMPRESSORS

Inlet or suction pressure is the total pressure measure at the compressor cylinder inlet flange. Normally expressed as gauge pressure but may be expressed as absolute pressure, which is gauge pressure plus atmospheric pressure (14.696). It is expressed in pounds per square inch, psig (gauge) or psia (absolute).

Discharge pressure is the total pressure measured at the discharge flange of the compressor. It is expressed the same as the suction pressure, psig or psia.

VACUUM

Vacuum is a type of pressure. A gas is said to be under vacuum when its pressure is below atmospheric. There are two methods of stating this pressure, only one of which is accurate in itself.

Vacuum is usually measured by a differential gauge that shows the difference in pressure between that of the system and atmospheric pressure. This measurement is expressed, for example, as

Millimeters of Hg vacuum = (mm Hg Vac)
Inches of Hg vacuum = (in. Hg Vac)
Inches of water vacuum = (in. H_2O Vac)

Unless the barometric equivalent of atmospheric pressure is also given, these expressions do not give an accurate specification of pressure. See Figure 1-6.

Subtracting the vacuum reading from the atmospheric pressure will give an accurate absolute pressure. This may be expressed as

Inches of Hg absolute = (in. Hg abs)
Millimeters of Hg absolute = (mm Hg Abs)
Pounds/sq in absolute = (psia)

The word *absolute* should never be omitted; otherwise, one is never sure whether a vacuum is expressed in differential or absolute terms.

THEORY OF RECIPROCATING COMPRESSORS

Reciprocating compressors are the best known and most widely used compressors of the positive displacement type. They operate on the same principle as the old, familiar bicycle pump, that is, by means of a piston in a cylinder. As the piston moves forward in the cylinder, it compresses the air or gas into a smaller space, thus raising its pressure.

The basic reciprocating compression element is a single cylinder compressing on one side of the piston (single-acting). A unit compressing on both sides of the piston (double-acting) consists of two basic single-acting elements operating in parallel in one casting. Most of the compressors in use are of the double-acting type.

Figure 1-7 shows a cross section of another variant—a V-oriented, two-stage, double-acting water-cooled compressor.

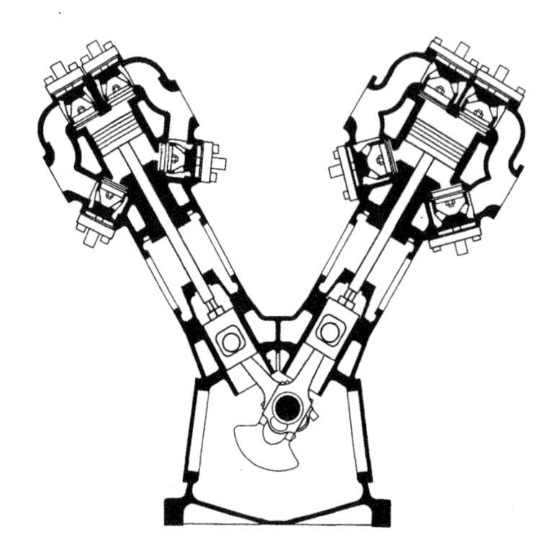

FIGURE 1-7. Multistage, double-acting reciprocating compressor in V-arrangement (*Source: Sulzer-Burkhardt, Winterthur, Switzerland*).

Rotary motion provided at the compressor shaft is converted to reciprocating (linear) motion by use of a crankshaft, crosshead, and a connecting rod between the two.

One end of the connecting rod is secured by the crankpin to the crankshaft, and the other by crosshead pin to the crosshead which, as the crankshaft turns, reciprocates in a linear motion.

Intake (suction) and discharge valves are located in the top and bottom of the cylinder. (Sometimes they may be located in the cylinder barrel.) These are basically check valves, permitting gas to flow in one direction only.

The movement of the piston to the top of the cylinder creates a partial vacuum in the lower end of the cylinder; the pressure differential between intake pressure and this vacuum across the intake valve then causes the valves to open, allowing air to flow into the cylinder from the intake line.

On the return stroke, when the pressure in the cylinder exceeds the pressure in the discharge line, the discharge valve opens, permitting air at that pressure to be discharged from the cylinder into the discharge or system line.

This action, when on one side of the piston only, is called "single-acting" compression; when on both sides of the piston, it is called "double-acting" compression.

Compressor Capacity

Determining compressor capacity would be relatively simple if a non-compressible, non-expandable fluid were handled. The quantity into the discharge line would be practically equal to the volume swept by the piston.

However, since air or gas is elastic, compressor capacity varies widely as pressure conditions change. For instance, with a given intake pressure, machine capacity is considerably less when discharging at 100 psi than at 50 psi. This makes it impossible to rate a given compressor for a given capacity. The only practicable rating is in terms of piston displacement—volume swept by the moving piston during one minute.

PISTON DISPLACEMENT

The piston displacement is the net volume actually displaced by the compressor piston at rated machine speed, as the piston travels the length of its stroke from bottom dead center to top dead center.

In Figure 1-8, the entire stroke, and thus the piston displacement, is represented by the travel of the piston from points B-H.

This volume is usually expressed in cubic feet per minute. For multistage units, the piston displacement of the first stage alone is commonly stated as that of the entire machine.

In the case of a double-acting cylinder, the displacement of the crank end of the cylinder is also included. The crank end displacement is, of course, less than the head end displacement by the amount that the piston rod displaces.

The piston displacement (PD) for a single-acting unit is readily computed by the following formulas:

1. Calculating PD for a single-acting cylinder:

$$PD = AHE \times \frac{S}{12} \times rpm$$

Where AHE = area of head end of piston in square feet
S = stroke in inches
rpm = revolutions per minute
PD = piston D is displacement in cubic feet per minute

2. Calculating PD for a double-acting cylinder:

$$PD = AHE \times \frac{S}{12} \times rpm + ACE \times \frac{S}{12} \times rpm$$

Where ACE = area of crank end of piston in square feet
This can be approximated by the expression:

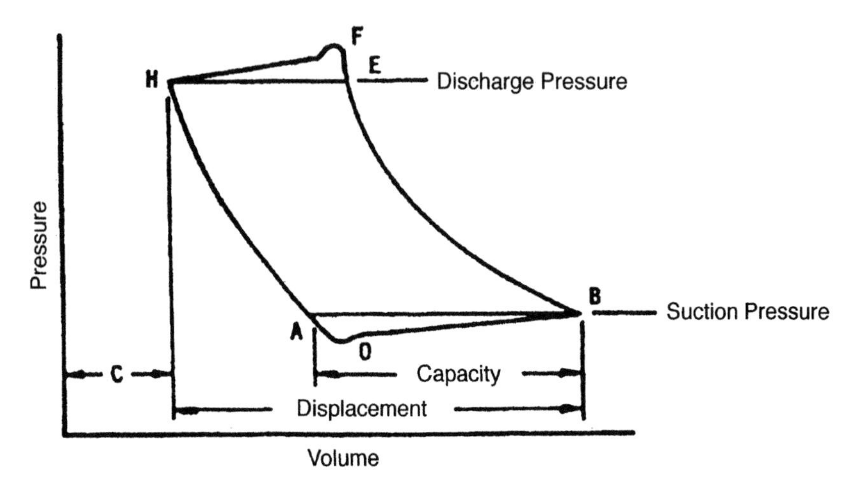

FIGURE 1-8. Actual compressor indicator card.

$$PD = 2\,(AHE - AR) \times \frac{S}{12} \times rpm$$

Where AR = area of rod in square feet

These are handy equations because any particular compressor unit has a standard stroke, speed and rod size. Therefore, these equations can be set up with constants for any specific unit, and only the AHE must be added to the equation to find any unknown PD for either the single- or double-acting cylinder.

A more practical card would be AOBFH. Here the area AOB, fluid losses through the inlet ports and valves; and the area EFH, fluid losses through the discharge valves and ports, are included in the card area. The larger area, by reason of fluid loss inclusion, means greater horsepower demand.

There are other considerations. During compression, represented graphically by BF, a small portion of the gas continually slips past the piston rings and suction valves. Work has been done on this gas, yet it is not delivered to the discharge system. Also, slippage past the discharge valves allows gas which has already been delivered to the discharge system to return to the cylinder. Re-compression and re-delivery take place.

Unless leakage is abnormal, the theoretical location of point E is not appreciably altered. Yet an overall loss has occurred: first, because more gas must be taken into the cylinder to compensate for piston ring and suction valve slip; secondly, because work is performed on this lost capacity; and finally, because leakages back through the discharge valves must be recompressed and re-delivered to the discharge system.

There is still another factor: the cooling effect of cylinder jacketing. Removal of heat by the jacket water would shrink the volume during compression and would tend to move point F to the left, reducing the power required. This is a true saving in power expended. Unfortunately, it is not of any great significance, except in small cylinders handling rather low density gas through high ratio of compression, where the jacket surface is large in proportion to the amount of work performed and heat generated.

These fluid losses are indicated on the indicator card AOBF, Figure 1-8. This figure represents a typical actual indicator card such as might be taken on a machine in the shop or in the field.

Brake Horsepower

The *actual indicated horsepower* is built upon the base of ideal horsepower and includes the thermodynamic losses in the cylinder. These thermodynamic losses (fluid losses) are summed up under the general term *compression efficiency.*

The major factor involved in determining the compression efficiency is the valve loss or pressure drop through the inlet and discharge valves. These fluid losses are a function of gas density and valve velocity. The suction and discharge pressures and the molecular weight establish the density. The valve velocity is fixed by the valve area available in the selected cylinders and by the piston speed. Valve velocity is normally stated in feet per minute; it is the ratio of piston area to valve area per cylinder end, multiplied by feet per minute piston speed.

A better understanding of the losses involved in compression efficiency may be obtained by reference to the indicator diagram, Figure 1-8. If it were possible to get the gas into and out of the cylinder without fluid losses, the indicator card ABEH could be realized.

This card may be said to represent the ideal or theoretical horsepower requirements. But fluid losses are present. Therefore, the actual inlet pressure in the cylinder is below that at the cylinder inlet flange. Likewise, the pressure in the cylinder during delivery interval EH is above that at the cylinder discharge flange.

VOLUMETRIC EFFICIENCY

A volumetric efficiency, which varies for different compression ratios, must then be applied to the piston displacement to determine actual free-air capacity. Volumetric efficiency also varies to some extent with the "n" value, and molecular weight, of the gas being compressed.

Greatest volumetric loss occurs because of clearance within the compressor cylinders. However, other losses, while of lesser importance, also affect compressor capacity.

CLEARANCE LOSS

When the compressor reaches the end of its stroke and has discharged all the gas it can, a small amount remains in the valve pockets and in the clearance space between piston and cylinder head.

When the piston starts its return stroke, this clearance gas at discharge pressure must expand to intake pressure before inlet valves can open; thus, no air enters the cylinder for that portion of the stroke, which reduces the intake volume by that amount.

Since the volume for this clearance gas, expanded to intake pressure, varies with the compression ratio, it follows that compressor volumetric efficiency, and hence its actual capacity, varies with compression ratio instead of with pressure.

Cylinder clearance cannot be completely eliminated. Normal clearance is the minimum obtainable in a given cylinder and will vary between 4% and 16% for most standard cylinders. Some special low-ratio cylinders have normal clearance much greater than this. Normal clearance does not include volume that may have been added for other purposes, such as capacity control.

Although clearance is of little importance to the average user (guarantees are made on actual delivered capacity), its effect on capacity should be understood because of the wide application of a variation in clearance for control and other purposes. There are many cases where extra clearance is added to a cylinder:

1. To reduce capacity at fixed pressure conditions.
2. To prevent driver overload under variable operating pressure conditions by reducing capacity as compression ratio changes.

If a compressor is designed for a given capacity at a given condition, the amount of normal clearance in the cylinder or cylinders has no effect on power.

When a piston has completed the compression and delivery stroke and is ready to reverse its movement, gas at discharge pressure is trapped in the clearance space.

This gas expands on the return stroke until its pressure is sufficiently below intake pressure to cause the suction valves to open. On a pV-diagram, Figure 1-9 shows the effect of this re-expansion on the quantity of fresh air or gas drawn in.

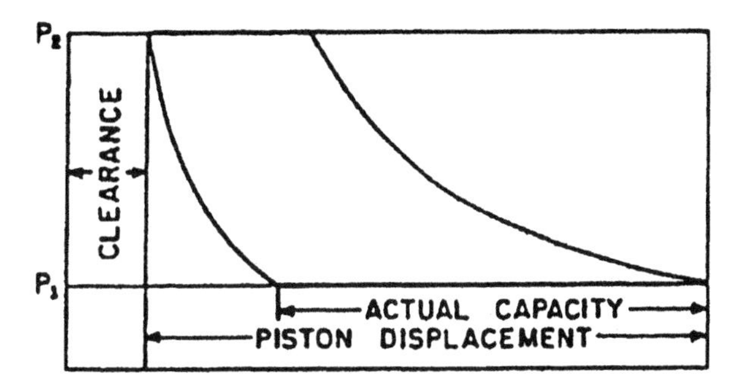

FIGURE 1-9. Effect of clearance on the capacity of a reciprocating compressor.

Figure 1-10 shows a series of theoretical pV diagrams based on a pressure ratio of 4.0 and clearances of 7, 14, and 21%. The effect of clearance is clearly indicated.

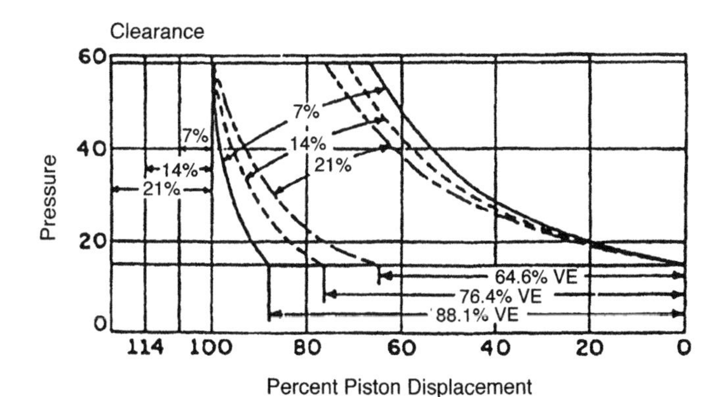

FIGURE 1-10. Quantitative effect of various cylinder clearances on the volumetric efficiency at constant compression ratio.

It can now be seen that volumetric efficiency decreases as

1. The clearance increases.
2. The compression ratio increases.

Figure 1-11 illustrates the effect of clearance at moderate and high compression ratio condition. A theoretical pV diagram for a ratio of 7 is superimposed on a diagram for a ratio of 4, all else being the same. A relatively high clearance (14%) has been used for illustrative purposes. The clearance for a commercial compressor designed for a ratio of 7 would be less than 14%.

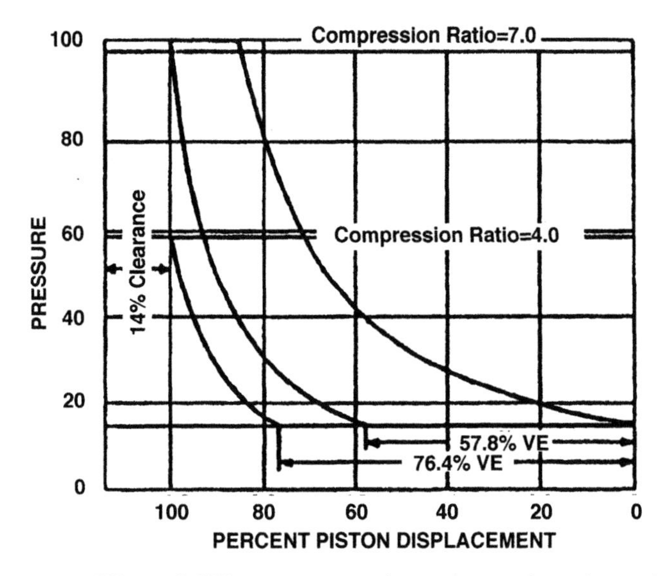

FIGURE 1-11. Effect of different compression ratios on the volumetric efficiency of a given cylinder.

Just as clearance in a cylinder has predominant control over volumetric efficiency, the valve area has predominant control over compression efficiency.

To obtain low clearance and a high volumetric efficiency, it is necessary to limit the size and number of valves. This may tend to lower the efficiency of compression and raise the horsepower. Both factors must be evaluated and compromises made.

Piston Ring Leakage

This leakage allows gas from the compression chamber to escape past the piston into the other end of the cylinder, which is taking suction with the inlet valve open. Capacity is reduced because this hot leakage gas heats up the incoming gas in that end of the cylinder.

Naturally, maximum piston leakage occurs as the piston approaches the end of its stroke because differential pressure across the rings and the time element are the greatest at this point. This leakage causes both a volumetric and a horsepower loss as evidenced by an increase in discharge temperatures.

Valve Slip

Valve slip means reversed gas flow through the valves before they have had time to seat at the end of the piston stroke. Obviously, this volume loss can occur through both intake and discharge valves. Minimum slippage occurs in a responsive valve; one that has minimum inertia so that the moving element can easily be controlled by air flow.

Slippage is usually much less through intake valves than through discharge valves. In the latter, differential pressure across the valve increases rapidly as the piston reaches dead center, so that if the valve does not respond instantaneously, high pressure gas naturally returns through the valve before it seats.

Effects of Multistaging

Multistaging has a marked effect on volumetric efficiency. Here, the low pressure cylinder largely determines the entire machine's volumetric efficiency because whatever volume this cylinder delivers to succeeding stages must be discharged, with the exception of slight leakage that occurs through packing boxes.

In other words, volumetric efficiency of a two-stage machine is the same as if the low pressure cylinder were a single-stage compressor delivering gas at intercooler pressure.

Figure 1-12 shows the pV combined diagram of a two-stage 100 psig air compressor. Further stages are added in the same manner. In a recip-

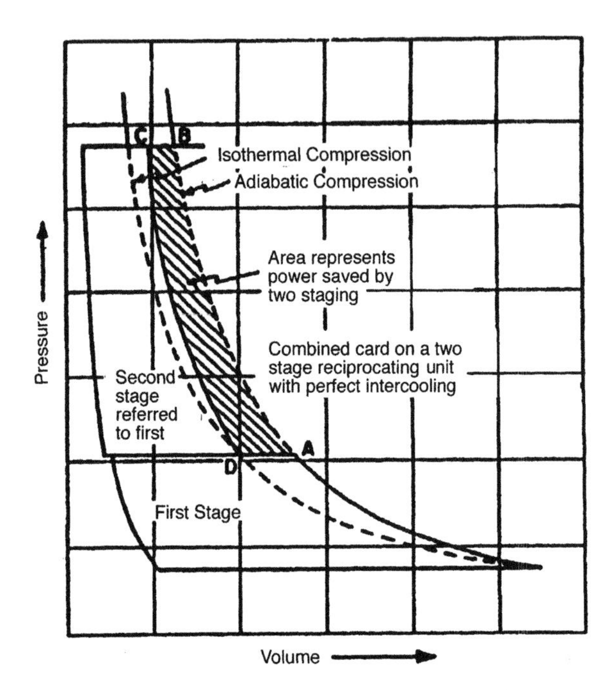

FIGURE 1-12. Theoretical combined indicator card of a two-stage positive displacement compressor with perfect intercooling.

rocating unit, all cylinders are commonly combined into one unit assembly and driven from a single crankshaft.

For reciprocating compressors, multistaging is used

1. To save power.
2. To limit gas discharge temperature.
3. To limit pressure differential.

Power saving has been demonstrated by the indicator card in Figure 1-8. Because there is intercooling between stages, there is a reduction in the maximum gas discharge temperature. Limitation of maximum discharge temperature is particularly important for safety when holding air in large, high pressure compressors where distortion of cylinder parts may be a problem. This is true even though the gas may not become appreciably more hazardous when heated.

The limitations imposed by high pressure differentials involve avoidance of excess strain in the frame, running gear, and other parts. This is a complex question to which designers must give thorough consideration. A problem of this nature is occasionally solved by increasing the number of compression stages.

Figure 1-13 shows the theoretical effect of two- and three-staging on the discharge temperature per stage.

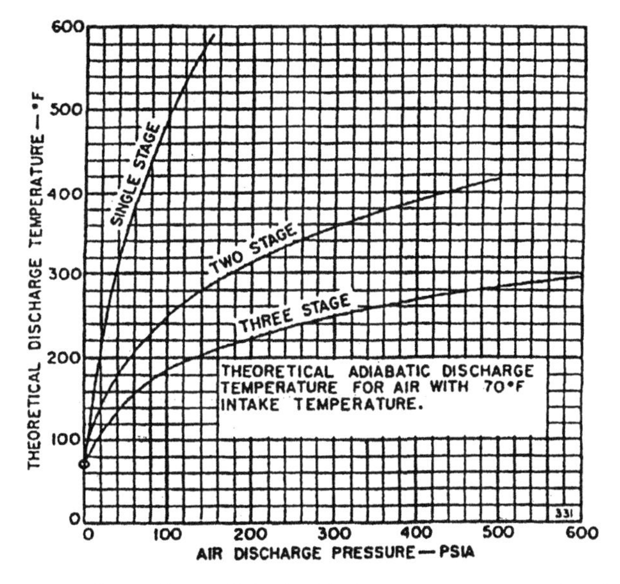

FIGURE 1-13. Theoretical adiabatic discharge temperature for air with 70°F intake temperature.

Figure 1-14 shows the effect of staging on power requirements.

In both figures, the compressors are handling normal air at 14.7 psia suction pressure. The data are theoretical, with intercooling to suction temperature between stages (perfect intercooling) and equal ratios for all stages, and are based on 70°F suction temperature.

Power savings are obvious. Percentages are shown in Figure 1-14. The importance to be placed upon power savings in unit selection will depend to a large degree upon load factor (percentage of total time a unit actually operates) and the size of the compressor. In actual practice, when compression stages exceed four, power savings are frequently slight through adding an extra stage, because of the greater gas friction losses through valves, piping, and coolers. There are often other practical advantages, however.

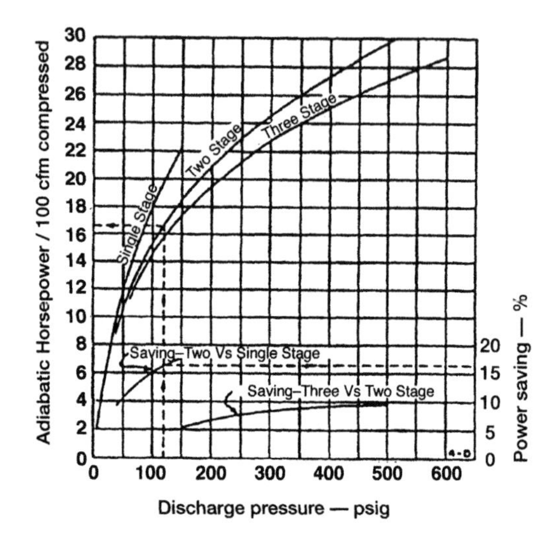

FIGURE 1-14. Comparative theoretical adiabatic horsepower per 100 cfm required for single- two-, and three-stage compression.

The desirability of imposing a maximum temperature limitation is not always fully appreciated. This applies particularly to air compressors where oxidizing atmospheres exist and where lubricating oil decomposition accelerates as temperatures rise. Actual discharge temperature will vary to some degree from the theoretical adiabatic, depending upon compressor size, design, method of cooling, and compression ratio. No rules can be set, but the deviation is not apt to be serious, and the theoretical limitation is an excellent guide.

A compressor in continuous heavy-duty service should definitely be designed more conservatively regarding discharge temperature than one operating on a relatively light or intermittent cycle.

As discharge temperatures go up, downtime and relative maintenance costs will certainly increase.

As a guide, for medium and large heavy-duty compressors (around 150 bhp and larger) handling air or any other oxidizing gas, the maximum discharge temperature should not exceed 350°F. For pressures over 300 psig, this temperature should be further scaled down.

INTERSTAGE PRESSURES

Actual interstage pressure readings are valuable indicators of the relative tightness of valves and piston rings and should be checked several times daily. Any variation from normal operating pressures is cause for immediate concern and investigation.

Since cylinder sizes for any multistage compressor are proportioned for definite intake and discharge temperature and pressure conditions, variations from design first-stage inlet pressure and temperature, as well as changes in final discharge pressure and in cooling water temperature, will cause interstage pressures to vary slightly.

In some cases, theoretical approximations are helpful. The following may be used:

Two Stage

P_1 is first-stage intake—psia
P_2 is intercooler—psia
P_3 is second-stage discharge—psia

$$P_2 = \sqrt{P_1 P_3}$$

Three Stage

P_1 is first-stage intake—psia
P_2 is first intercooler—psia
P_3 is second intercooler—psia
P_4 is third-stage discharge—psia

$$P_2 = \sqrt[3]{P_1^2 P_4}$$

$$P_3 = \sqrt[3]{P_1 P_4^2}$$

Because of variation in cylinder proportions and clearances in an actual compressor, the theoretical approach, with its equal compression ratios per stage, cannot be considered to be completely accurate. Actual readings from the specific machine, when in good condition and operating at design conditions, should be considered the standard for reference.

Effect of Altitude

The altitude at which a compressor is installed must always be given consideration. As altitude above sea level increases, the weight of the earth's atmosphere decreases. This is reflected in the barometer and in absolute intake pressure, which decreases with altitude. This fact is well understood and allowed for with process compressors.

At higher altitudes, the low-pressure cylinder size is increased to provide greater inlet capacity and to bring the power imposed on the frame and running gear closer to normal values.

Single-stage reciprocating and other positive displacement compressors are limited somewhat by the allowable compression ratio and discharge temperature. Frequently, they must be materially derated for altitude operation.

Although the power required by a given compressor decreases as the altitude increases, the ability of engines and electric motors to safely develop this power usually decreases even more rapidly.

Brake Horsepower

Reciprocating units are calculated on the basis of theoretical adiabatic horsepower modified by compression and mechanical efficiencies which result in the brake horsepower (bhp). Compression efficiency depends on many factors—effectiveness of valving, compression ratio, gas composition, compressor size, etc. Mechanical efficiency varies with machine type and size.

For preliminary estimation of sea-level air compressors for general power services, the data shown in Figure 1-15 are reasonable but subject to confirmation by the manufacturer. Information is based on 100 cfm actually delivered intake air and heavy-duty water-cooled compressors.

For altitude installation, the performance will differ. Figure 1-16 also gives approximate altitude correction factors for bhp/100.

Multistage machines may be approximated by using equal compression ratios per stage and multiplying the single-stage bhp/million by the number of stages. A compression ratio per stage of over 3.5 should not normally be used, although there will be exceptions. If involved, compressibility must be allowed for separately, stage by stage. Interstage pressure drop, imperfect intercooling, and vapor condensation between stages that reduces the volume handled, must also be allowed for in this manner.

Characteristics of Reciprocating Compressors

Reciprocating compressors are the most widely used of all compression equipment and also provide the widest range of sizes and types. Rat-

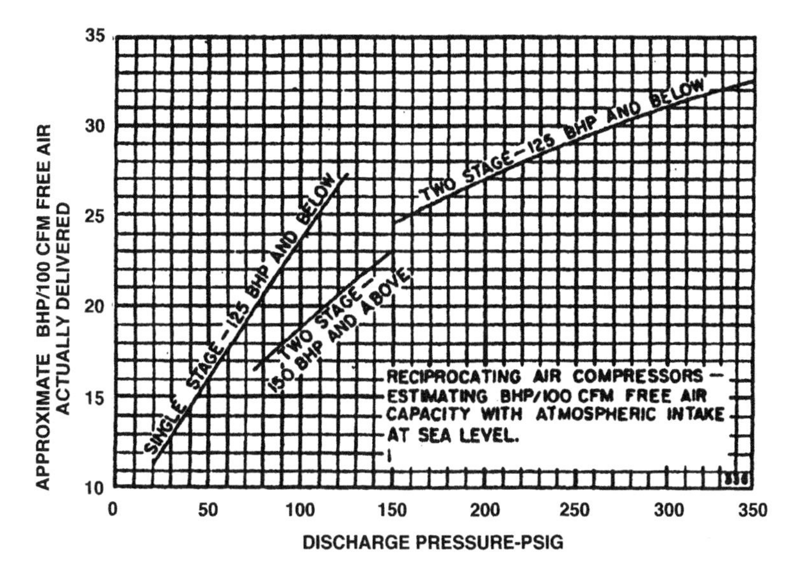

FIGURE 1-15. Power vs. capacity and pressure for various air compressors (general approximations).

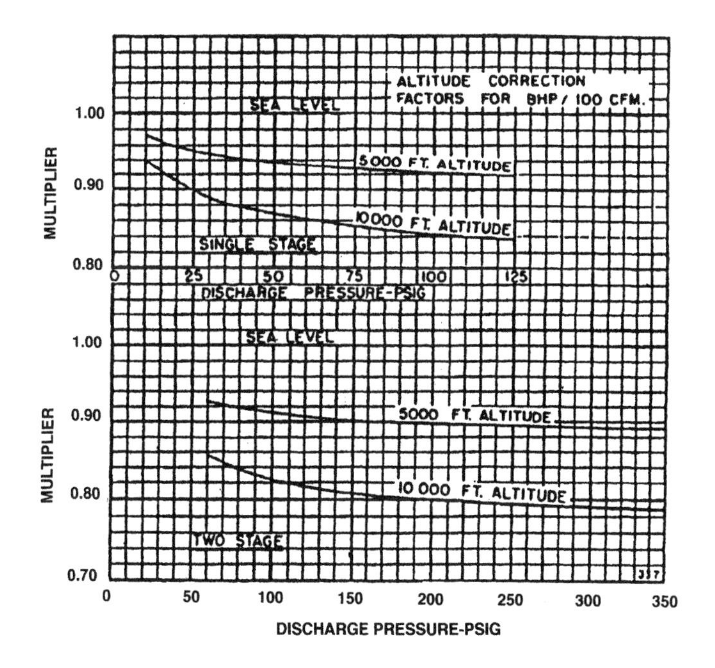

FIGURE 1-16. Altitude correction factors for air compressors applied to Bhp/100 cfm values.

ings vary from fractions to more than 20,000 HP per unit. Pressures range from low vacuum (at intake) to special process compressors for 65,000 psig or higher. For industrial plant air service, they are used from fractional HP to 2000–3000 HP and in pressure ranges from low vacuum to perhaps 500 psi for soot blowing.

In common with all positive displacement compressors, the reciprocating compressor is classified as a "constant-volume variable pressure" machine.

For most applications, they are the most efficient built today. They can be fitted with capacity control devices to closely maintain their efficiency at partial loads (reduced capacity output). They can be built to handle almost any commercial gas, provided corrosion problems in some extreme cases can be solved. Gas cylinders are generally lubricated, although a non-lubricated design is available when warranted.

Because of the reciprocating pistons and other parts, as well as some unbalanced rotating parts, inertia forces are set up that tend to shake the unit. It is necessary to provide a mounting that will stabilize the installation. The extent of this requirement depends on the type and size of compressor involved. These machines are normally designed to be installed in a building, but can be fitted for outdoor installation.

Reciprocating compressors should be supplied with clean gas. Inlet filters are recommended on air compressors. These compressors cannot satisfactorily handle liquids that may be entrained in the gas, although vapors are no problem if condensation within the cylinder does not take place. Liquids tend to destroy lubrication and cause excessive wear.

Reciprocating compressors deliver a pulsating flow of gas. This is sometimes a disadvantage, but pulsation dampeners can usually eliminate the problem.

Reciprocating compressors, like the rotary sliding vane and helical lobe screw machines, are positive displacement compressors. This means that gas is compressed by trapping a charge of gas and then reducing the confining space, causing a build-up in pressure.

The reciprocating compressors, more commonly called "piston compressors," compress gas by use of a piston, cylinder, and valve arrangement. Figure 1-17 shows volume reduction and subsequent increase in pressure, as a piston moves in a cylinder.

Compressor Classifications

Manufacturers design compressors to fill definite user needs. These compressors fall into two general groups, moderate duty machines and

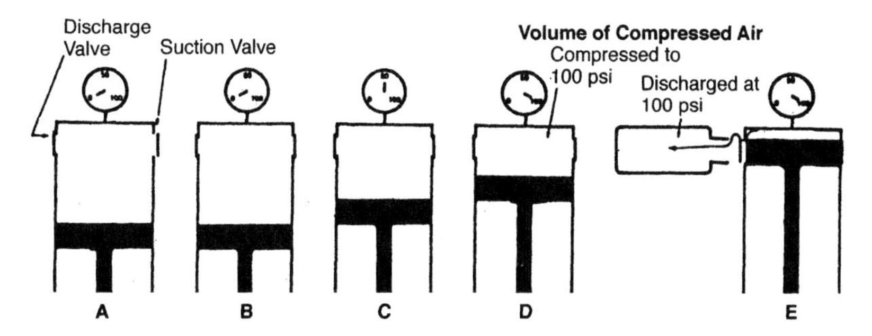

FIGURE 1-17. Volume reduction in a compressor cylinder relative to piston stroke position. (A) Cylinder full of air at atmospheric pressure, gauge reading "0" psi. (B) Same but with intake valve opening closed. (C) Volume reduced to about one-half the original, gauge reading "50" psi. (D) Volume reduced to about one-eighth the original, gauge reading "100" psi. (E) Piston at top dead center after having compressed and discharged the volume of compressed air in D.

heavy duty units. (Light duty, the fractional horsepower to 3–5 HP, will not be discussed.)

Moderate Duty Compressors

Moderate duty compressors are designed for reliable operation over a reasonable service life but should not be installed where continuous full-load, long-time operation is required. This does not mean that these units will not operate for long full-load periods. It does mean that maintenance will be greater than normal.

Generally, moderate duty compressors are of single-acting cylinder design. Usually, these compressors are air-cooled. However, they are also offered as water-cooled designs in horsepower ranges of 30 to 125. Maximum rating is 125 horsepower in either version.

They are built as single-stage units for pressure ratings up to 50 psig and as two-stage units up to 250 psig.

Cooling Arrangements

1. Air-cooled compressors have fins cast as part of the cylinder to dissipate some of the heat generated by the compression of the gas. Most have vanes cast as part of the flywheel or sheave to act as a fan to help remove the heat from the cylinder surface.
2. Combination air/water-cooled compressors have fins forming part of the cylinder casting and have cooling water circulating in the heads

which contain the discharge valves. This water is circulated through a radiator/fan arrangement identical to that in an automobile engine.

3. Water-cooled compressors have jackets cast as an integral part of the cylinder and heads to remove the heat of compression. These compressors do not depend on radiation to dissipate the heat and can therefore operate at higher ratios of compression.

Trunk Piston Design

The moderate duty compressor, whether air-cooled or water-cooled, is built similarly to the automotive engine and uses the length of the piston to guide it in the cylinder. This is called a "trunk piston" design.

In the trunk piston design, the sides of the piston thus act as the guiding surface in the cylinder bore. For this reason the piston must be fairly long in relation to its diameter. With the trunk piston design, compression is possible on top of the piston only; it is called single-acting.

Trunk piston design compressors allow for higher rotation speeds than those designed with a crosshead.

Generally, the bearings, wrist pin and piston rings are lubricated by oil thrown from the crank, a process known as "splash" lubrication. In some cases an oil pump is added, and the wrist pin, crank pins, and bearings are pressure lubricated.

Heavy Duty Compressors

It is generally agreed that for a reciprocating compressor to be considered heavy duty or "continuous duty," it must be both water-cooled and double-acting.

Water-Cooled

We have seen that whenever gas is compressed, heat is generated. Proper cooling of the internal parts of the compressor in order to maintain the coolest possible temperatures at critical points is a basic part of the design.

In the case of the water-cooled reciprocating compressor, the cylinders and cylinder heads are surrounded by water jackets, and the heat transfers through the metal to the water much more efficiently than heat transfer through metal to air. Water-cooled reciprocating units handle cooling more efficiently than comparable air-cooled units. This allows for continuous duty (operation at 100% load and 24-hour days) with low maintenance.

Crosshead Design

In addition to being water-cooled, heavy duty reciprocating compressors are designed with a separate crosshead to guide the piston in the cylinder bore instead of depending on the piston skirt to do the guiding.

- Allows the use of a narrow piston and larger valve area for greater efficiency.
- Permits a longer stroke and greater capacity.
- Separates crankcase from cylinder, allowing control of oil carryover into the cylinder.
- Gives greater stability to piston, eliminating piston "slap" and reducing ring wear.
- Permits stronger piston design and higher operating pressures.
- Makes possible a piston design which allows "pumping" or compression at both ends of the stroke. "Double-acting" results in twice the capacity for every revolution, as opposed to the single-acting design.

The heavy duty, water-cooled, crosshead type reciprocating compressor has a relatively slow rotative speed (180–900 rpm). These machines are conservatively designed for long service life at low maintenance. It is not unusual for this type of machine to be in operation for 50 to 55 years after installation.

Because they are conservatively designed with crossheads and water-cooled cylinders, they are heavier, more expensive to manufacture, and more expensive to install than the other type compressors.

They are also the most efficient of all compressors on a BHP/CFM basis, particularly at part loads, due to their ability to be controlled effectively at part loads. All other types of compressors are compared to the heavy duty, water-cooled compressor when efficiency is discussed.

AUTOMATIC VALVES

The reciprocating compressor uses automatic spring-loaded valves that open only when the proper differential pressure exists across the valve. Inlet valves open when the pressure in the cylinder is slightly below the intake pressure. Discharge valves open when the pressure in the cylinder is slightly above the discharge pressure.

Figure 1-18 shows a typical ideal pressure-volume (pV) diagram (idealized indicator card) of a single-stage compressor with corresponding compressor piston locations.

At position 1, the piston is at the start of the compression stroke. At this point, the cylinder has a full charge of air or gas at suction pressure and begins compressing the gas along line 1-2. As compression begins, the suction valves immediately close, acting as check valves to shut off the cylinder from the suction line.

At point 2, the pressure in the cylinder is slightly higher than that existing in the discharge line, and the discharge valves open, allowing the piston to push the compressed air out of the cylinder into the discharge system (line 2-3).

At point 3, the piston has completed the discharge stroke. As soon as it starts its return stroke, the pressure in the cylinder drops, closing the discharge valve. Notice the volume of gas that is trapped between the end of the piston and the end of the cylinder (Volume C). This is known as "clearance volume."

As the piston begins making its return stroke, this clearance volume gas expands along line 3-4, until the pressure in the cylinder is slightly lower than the suction line. This condition occurs at point 4 and thus, at point 4, the suction valves open, and the cylinder starts to take gas from the suction line. The intake stroke occurs along line 4-1 on the card.

In a double-acting cylinder, the same cycle occurs on the opposite side (crank end) of the piston, 180° out of phase with the head end; this is shown on the diagram as 1′, 2′, 3′, 4′.

COMPRESSOR TERMINOLOGY

Compressors are machines designed for compressing air or other gases from an initial intake at approximately atmospheric pressure to a higher discharge pressure.

Booster compressors are machines for compressing gases from an initial pressure, which is considerably above atmospheric pressure, to a still higher pressure.

Vacuum pumps are machines for compressing gas from an initial pressure which is below atmospheric, to a final pressure which is near atmospheric.

Reciprocating compressors are those in which each compressing element consists of a piston moving back and forth in a cylinder.

Single-acting compressors are those in which compression takes place on one side of the piston, usually on the side away from the crankshaft.

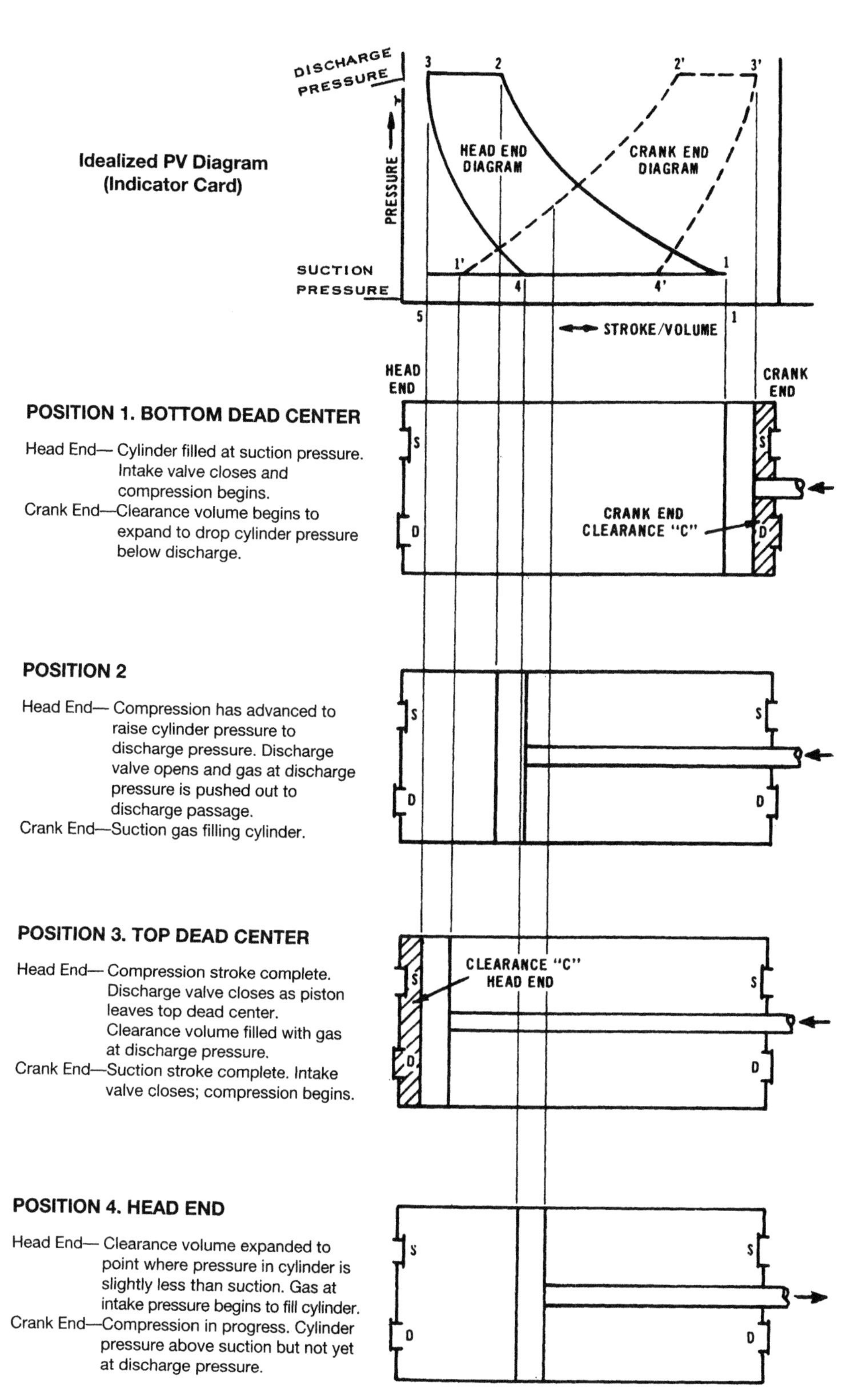

FIGURE 1-18. Ideal compression cycle, single-stage, double-acting.

Double-acting compressors are those in which compression takes place on both sides of the piston, with intake and discharge valves at both ends of the cylinder.

Single-stage compressors are those in which compression from initial to final pressure is completed in a single step or stage.

Multi-stage compressors are those in which compression from initial to final pressure is completed in two or more distinct steps or stages.

Two-stage compressors are those in which compression from initial to final pressure is completed in two distinct steps or stages.

Portable compressors are those consisting of compressor or driver so mounted that they may readily be moved as a unit.

Intercoolers are devices for removing the heat of compression of the gas between consecutive stages of multi-stage compressors.

Aftercoolers are devices for removing the heat of compression from the gas after the last stage of compression is completed. Aftercoolers are one of the most effective means of removing the major amount of moisture from the air, provided they have the capability of cooling the air down to less than 100°F.

Moisture separators are devices for removing moisture precipitated from gas during the process of cooling, allowing the condensate to collect and then drain via either a manual or an automatic condensate trap.

Air receivers are pressure vessels that must be built in accordance with the ASME code. Air receivers take the discharge of a compressor (or compressors) piped into them, preferably after the air has been cooled, for the purpose of storing the air and simultaneously reducing the rate of pressure fluctuation in the air system. This will reduce the frequency at which the compressor will either load or unload, or start and stop, as the case may be.

COMPRESSOR TYPE SELECTIONS

CONSIDERATIONS IN SELECTION

Before choosing a specific type of compressor, consideration must be given to a number of important factors.

1. Discharge pressure required.
2. Capacity required.
3. Power supply characteristics.
4. Availability and cost of cooling water.
5. Space required for compressor.
6. Compressor weight.
7. Type and size of foundation required.
8. Type of control required.
9. Maintenance costs.

The initial choice is between the two basic types, positive displacement and dynamic. Once this decision is made, you can make a further study of the characteristics of various compressor types to see which one will be best for the job. Keep in mind that the selection of a compressor is an engineering decision in which every factor should be considered. Remember, the average life of a compressor is usually 20 years or more, so every decision made that affects the operating cost will be in effect for a long time.

REVIEW OF SELECTION POSSIBILITIES

For a quick review of possibilities, it is best to consider only horsepower and pressure. The upper limits for each type are summarized from previous references. These columns do not usually apply simultaneously.

Figure 1-19 shows the approximate application ranges for reciprocating, centrifugal, and axial flow compressors.

Compressor Type	Approx. Max.* BHP	Power KW	Approx. Max. psig
Reciprocating	20,000	15,000	100,000
Vane Type Rotary	860 Twin Unit	640	400
Helical Lobe Rotary	8,000	6,000	250
Centrifugal Dynamic	60,000	45,000	10,000
Axial Flow Dynamic	100,000	74,600	500

**These maximums are subject to certain limitations imposed by other factors and under certain conditions can be exceeded.*

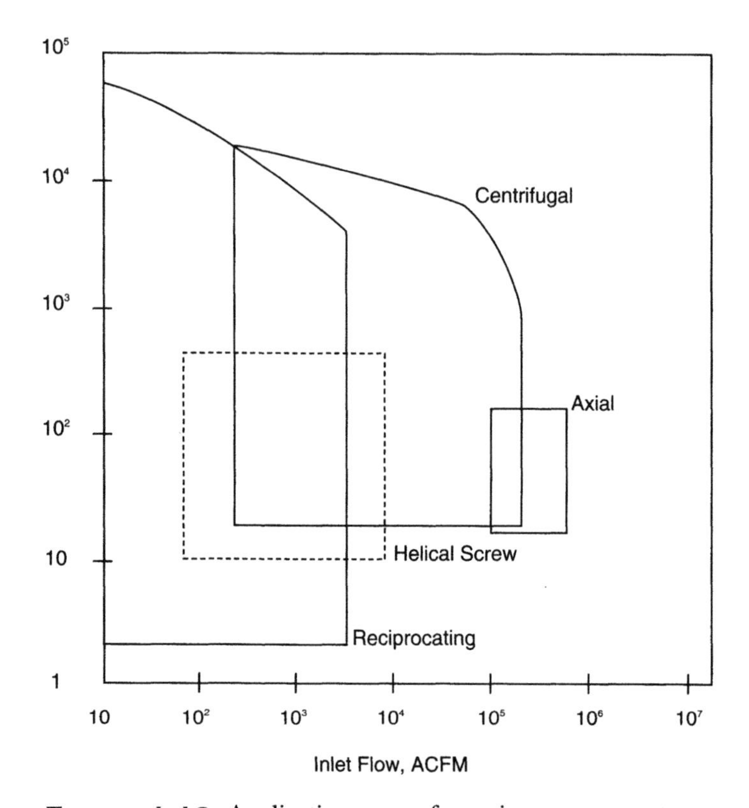

FIGURE 1-19. Application ranges for major compressor types.

CONSIDERATIONS IN SELECTION OF AIR COMPRESSORS

Load Factor

Load factor is a consideration largely in the smaller installations where only one or two compressors are to be installed.

Load factor is the ratio of actual compressed gas output (while the unit is operating) to the rated full-load output during this same period. It should never be 100%, a good rule being to select an installation for from 50 to 80% load factor, dependent upon the size, type and number of units involved. Proper use of load factor results in

1. More uniform pressure, even during peak demand periods.
2. A cooling-off period (extremely desirable for air-cooled units).

3. Less maintenance.
4. Ability to increase use of air without immediately increasing plant size.

Load factor is particularly important with air-cooled machines where sustained full-load operation results in an early build-up of deposit on valves and other parts, thereby adding to maintenance. Intermittent operation is always recommended for these units; the degree depends upon the size and operating pressure. Air-cooled units for higher than 200 psig pressure are usually rated by a rule that states that "pump-up" (compressing) time shall not ordinarily exceed 30 minutes, nor be less than 10 minutes. Shutdown or unloaded time should be at least equal to compression time. This means 50% load factor.

Multistaging

Any air-cooled compressor for 80 psig or higher pressure should have two or more stages of compression, unless it is very small. A two-stage unit for 100 to 200 psig will run 100° to 150°F cooler than a single-stage unit, thus reducing deposit formation and the need for cleaning valves. Two-stage compression of air to 100 psig also saves 10 to 15% in power over single-stage compression.

Heavy duty water-cooled units for the same service are more economical to operate than the air-cooled units and normally run at considerably lower speed and temperatures. Hence, maintenance will consistently be less. They are universally two-staged for 100 psig air service above approximately 125 BHP.

Floor Space

Floor space and its shape will, at times, influence selection. The opportunity for selection will be broadened if the exact dimensions of the space available are given to all manufacturers. Design adjustments or alternate arrangements are often available.

Foundation Needs

Foundation requirements for rotating compression machinery will almost always be less than for an equivalent reciprocating compressor, unless one can utilize the smaller moderate duty unit. The latter require (generally) limited foundations.

Capacity Control

Variation in demand for gas must be considered; it usually ranges from full compressor capacity to zero. The different types have widely varying abilities to handle this range, some being much more economical at part load than others. The reciprocating compressor is particularly favored in this area.

Variety of Choice

Within the four general divisions of compressors, there is a certain freedom of choice between arrangements. Similarly, many selection problems can be solved by considering more than one of the available reciprocating compressor designs.

Oil Free Air

Any requirement for air completely free of oil will rule out the rotary vane and most of the helical-lobe (screw) type compressors. There are, however, some screw compressor models which are offered as oil free and thus do not operate with oil in the rotor chamber.

Cost of Air

Compressed air is never "free." It costs a considerable sum, and the selection of a compressor should always consider this factor, particularly for heavy duty service. Two facts should be mentioned:

1. Power cost is by far the largest item in over-all air cost.
2. Power cost over the service life may be many times the first cost of any compressor.

SELECTION OF PROCESS COMPRESSORS

The process gas compressor is called upon to handle many diverse types of gas. Its capacity control requirements have a lower range (50%

being a quite frequent minimum even with reciprocating units where control possibilities are the greatest), and they are practically never "spared." Twenty-four hours a day, seven days per week operation is the general rule for months at a time. On-the-line continuity, or "availability" of operation, is highly important.

Gas Characteristics

Gas composition and characteristics can have a decided influence on compressor type. A low gas inlet density, for example, will usually affect the centrifugal compressor to a greater degree than it will affect the positive displacement machine. A centrifugal machine handling low density gas will require many more stages (a larger unit) than when handling a high molecular weight or specific volume inlet gas. Reciprocating and other positive displacement compressors are not seriously affected by the gas, molecular weight, specific volume, or inlet density.

Process Conditions

In any compression problem, for a given compression ratio, the flow rate to be handled establishes the physical size of the equipment under consideration.

If 80,000 CFM of gas must be compressed at near atmospheric suction conditions, a centrifugal compressor would almost certainly be used for the lower stages of compression.

In centrifugal compressors, if suction pressure is raised, the discharge pressure will exceed the design point, the horsepower will increase, and excess pressure may have to be throttled down. If the suction pressure is lowered, the centrifugal will not compress to the desired discharge pressure.

Therefore, it is most important that the suction and discharge pressures, their variation, and resulting influences be accurately evaluated.

Low compression ratios with reasonable capacities favor the centrifugal compressor. High compression ratios and higher pressures favor the reciprocating machine. One could not attempt to define the borderline in pressure between reciprocating and centrifugal compressors because there are too many other factors to be considered.

In some multistage process compressors, the interstage pressure is set by the process. This may be for washing out undesirable elements, adding gas, or carrying out chemical reactions that change the nature of the gas between stages. It is important that such interstage pressure restrictions be considered, including the pressure drop involved.

Driver

Generally speaking, if choice of driver is dictated by the available source of power, the heat balance, waste gas use, or any other factors in the plant or process, the compressor that best fits the driver should receive first consideration. In other words, the centrifugal compressor always receives first consideration if the driver must be a turbine. The reciprocating compressor should probably always receive first consideration if the driver is to be an electric motor. This is because the majority of motor-driven centrifugal compressors require a speed-increasing gear. The converse is true if turbines are used.

Oil Contamination

Oil contamination can have a bearing on compressor type selected. The degree of importance attributed to oil contamination will also influence decisions. Partially lubricated or non-lubricated reciprocating compressor designs may be more suitable than the small centrifugal (or common helical lobe units with inexpensive seal systems).

Foundation and Floor Space

Soil and foundation conditions must be considered in compressor selection. In a few instances, this dictates the type of machine. A centrifugal compressor operates without producing unbalanced forces. The foundation need only support the dead-weight load with adequate stiffness to maintain alignment.

The type of reciprocating compressor most widely applied today in chemical and process plants is designed specifically for a minimum of unbalanced forces. Generally, a foundation that will support the dead-weight load of the compressor and maintain alignment of compressor and driver will have adequate mass to absorb the small unbalanced forces that may be present.

The centrifugal machine will usually be favored from both the foundation and floor space standpoint if this is the only consideration.

Multiple-Service Units

For sound economic reasons, process plants have been installing fewer compressors to do a given job, using larger sizes and no spares. When a process requires handling of many streams, a common occurrence in some types of plants, machines can be so configured that several streams may be handled by a single driver.

It is rare that more than two services are handled in a combination of centrifugal casings having a common driver. There are problems of capacity and pressure control, as well as other factors.

With reciprocating compressors, greater flexibility is possible in the number and size of streams that can be handled on a single driver. Desired capacity control is usually more easily attained as well. As many as six separate streams have been handled on large existing compressors.

First Cost

There is no formula for establishing the relative cost of the centrifugal versus the reciprocating compressor. If volume, pressure, k factor, and all other factors are the same, then gas specific gravity can influence the cost of the centrifugal, but it will have little effect on the reciprocating compressor.

Power Cost

As has been mentioned earlier, power cost throughout the service life of a compressor is many times the first cost. While it is possible that the more efficient machine may be higher in original installed cost, the power savings over a period of years will usually quickly pay off the differential and return a profit for the remaining life.

Except at very low ratios of compression, the centrifugal compressor is inherently less efficient than the reciprocating compressor.

Very large volumes, low ratios of compression, and low final pressures favor the centrifugal compressor. No one today would think of using the reciprocating compressor for blast furnace blowers. Conversely, higher ratios of compression and higher terminal pressures favor the reciprocating compressor in other services.

RECIPROCATING COMPRESSOR CYLINDER ARRANGEMENTS

By arranging the cylinders of the reciprocating compressor in various combinations, greater capacity and output can be achieved.

Some of the many possible combinations of cylinders and their arrangements are shown in Figure 1-20.

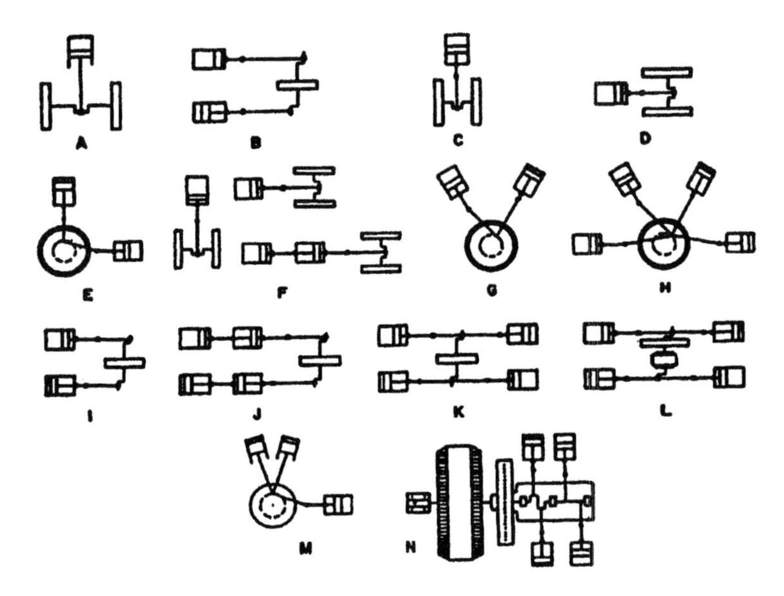

FIGURE 1-20. Cylinder combinations and available arrangements.

Single - Frame straight-line reciprocating compressors, Figures A, C, D, and F, are horizontal or vertical double-acting compressors with one or more cylinders in line on a single frame having one crank throw and one connecting rod and crosshead. They may be belted or direct-connected, motor-driven, or steam-driven with the steam and air cylinder in tandem.

V- or Y-type reciprocating compressors are two-cylinder, vertical double-acting machines in which the compressor cylinders are arranged at some angle, using 45° from the vertical, and are driven from a single crank (Figure G).

Semi-Radial reciprocating compressors are similar to the V- or Y-type except that, in addition to the double-acting compressor cylinders arranged at an angle from the vertical, horizontal double-acting cylinders are also arranged on each side, all operated from a single crankpin (Figure H).

Duplex reciprocating compressors are machines with cylinders mounted on two parallel frames connected through a common crankshaft (Figure I).

Duplex Four-Cornered Steam-Driven reciprocating compressors are of the duplex type and have steam and compressor cylinders on opposite ends of each frame (Figure J).

Four-Cornered Motor-Driven reciprocating compressors are of the duplex type with one or more compressing cylinders on each end of each frame. The driving motor is mounted on the shaft between the frames (Figure L).

Angle or L-type Integral gas or oil-driven compressors have the power cylinders in a vertical or vertical-V arrangement and the compressing cylinders in a horizontal plane (Figure M).

Horizontal Opposed reciprocating compressors are multi-cylinder machines on opposite sides of the crankcase; they employ a multi-throw type of shaft with one crank throw per cylinder. Power is applied at the end of the shaft (Figure N).

CHAPTER 2

Design and Materials for Reciprocating Compressor Components

Today's modern, heavy-duty, continuous-service reciprocating compressor cylinders can be attached to a single horizontal or vertical frame. When more than one cylinder is used, the various configurations are almost endless. As previously seen, these machines range from simple one-cylinder, single-stage compressors for air service to multi-stage, multi-service process gas compressors.

Regardless of the service, from the smallest to the largest, reciprocating compressors share both the same principles of operation and the same basic design features. Figure 2-1 illustrates a modern two-cylinder, two-stage compressor with intercooling between stages.

Some type of driver rotates a crankshaft, which converts rotary motion into reciprocating motion. The crankshaft is usually made from a steel forging and is supported by at least two main bearings.

The number of main bearings increases with the number of throws on the crankshaft that are required for the number of cylinders used. A typical crankshaft is counter-weighted, either integral with the shaft or separate from and bolted to it, in order to offset the effect of the unbalanced forces associated with a reciprocating compressor.

Most modern compressors use anti-friction or sleeve main bearings, although there are many older design compressors still in service that use two- or three-piece, adjustable babbitted bearings. This practice was

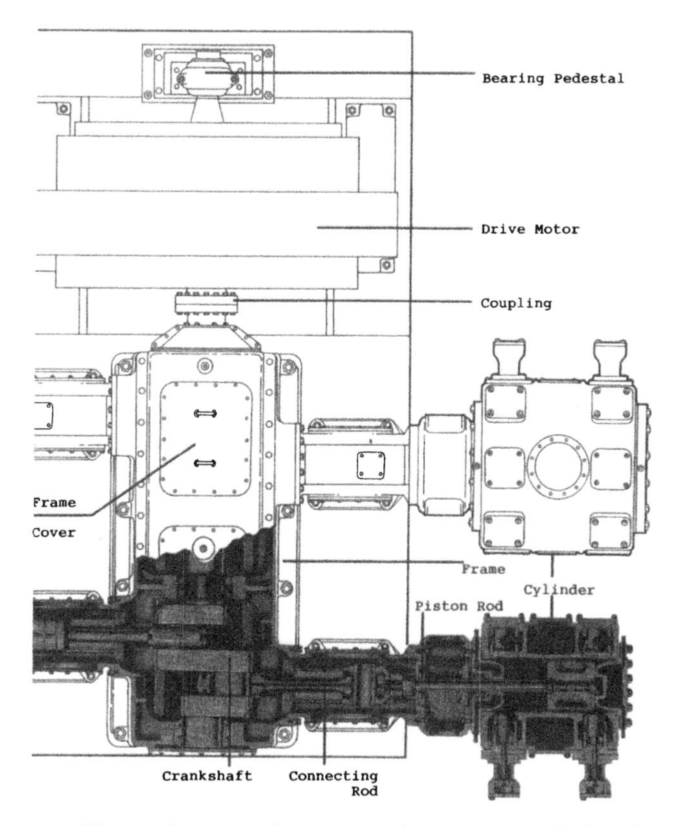

FIGURE 2-1. Heavy duty, continuous-service, water-cooled reciprocating compressor.

common in compressors designed with dual parallel frames, like the horizontal duplex types built as recently as the 1960s.

Modern designs take advantage of pressure lubrication systems from a gear-type oil pump, while some smaller units rely on splash to lubricate the bearings and crossheads.

A connecting rod is fastened to the crank throws; this component uses a two-piece sleeve bearing liner. At the opposite end of the connecting rod and secured to it by a pin through a bushing is the crosshead. This crosshead converts the rotary motion of the crank into reciprocating motion, and the piston rod is fastened to it.

The crosshead rides in a guide and is supported by shoes or slippers. These shoes are made from a bearing material of aluminum or babbitt.

The guides are machined in the frame at 90 degrees to the crankshaft and may be separate pieces bolted to the frame.

Crankcase lubricating oil is supplied to all bearing surfaces from the main lube oil pump. A segmental metallic seal called an oil scraper is mounted in the crankcase where the piston rod passes through the distance piece. This device scrapes the oil from the piston rod during the outward stroke of the piston. The oil drains back to the crankcase.

Figure 2-2 shows the principal components of a horizontal reciprocating compressor.

FIGURE 2-2. Components of a small reciprocating compressor (*Source: Pennsylvania Process Compressors, Easton, Pennsylvania*).

MATERIALS OF CONSTRUCTION

A knowledge of the materials used in the construction of the major components of the heavy-duty water-cooled compressor will permit the proper maintenance and repair of those components.

Frame or crankcase is a high-grade cast-iron casting similar to ASTM class 40 or 50; it is designed with suitable supports or ribbing to mount the compressor cylinders, crankshaft, and other running gear parts and hold them in accurate alignment under the stresses imposed during operation.

Crankshafts are made from carbon steel forgings or nodular iron (ductile) castings. Most forgings are carbon steel conforming to AISI 1020, ASTM 668 for small shafts and AISI 1045, ASTM 668 class F for large shafts. Nodular iron crankshafts conform to ASTM A-536 Grade 80-55-06. No hard surface treatment is used, and the shafts are not dynamically balanced, unless they are used on compressors with rotative speeds of 900 rpm and above.

Main bearings may be horizontally split shells made of steel or cast iron, with a lining of babbitt. Some are made of aluminum without babbitt or of a steel/bronze/babbitt tri-metal construction. The shells may have a laminated shim placed between the halves to permit adjustment for wear. On some smaller compressors, anti-friction roller bearings are used.

Connecting rod as shown in Figure 2-3 is a semi-marine type, made from a low carbon steel forging. Oil under pressure is conducted from the

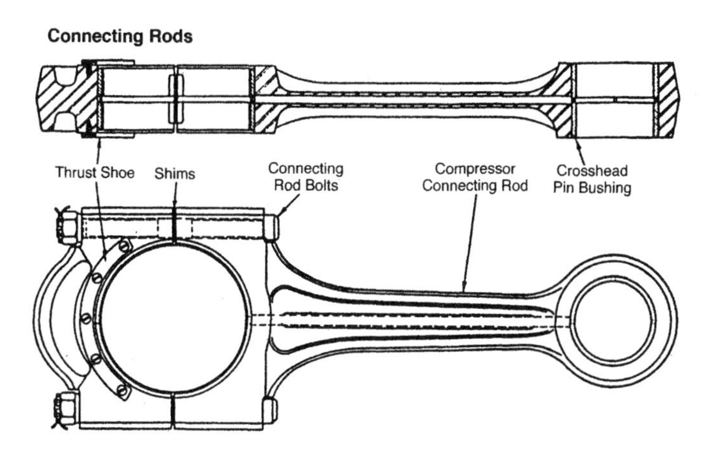

FIGURE 2-3. Connecting rods (*Source: Dresser-Rand, Painted Post, New York*).

crankpins to the crosshead pin. Similar to the main bearing, crankpin bearings are babbitt-lined steel or cast-iron shells. They may be shim-adjustable or shimless. In other designs, bearings are made from aluminum, bronze, or tri-metal construction.

The **crosshead pin bushings** are bronze, aluminum or babbitt-lined on steel/cast-iron backing.

Crossheads, as illustrated in Figure 2-4, are generally made of cast grey iron or nodular iron, but in some older designs cast steel is used. Shoes or slippers of cast iron with babbitt overlay or shoes of aluminum are bolted to the crosshead.

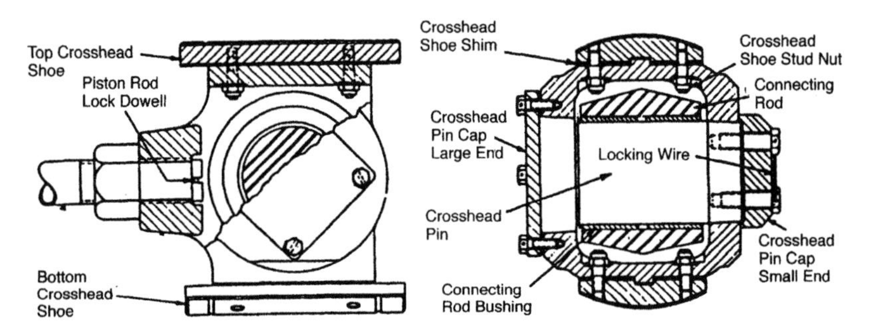

FIGURE 2-4. Crosshead with replaceable shoes (*Source: Dresser-Rand, Painted Post, New York*).

The crosshead pin is made of steel and hardened to approximately 50 Rc; it is tapered at the ends and held in place with caps. Other designs have no taper and are free to rotate or "float," retained by lock rings.

The piston rod is threaded into the crosshead and is locked with a dowel or set screws.

Other crosshead designs, such as the one in Figure 2-5, do not have replaceable shoes or slippers but have the faces babbitted and machined for a bearing surface. The crosshead pin is "floating"—it is not locked so rotation is allowed, but it is restrained in the lateral direction by retaining rings inserted in grooves in the pin. The total lateral movement of the pin is .060″ to .063″.

Figure 2-6 shows a connecting rod, crosshead assembly, and all the component parts of the connecting rod and the crosshead.

FIGURE 2-5. Crosshead design with non-replaceable shoes (*Source: Dresser-Rand, Painted Post, New York*).

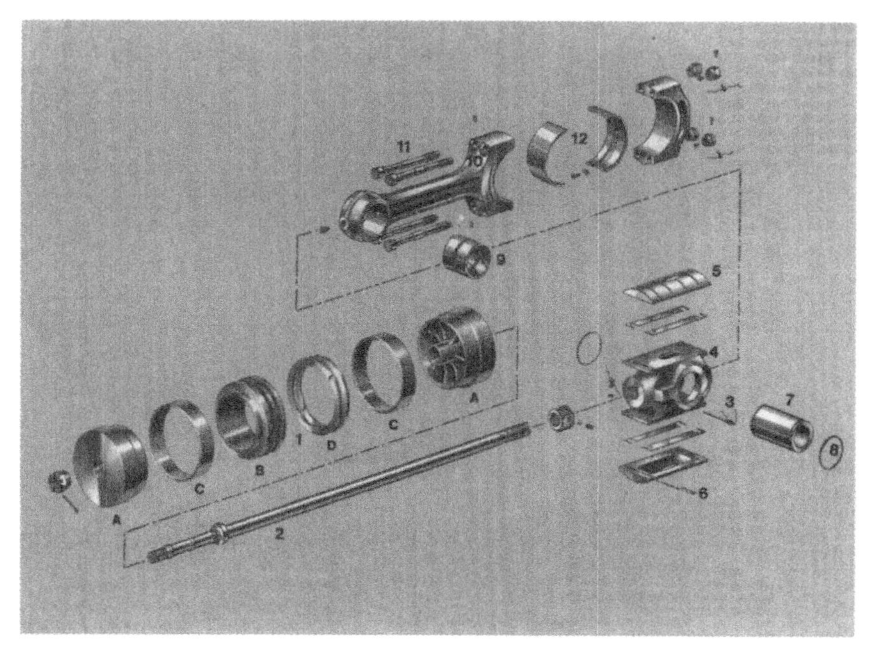

FIGURE 2-6. Connecting rod, crosshead assembly, and associated parts (*Source: Dresser-Rand, Painted Post, New York*).

CYLINDERS

A typical double-acting cylinder, like that shown in Figure 2-7, consists of a barrel, usually water-jacketed, with a front and rear head. In some designs the rear head is an integral part of the cylinder barrel; in others, it is a separate, bolted piece as shown. These heads are also water-cooled to remove heat of compression.

Provision is made in the rear head for the pressure packing. Valves are installed around the barrel, but in some designs valves are installed in the heads.

Cylinders may be double-acting, that is, compressing on both sides of the piston, or single-acting, compressing at either the head or crank end, but not both.

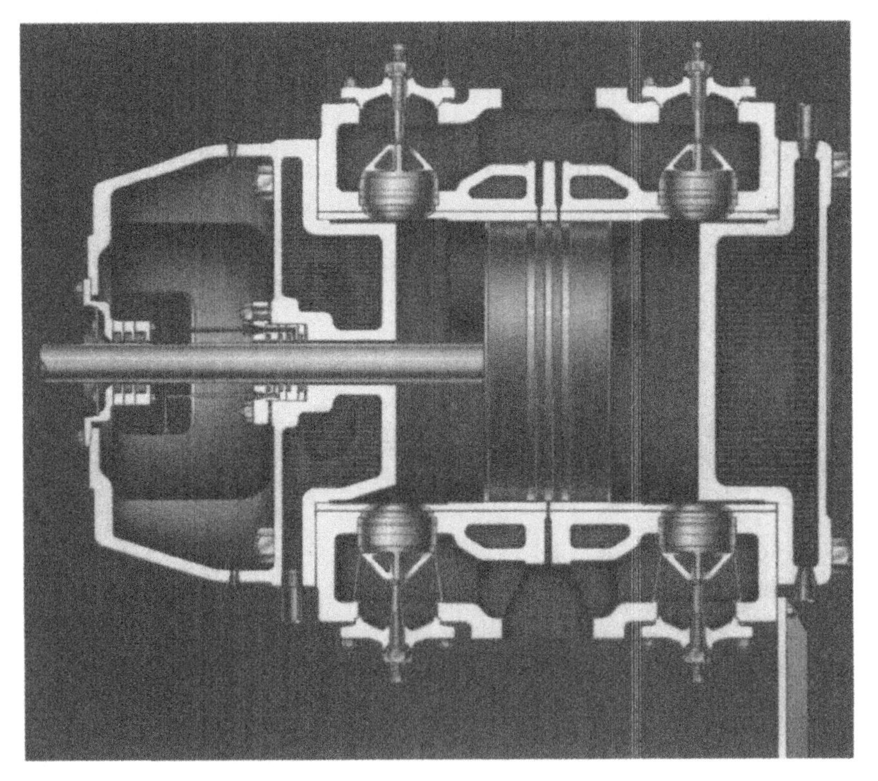

FIGURE 2-7. Typical double-acting compressor cylinder (*Source: Dresser-Rand, Painted Post, New York*).

All cylinders can be modified to provide capacity control or to provide openings for clearance pockets and internal plug unloaders.

Cylinder Materials

Cylinders are made of material selected for the particular pressure and gas being handled. Variables which must be considered in the selection of materials include bore diameter, pressure differential, and the type of gas to be handled.

Low and Medium Pressure Cylinders

These are normally made of cast iron but are also available in nodular iron or cast steel depending on application. The cylinder shown in Figure 2-7 has a solid bore, but in most designs the cylinders may be furnished with sleeves or liners. These cylinders normally have generous water cooling around the bore and in both heads. Pressures can be up to 2000 psi depending on the bore diameter.

Medium to High Pressure Cylinders

This cylinder, Figure 2-8, has heavier wall sections and smaller bores than the cylinder shown in Figure 2-7. The material used would normally be nodular iron, but steel has been used in the past. Cooling and control

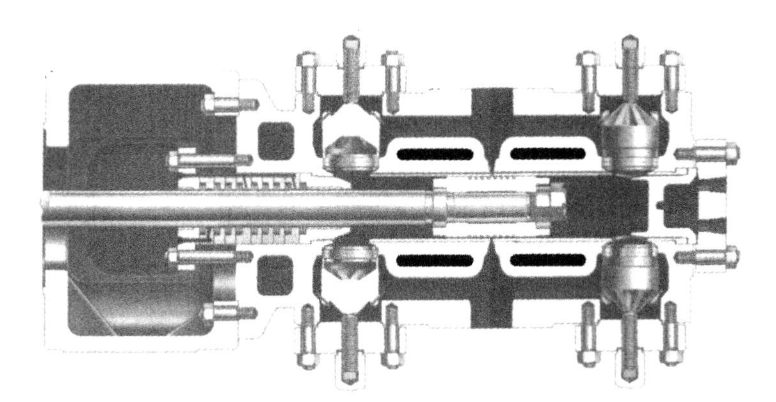

FIGURE 2-8. Medium- to high-pressure compressor cylinder (*Source: Dresser-Rand, Painted Post, New York*).

options are similar to those in low and medium pressure cylinders. Pressure range is from 1000 psi to 2500 psi, depending on the bore sizes.

High-Pressure Cylinders

High-pressure cylinders, as shown in Figure 2-9, are made from a steel forging, with only nominal water cooling as compared to the other cylinders. Because the smallest possible number of openings in the forging is desired, capacity control mechanisms are not normally provided. Tie bolts may be installed perpendicular to the bore to pre-stress the forging, decreasing the maximum tensile stress induced by the gas pressure. Valve ports are finished to a highly polished surface, which reduces the maximum tensile stresses. The pressure range for these cylinders is approximately 2000 psi to 8000 psi.

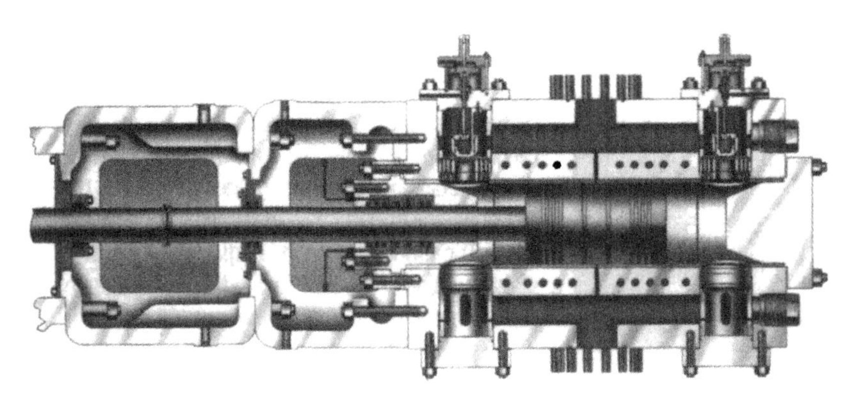

FIGURE 2-9. Forged steel high-pressure cylinder (*Source: Dresser-Rand, Painted Post, New York*).

High-Pressure, Low-Differential Pressures

This type of cylinder, Figure 2-10, is commonly called a recycle cylinder. It generally uses a "tail rod" because it is normally difficult to obtain reversible frame loading with a conventional piston rod design. An additional packing assembly is required. Made from steel forging, the cylinder has a cover pressure range from 2000 psi to 8000 psi.

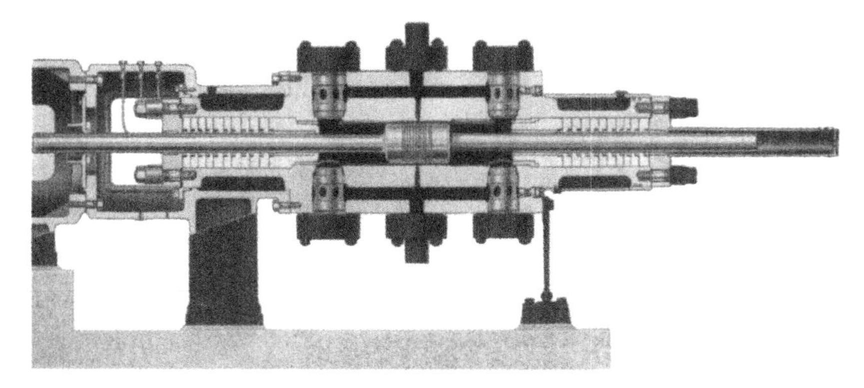

FIGURE 2-10. High-pressure recycle, "tail rod" cylinder (*Source: Dresser-Rand, Painted Post, New York*).

High-Pressure, Opposed Plunger, Conventional Running Gear

This design, as shown in Figure 2-11, is used in very high-pressure cylinders. Both single-acting ends may be identical in bore diameter and can be used as single stages with both compressing the gas at identical pressures and temperatures. An alternate design contains two stages with different bore sizes, and operating pressures with resultant similar frame loads. Each cylinder has one valve assembly, with the cylinder ported to serve both suction and discharge valves. Plungers may be internally oil-cooled. Pressures range from 5000 psi to 60,000 psi.

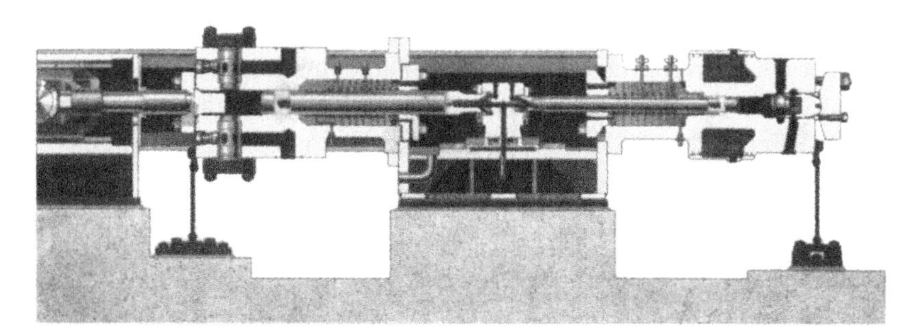

FIGURE 2-11. High-pressure, opposed plunger, "hyper" cylinder (*Source: Dresser-Rand, Painted Post, New York*).

High-Pressure, Opposed Plunger, "Hyper Compressor" Running Gear

Special designs intended for high-pressure ethylene service in LDPE (low density polyethylene) plants often employ the construction principles illustrated in Figures 2-12 and 2-13. Here, the plungers for two opposing cylinders are operating on the same centerline. These plungers

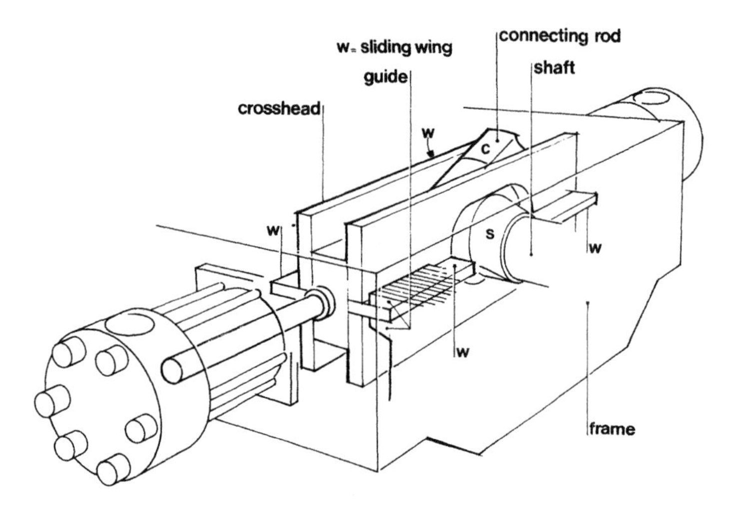

FIGURE 2-12. Principal construction features of a hyper compressor for high-pressure ethylene service in the production of LDPE (low density polyethylene) (*Source: Nuovo Pignone, Florence, Italy*).

are attached to a massive crosshead, and the crosshead is equipped with wings that slide on a suitably sized wing guides (Fig. 2-12). A number of different cylinder configurations are available, and Figure 2-13 depicts a typical execution popular in Nuovo Pignone hyper compressors. Note how it differs from the liner configuration that is customary in conventional compressor cylinders, shown in Figure 2-14.

LINERS

Liners and sleeves are used in compressor cylinders either to form the cylinder wall or to be a removable part of the cylinder in case of accidental scoring or wear over a long period of service. Refer to Figure 2-14.

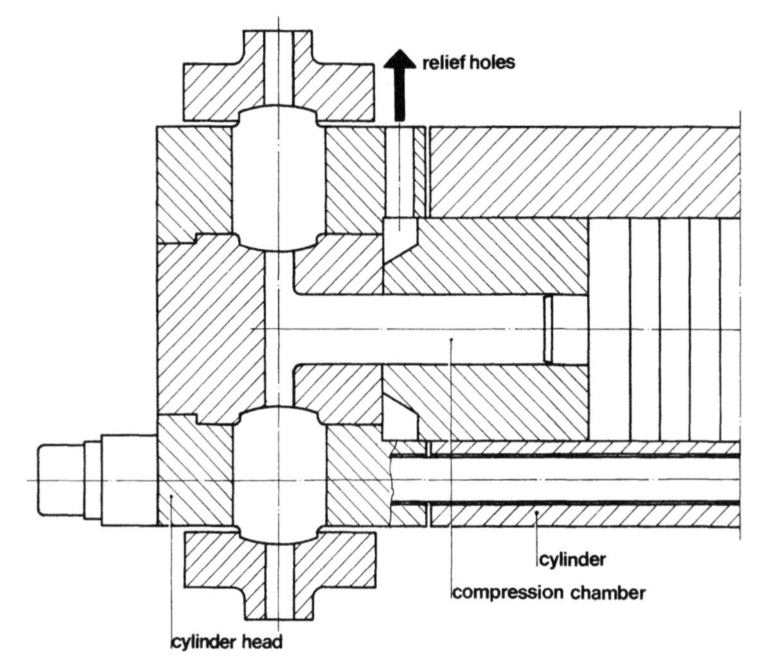

FIGURE 2-13. Typical hyper compressor cylinder for high pressure ethylene service (*Source: Nuovo Pignone, Florence, Italy*).

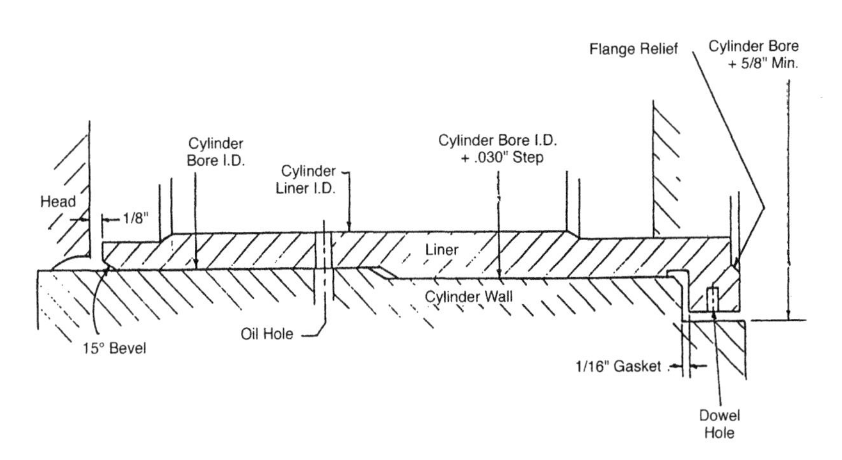

FIGURE 2-14. Conventional liner.

In general, liners are not used in low- or medium-pressure cylinders where the gas being handled is non-corrosive, such as air. Liners are almost always used in medium- to high-pressure cylinders where the gas being handled is corrosive.

Certain industry users insist on liners in their compressors. Liners have become standard design for certain industry applications, even though the gas involved may be non-corrosive and of medium pressure, or with low differentials. Such a standard exists in natural gas pipeline transmission service. The API 618 specifications for reciprocating compressors in refinery service also reflect this standard.

It follows that a liner which increases the initial cost of the compressor cylinder is more economical to replace than the complex cylinder casting.

Another application of the liner is to reduce the cylinder bore size to meet certain capacity conditions. By installing liners with different bore sizes, the same cylinder casting can be used to accommodate a range of capacity and pressure conditions.

Cast iron is generally used for liners unless the cylinder is high pressure—above 3000 psi on lubricated machines and above 500 psi on non-lubricated compressors. On these, hardened cast steel may be used.

For corrosive applications a "Ni-Resist" cast iron which contains 20% nickel is used. In very high pressures where high strength is required, forged steel, hardened by nitriding, is used.

PISTONS

The design and materials used for compressor pistons will vary with the make, type, and application of the compressor. They are designed to take into account a number of conditions:

- Cylinder bore diameter
- Discharge pressure
- Compressor rotative speed
- Compressor stroke
- Required piston weight

Compressor pistons are typically designed as one of three types:

One piece, either solid cast iron or steel, for small bores and high pressure differential applications, or one piece hollow-cored cast iron or aluminum, for large diameter and lower pressures (Figure 2-15A).

Two piece, aluminum or cast iron, which is split for ease of hollow casting and weight control. These are generally used above 10″ bore diameters. Aluminum is used when the reciprocating weight must be reduced (Figure 2-15B).

Three piece, in which a ring carrier is added to permit band-type rider rings to be installed directly into the piston grooves. While this design adds a part, it allows thicker rings to be used since the ring does not have to be stretched over the outside diameter. It is also used as a carrier for the piston rings on large diameter pistons, where metallic rings are used which might wear into an aluminum piston. Aluminum is used when weight reduction is required. A cast iron ring is used for its superior ring groove wall wearing qualities (Figure 2-15 C).

Piston Materials

Material selection for pistons is very important, and many factors must be considered. Some of these include

- Weight
- Strength, for differential pressures and inertia forces
- Corrosion resistance
- Compression and rider ring wall wear resistance
- Outside diameter wear resistance

COMPRESSOR PISTONS

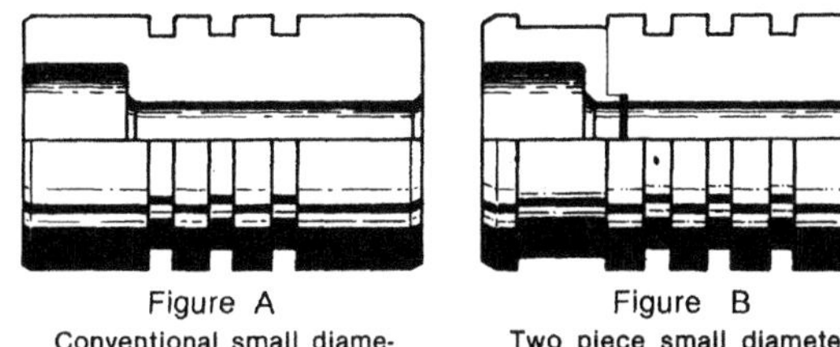

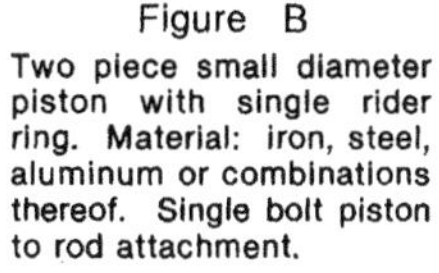

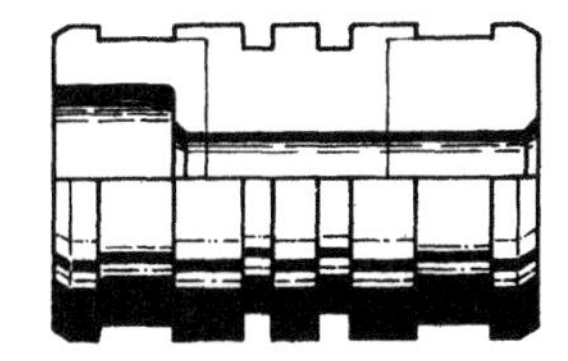

Figure A
Conventional small diameter piston without rider rings. Material: iron, steel or aluminum. Single bolt piston to rod attachment.

Figure B
Two piece small diameter piston with single rider ring. Material: iron, steel, aluminum or combinations thereof. Single bolt piston to rod attachment.

Figure C
Three piece piston with two rider rings. This design utilizes two rider ring carriers and one piston ring carrier. Materials as in B. Single bolt piston to rod attachment.

FIGURE 2-15. Typical piston designs (*Source: Anglo Compression, Mount Vernon, Ohio*).

Materials commonly used for compressor pistons are aluminum, cast iron, and steel.

Aluminum is used when lightweight pistons are required in order to balance reciprocating weights or to reduce inertia forces so they do not exceed rated frame load limits. The aluminum used is a special alloy with a tensile strength of 40,000 psi and a hardness of 100–110 Bhn. It may be given a surface anodizing treatment to achieve a hardness of 370–475 Bhn; this improves wear resistance. Applications are limited to approximately 200°F and a differential pressure of 125 psi for castings.

Cast Iron is the most common piston material due to its high strength and good wear and corrosion resistance. It is used in either the cast or solid form, conforming to ASTM A275, class 40.

Steels are used for small bore, high differential pistons when strength requirements are higher. They conform to ASTM A354 or A320. Steel is also used in fabricating built-up type pistons in some designs.

Rider bands may or may not be used on large diameter pistons for lubricated compressors. Generally, aluminum pistons do not use a rider band, while cast-iron pistons will have a rider band, usually a high lead bronze, such as "Allen Metal." In conventional non-lubricated or oil-free compressors, rider bands are always used. Figure 2-16 shows a segmental piston with a rider band.

Piston-to-Piston Rod Connections

There are several methods commonly used to fasten the piston to the piston rod.

Taper Fit. The end of the piston rod is machined with a tapered bore in the piston. The end of the piston rod is threaded, and a nut draws the piston to the rod and holds it securely (Figure 2-17).

Interference Fit. The piston bore is machined to accept the piston rod with an interference. No threaded connection or nut is used to hold the piston to the rod, and the piston is "peened" at the rod end (Figure 2-18).

Single Nut. The piston bore is closely fitted to the piston rod, and the piston is held to a shoulder or collar machined on the rod. The piston is held to the rod by a nut which is torqued to the proper value dependent on material, diameter, and number of thread. This is the most common fastening arrangement (Figure 2-19).

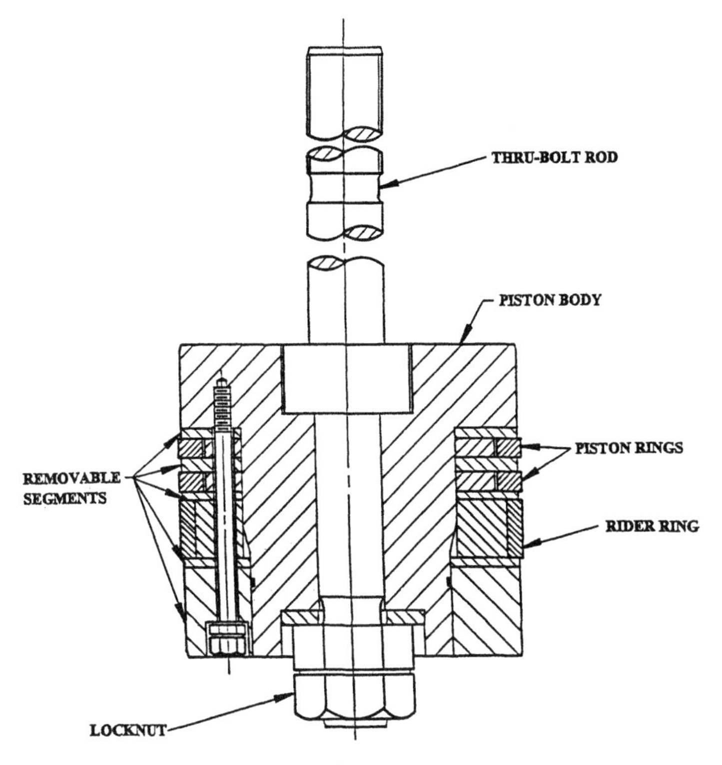

FIGURE 2-16. Segmental piston with rider band (*Source: Anglo Compression, Mount Vernon, Ohio*).

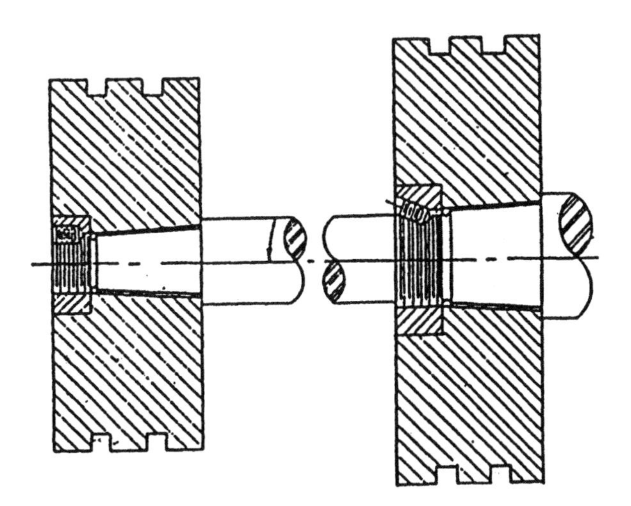

FIGURE 2-17. Taper fit piston rod connection (*Source: Dresser-Rand, Painted Post, New York*).

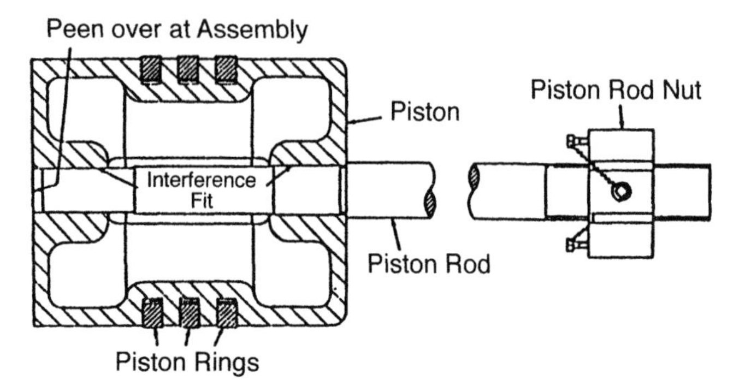

FIGURE 2-18. Interference fit piston rod connection for small compressors (*Source: Dresser-Rand, Painted Post, New York*).

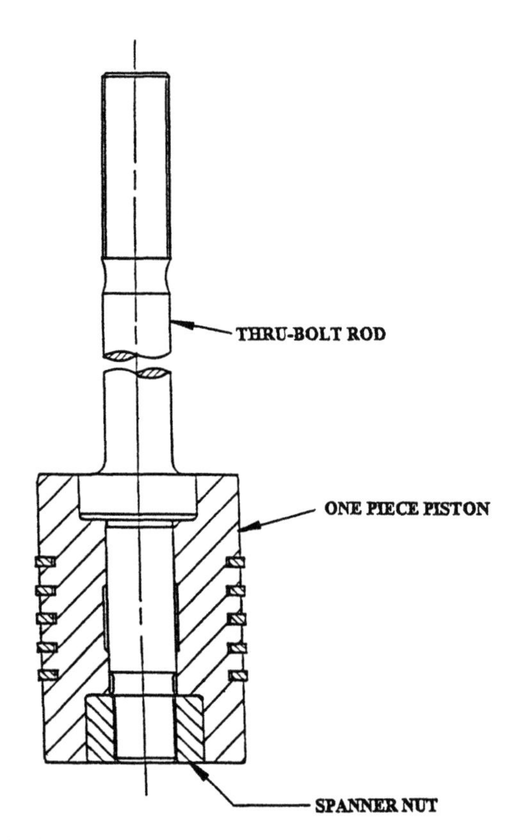

FIGURE 2-19. Single nut, one-piece piston arrangement (*Source: Anglo Compression, Mount Vernon, Ohio*).

Multi-Bolt. In this arrangement (see Figure 2-20) the piston rod is made with a flange at the piston end, which is drilled and tapped to accept studs. The piston is drilled to allow studs to pass through and locked to the rod with nuts. This arrangement has the advantage of allowing removal of the piston from the cylinder without having to unscrew the piston rod from the crosshead.

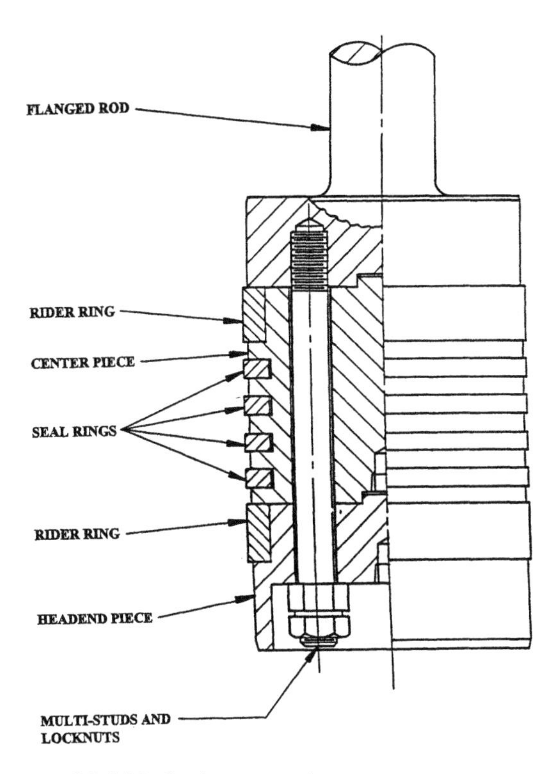

FIGURE 2-20. Multi-bolt piston attachment allowing removal of piston without having to unscrew the piston rod from the crosshead (*Source: Anglo Compression, Mount Vernon, Ohio*).

PISTON RODS

Like other components in the modern reciprocating compressor, the piston rod is designed for specific applications such as

- Operating pressures
- Gas composition

- Capacity
- Rotating speed

Piston rods incorporate diameter, length, material, composition, and fastening arrangements as dictated by both the operating conditions of the compressor and the design of the piston to which they are connected. Several different piston rods are shown in Figures 2-21 through 2-23.

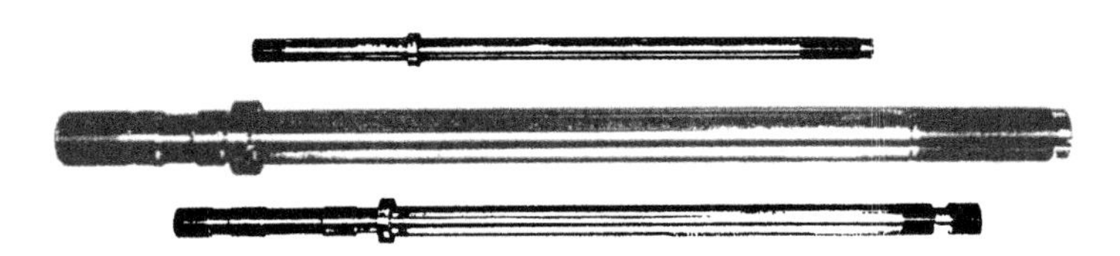

FIGURE 2-21. Different types of piston rods used in reciprocating compressors (*Source: Dresser-Rand, Painted Post, New York*).

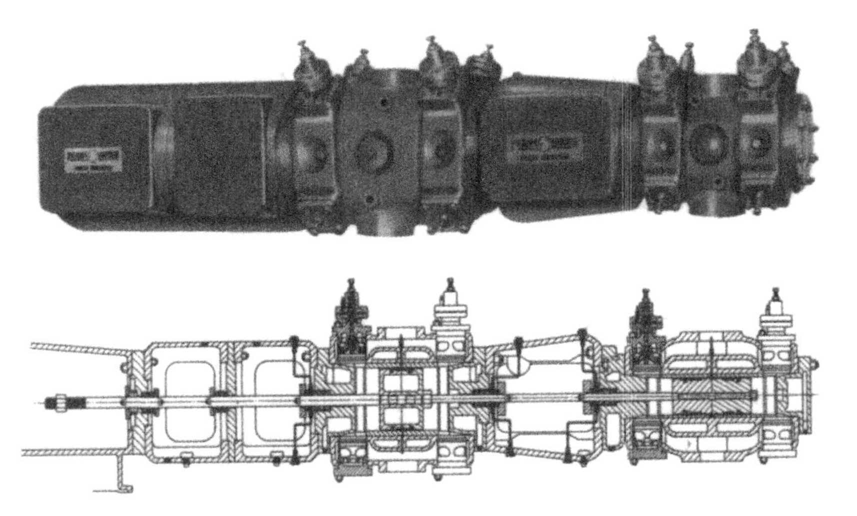

FIGURE 2-22. Tandem cylinders (*Source: Pennsylvania Process Compressors, Easton, Pennsylvania*).

Types of Piston Rods

Configurations vary with cylinder design.

Single Piston Rod. This is the most common type. It contains a single piston.

Tandem Piston Rod. This is a piston rod on which two or more pistons are mounted in tandem. This is used where loading is low so that the combined loading does not exceed the allowable frame load. Figure 2-22 shows a tandem rod in a tandem cylinder compressor.

Piston Rod for Truncated Cylinder. This is a special configuration which accommodates the truncated cylinders illustrated in Figure 2-23. A truncated, or stepped, piston compressor cylinder is designed with different bore sizes for head end and crank end. Two stages of compression can be handled—each stage single-acting. This type of compressor cylinder is custom-designed for each application and is typically available in forged steel and fabricated steel. Normally the void space between the

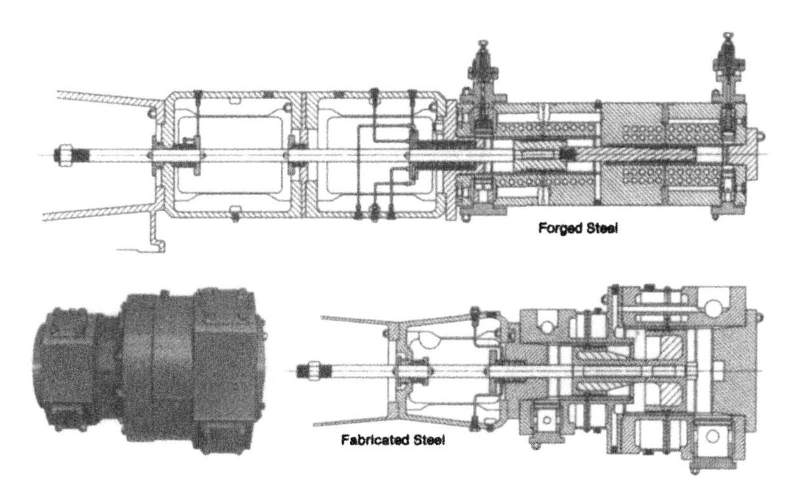

FIGURE 2-23. Truncated cylinders (*Source: Pennsylvania Process Compressors, Easton, Pennsylvania*).

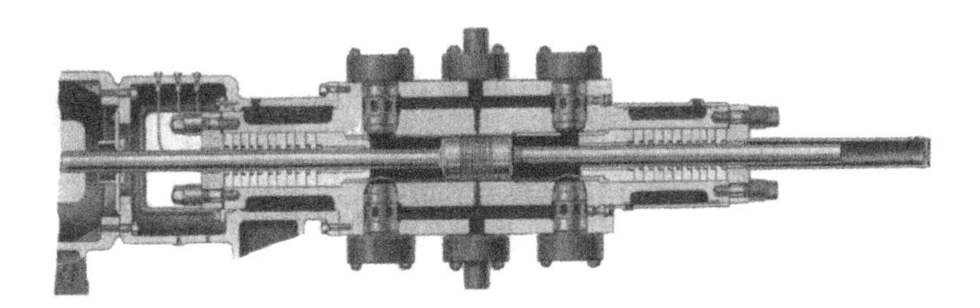

FIGURE 2-24. Compressor cylinder with tail rod arrangement (*Source: Dresser-Rand, Painted Post, New York*).

two pistons is connected to low-stage suction which acts to minimize the differential pressure across the piston rings.

The applications best handled in truncated forged cylinders involve small piston displacements for each stage and medium-to-high discharge pressures.

The large-bore crank end is used for the low stage and the small-bore head end for the high stage. By varying the bore sizes, an optimum design is achieved to minimize rod loads. With this arrangement it is not necessary to pack against second-stage discharge pressure.

Truncated cylinders of fabricated steel often find application at lower pressures where it is desired to avoid the maintenance problems of the tandem configuration. With the large-bore, low stage on the head end, it is possible to remove the two-stage piston and rod assembly with a minimum of disassembly.

Tail Rod. In this arrangement, the piston is in the center of the piston rod, and the rod is the same diameter on both sides of the piston. This prevents non-reversal loading due to equal areas and equal pressures on both sides of the piston. Refer to Figure 2-24.

Cooled Piston Rod. This rod is drilled through the center axis to allow pressure fed coolant, usually oil, to circulate up the rod core. Such circulation helps to remove the heat of compression from rod and piston.

Piston Rod Materials

Compressor piston rods are made from various types of steel depending on stress levels and the composition of the gas handled.

Low carbon steels, such as AISI 1037, and low alloy steels, such as AISI 4140, are commonly used. For certain corrosive gases, 410 stainless, K-Monel, or Hastalloy steels may be used.

The normal frictional forces acting between the piston rod packing rings and the piston rod increase with pressure. Experience has shown that piston rod packing wear becomes excessive with pressures above 1000 psi. Figure 2-25 shows the friction forces which act on a piston rod.

Therefore, a piston rod with a hardened surface in the packing area will have less wear than will one not so hardened. This hardening may take the form of carburizing, nitriding, induction hardening, plasma spray with chrome oxide, tungsten carbide, or flame hardening. This also implies that surface finish is important!

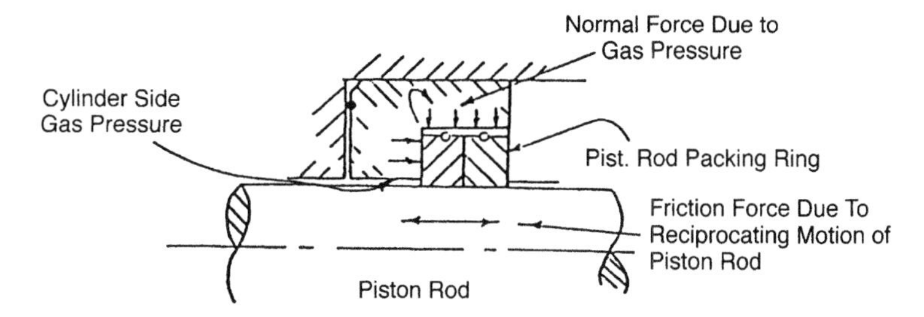

FIGURE 2-25. Friction forces acting on compressor piston rod.

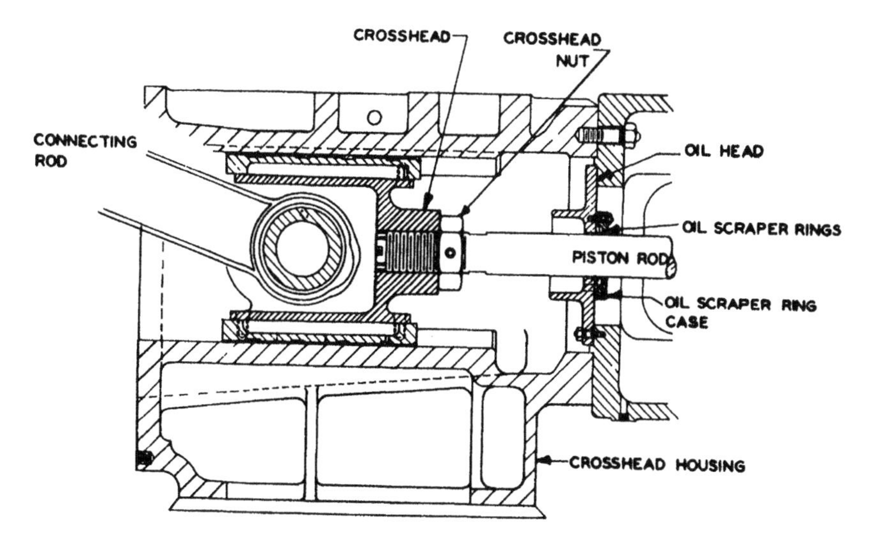

FIGURE 2-26. Screwed piston rod—crosshead connection (*Source: Dresser-Rand, Painted Post, New York*).

Piston Rod to Crosshead Connections

The piston rod is always screwed into the crosshead and locked with a nut except in the case of the floating coupling arrangement to be described. This discussion will concern only the arrangement which is screwed into the crosshead. Refer to Figure 2-26 for a section through the crosshead frame.

Castellated Piston Rod. The end of the piston rod is machined with castellations, and a pin or lock dowel is inserted through the crosshead to prevent rod rotation. This method also uses a nut against the face of the crosshead for locking, as in Figure 2-27.

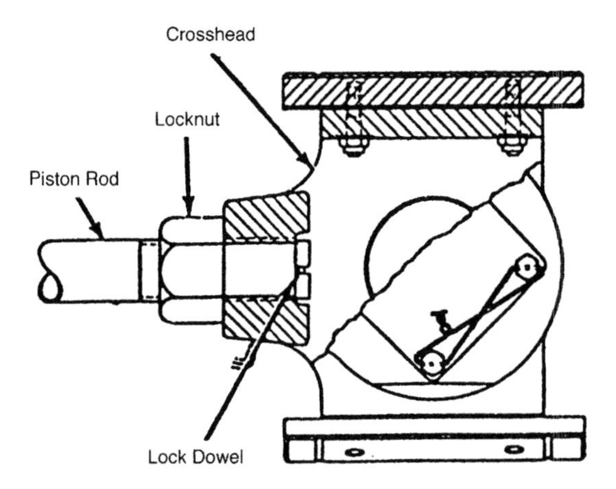

FIGURE 2-27. Castellated rod end with locking dowel (*Source: Dresser-Rand, Painted Post, New York*).

Set Screw Lock. The crosshead is drilled and tapped for a set screw at the threads of the piston rod. A copper disc is inserted into the hole to prevent damage of the rod threads by the set screw. The crosshead nuts are also locked in a similar manner with the set screws and copper discs (Figure 2-28).

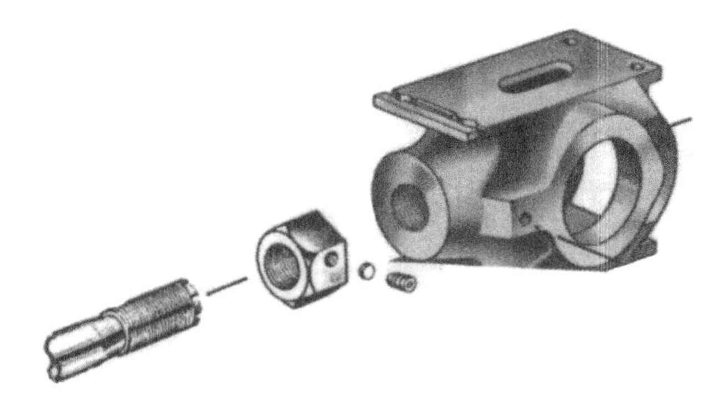

FIGURE 2-28. Set screw locking approach used on small compressors (*Source: Dresser-Rand, Painted Post, New York*).

Jam Screw Lock to Crosshead. This arrangement uses several set screws through the nut which bears against the face of the crosshead. It also uses a set screw and copper disc through the nut and bears against the piston rod threads (Figure 2-29A). A more modern version is shown in Figure 2-29B.

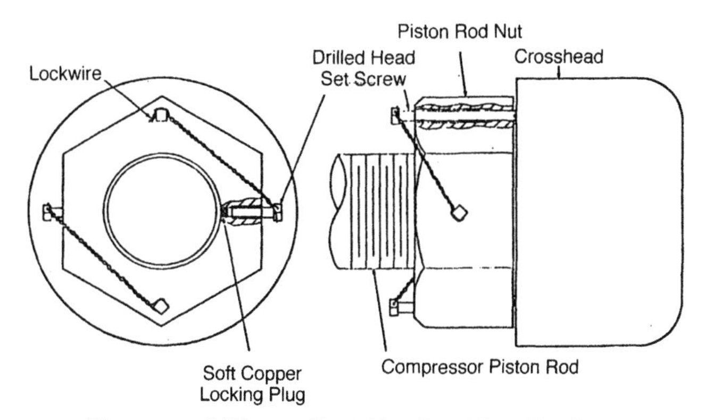

FIGURE 2-29A. Jam screw lock to crosshead (*Source: Dresser-Rand, Painted Post, New York*).

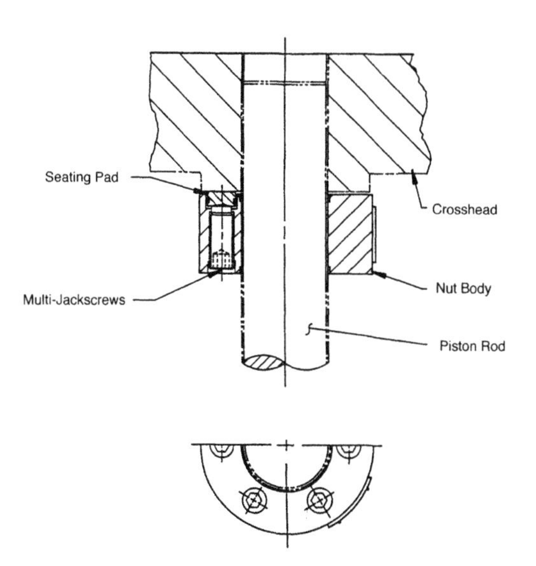

FIGURE 2-29B. Multi-jackscrew ("Extractorque"™) rod-to-crosshead lock nut (*Source: Anglo Compression, Mount Vernon, Ohio*).

Multi-Bolt Lock. In this arrangement a circular nut, instead of a hex nut, is used at the crosshead. Torque of the nut is obtained by 4 to 8 bolts which fasten the nut to the face of the crosshead. Allowing a clearance of .020″ to .040″ between nut and crosshead faces and tightening the bolts on the nut effectively locks the piston rod (Figure 2-30).

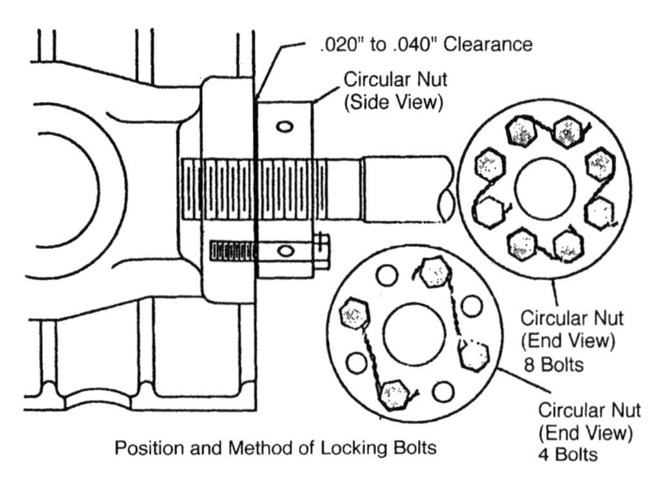

FIGURE 2-30. Multi-bolt piston rod locking device (*Source: Dresser-Rand, Painted Post, New York*).

Floating Coupling. This type of piston rod is not threaded into the crosshead. Instead, it is connected by a three-piece coupling arrangement which allows the rod and piston to continuously align with the cylinder bore. While typically used on high-pressure small bore cylinders, it is sometimes found on European-designed compressors as well (Figure 2-31A).

Again, a more modern variant is depicted in Figures 2-31B and 2-31C. The pre-stressed connection between piston rod and crosshead contains only one connecting element comprising an anti-fatigue shaft and a screw thread at the piston rod end. The optimized pre-stress transmits low alternating loads to the thread and can be adjusted entirely without any measuring devices. A simple distance ring is all that is required to join the two components properly. Pre-stressing is done hydraulically for most compressor sizes requiring no special hydraulic tools.

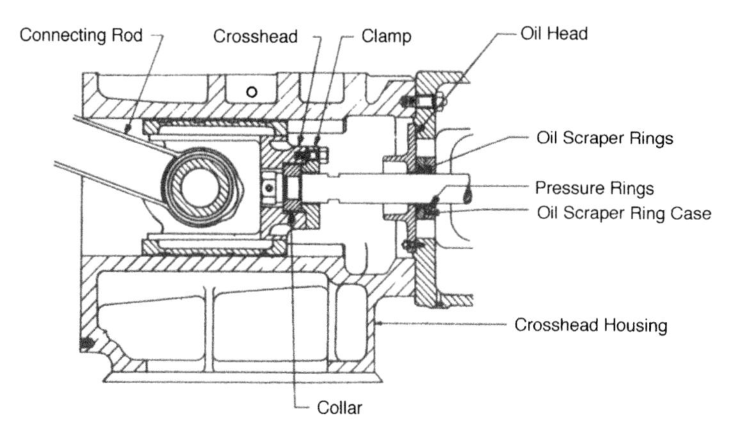

FIGURE 2-31A. Floating coupling piston rod locking device (*Source: Dresser-Rand, Painted Post, New York*).

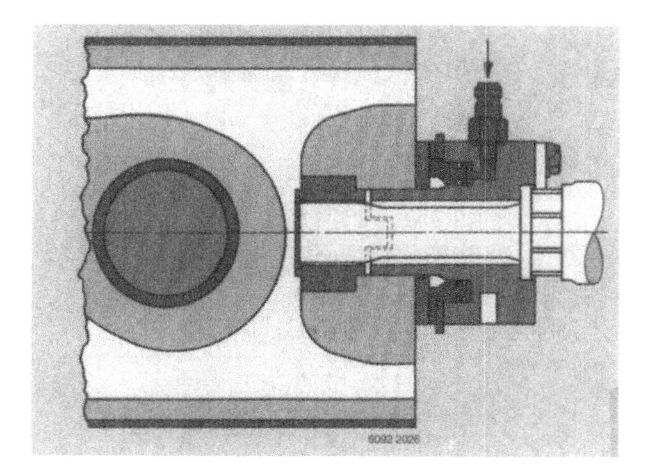

FIGURE 2-31B. Pre-stressed connection between piston rod and crosshead, cross-section view (*Source: Sulzer-Burckhardt, Winterthur, Switzerland*).

FIGURE 2-31C. Assembled view of pre-stressed connection between piston rod and crosshead (*Source: Sulzer-Burckhardt, Winterthur, Switzerland*).

Figure 2-32A shows a typical piston and piston rod assembly. This is a two-piece cast iron piston which fits against a collar machined on the piston rod and is held in place by a nut. It shows the threaded end, which screws into the crosshead and is locked by a nut. Also seen are the cotter pin locking the piston nut and the disc and set screws at the crosshead nut. A more advanced version is shown in Figure 2-32B.

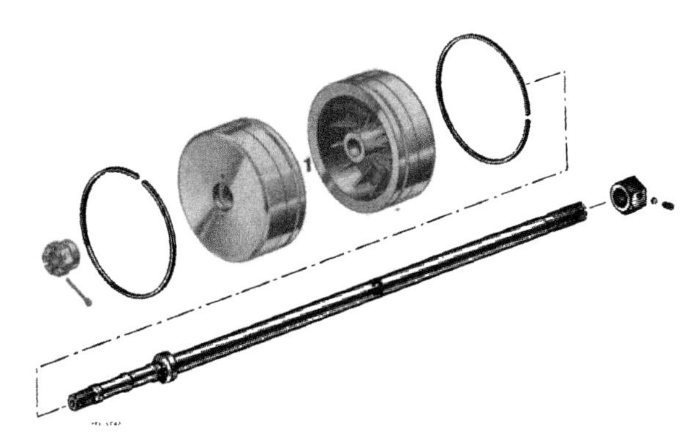

FIGURE 2-32A. Typical piston and piston rod assembly (*Source: Dresser-Rand, Painted Post, New York*).

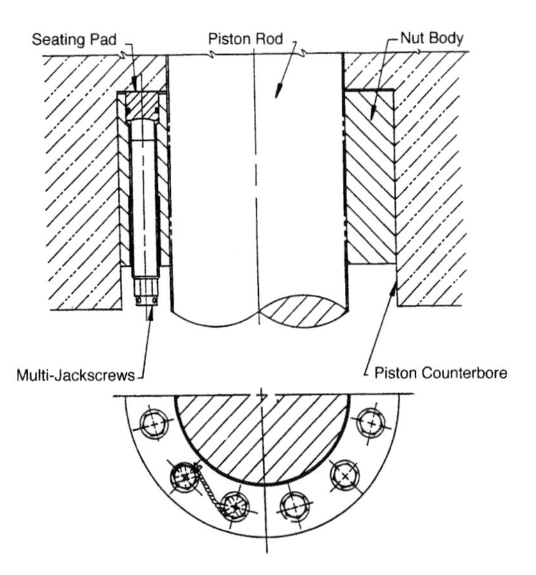

FIGURE 2-32B. Multi-jackscrew ("Extractorque"™) connection between piston and piston rod (*Source: Anglo Compression, Mount Vernon, Ohio*).

NON-LUBRICATED OR OIL-FREE CYLINDER CONSTRUCTION

There are many applications for industrial compressors in which oil in the gas stream cannot be tolerated. Oil-free compressed air is essential in industries such as the food industry, the brewing industry, and the packaging (pharmaceuticals) industry, as well as in some industrial air control systems.

But even in general industry or manufacturing, there may be reasons to consider reducing the amount of lubricating oil used in the compressor cylinders. Excess oil can build up in the discharge valve port areas, and even the best premium grades of compressor oil will oxidize when subjected to high temperatures. These oils may eventually form gummy or sludge-like deposits, which reduce the performance of a compressor and can, in some cases, lead to fires in the air system if they are allowed to build up. For these and other reasons, non-lubricated operation has become increasingly popular.

HISTORY

Before looking at the design details of the oil-free reciprocating compressor, it may be of interest to review what has been produced in the past and to examine the present "state of the art."

Oil-free cylinder designs were created in the early 1930s. These cylinder designs used water for lubrication and saw service in brewery applications. Soap and water lubrication was used for compressors pumping oxygen.

In about the mid 1930s, the first high-pressure, 2000 psi non-lubricated air compressor was made using carbon rings. In subsequent years, many single- and multi-stage compressors were made using carbon as the wearing material for the piston and rider bands. This carbon piston ring construction is shown in Figure 2-33.

This was a "non-floating" type piston, which meant that the carbon rings transferred the weight and load of the iron piston onto the cylinder bore. Piston rings with expanders were used to seal the gas.

Another type of construction was a "floating" piston, in which a tail rod was used with a small auxiliary crosshead. The tail rod supported the piston and prevented it from touching the bore. Carbon-rider rings were not used.

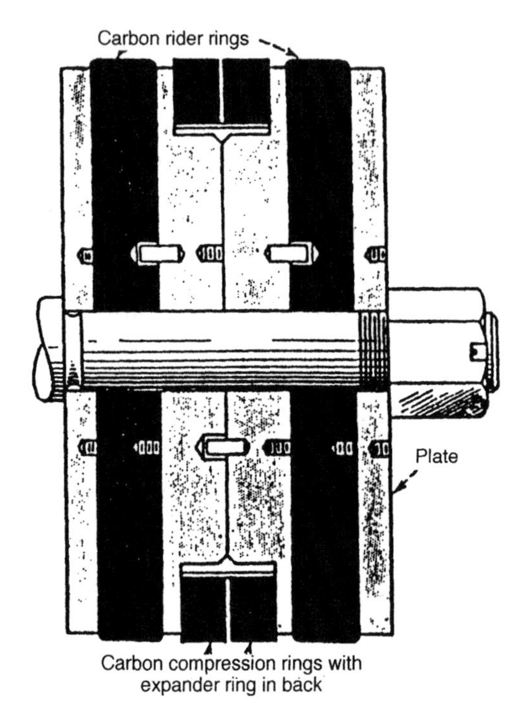

FIGURE 2-33. Early version of non-lubricated compressor piston.

The pressure packing was either a soft braided asbestos yarn, sometimes filled with animal fat lubricant, or rings made of graphite or segmented carbon.

Carbon has a great disadvantage; it is an extremely brittle material and requires extreme care when installing to prevent chipping and breakage.

The carbon dust generated as a result of wear is somewhat abrasive and accelerates ring wear. Ring slap, caused by the resulting excess side clearance of the rings in their grooves, tends to chip or fracture the rings. Plain carbon rings are thus banned from non-lubed reciprocating compressors. Prior to the advent of high performance polymers, the process industry had adopted a standard of ordinary Teflon construction. Piston and packing rings are often fabricated from a group of materials based on DuPont's polytetrafluroethylene (PTFE). Various fillers are used such as glass (fibre), carbon, bronze or graphite.

Compressors are often manufactured as either *oil-free* (totally non-lubricated) or *mini-lube,* where are a reduced amount of lubrication (usually 10% of the amount of lubrication used in a lubricated compressor) is used.

By far, the best non-lubricated compressor design is one embodying labyrinth pistons. These are primarily available from Sulzer-Burckhardt (Winterthur, Switzerland). They merit special consideration and are discussed at the end of this chapter.

DEFINITION OF NON-LUBRICATED CYLINDERS

Field references to non-lubricated compressors are often misleading, so a definition of cylinder terminology is in order.

Since the cylinder assemblies of reciprocating compressors must be designed relative to their lubrication, the nomenclature used to describe and classify the types of cylinder construction likewise refers to lubrication.

The classifications most commonly used are

1. **Lubricated Cylinder Construction**—The lubricated cylinder assembly is the conventional cylinder construction, which has a liquid lubricant introduced directly into the cylinder and piston rod packing in sufficient amounts to provide a lubrication film between the mated parts. Previous descriptions have been made of this general construction.

   The gas stream from the lubricated cylinder is contaminated with the lubricant, normally a hydro-carbon or a synthetic oil.
2. **Mini-Lube**—A partially lubricated cylinder construction with oil feed to the cylinders reduced to at least one-third of that for a lubricated cylinder. Teflon trim is used on the piston and for the pressure packing.

   The aims of Mini-Lube construction are to reduce the amount of oil carried within the exit gas stream and to reduce contamination of downstream systems. Reduction in the amount of oil used and reduction of load on downstream oil separation reduces costs.
3. **Micro-Lube**—No lubrication to the cylinder from conventional oil feed, but some oil enters the cylinder from migration along the piston rod.

   Teflon trim is used on the piston and for the pressure packing. The oil scraper rings are usually removed, which allows oil migration along the piston rod.

   The reasons for this construction are the same as for Mini-Lube, except that the system receives an even smaller amount of oil.
4. **Non-Lube or Oil-Free Cylinder Construction**—No lubrication reaches the cylinder. A longer distance piece is used to separate the

crosshead guide from the cylinder. This necessitates a longer piston rod on which a "collar" or oil deflector is installed. This collar prevents oil migration along the rod and into the cylinder.

A Teflon-containing material is used on the piston and in the pressure packing. Since the pistons of these non-lubricated compressors are fitted with rider bands that contact the cylinder walls, we call the compressors "conventional."

OPERATIONAL DIFFERENCES

A basic operational difference between the lubricated and non-lubricated cylinder should be explained. The piston works against pressure and should form a sliding seal so that it can compress the gas without leakage. So perhaps, in the lubricated cylinder, the simplest piston would be a plug piston with a very close fit to the cylinder bore. But, because of temperatures and other engineering and economic reasons, this is not practical. Piston rings are therefore used for sealing.

These piston rings have many variations, but all follow the basic principle of a thin metallic split ring fitted into a groove around the piston. The ring is made with "spring," or tension, which tends to push out against the cylinder wall and make a tight sliding fit.

It is important to note that piston rings float in the ring grooves of the piston and that they only seal. THEY DO NOT SUPPORT THE PISTON, nor is there any other device to support the piston. The piston is supported off the cylinder wall by the liquid lubrication film only.

In the non-lube or oil-free piston and piston ring assembly there is no oil film to support the piston, so the metallic piston must be kept off the cylinder bore by other means or serious damage will result. Note that this is the difference between lube and non-lube principle.

In the conventional non-lubricated compressor, the piston is kept off the cylinder wall by a guide ring which is referred to as a bull, wear, or rider ring. This rider is of a low friction material, such as carbon or Teflon, and of low unit loading relative to the piston weight.

The outside diameter of the piston is smaller than that of the piston in the lube compressor; this creates clearance between the piston outside diameter and the cylinder bore. This clearance allows for rider band wear before metal contact is achieved with the cylinder bore.

The rider ring is either a solid or a split configuration; its size is determined by piston assembly weight only and is independent of operating pressures.

Conventional Design for Non-Lube Service

It should be obvious that the non-lube cylinder is not a mere factory conversion of a lubricated cylinder. Simply changing out the piston rings and packing rings from metallic to carbon or teflon and shutting off the lubricating oil does not make a non-lube cylinder. Successful operation means that the non-lube cylinder must be engineered for non-lube service, using materials developed and thoroughly tested for this service.

Figure 2-34 shows a modern compressor cylinder designed for conventional non-lube service.

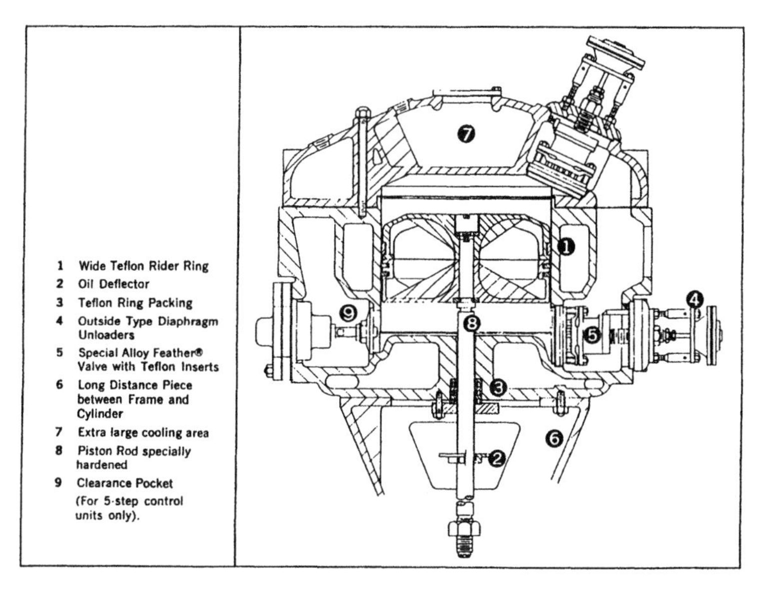

FIGURE 2-34. Conventional non-lubricated compressor cylinder (*Source: Dresser-Rand, Painted Post, New York*).

Let's examine the similarities and differences between oil-lubricated and non-lubricated compressors. Both have a conventional lubricated frame, of either vertical, horizontal, or "Y" configuration. Figure 2-35, a section of a vertical compressor, is used for reference.

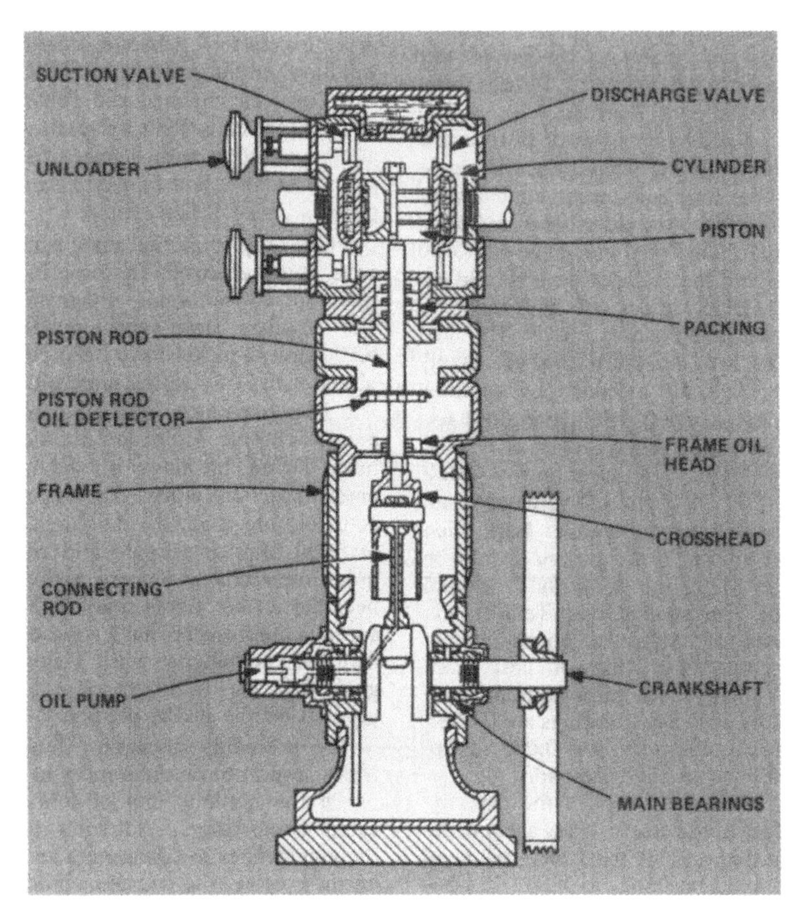

FIGURE 2-35. Conventional non-lubricated compressor, vertically, oriented (Worthington Type VBB) (*Source: Plant Engineering, October 18, 1979*).

The crankcase components are conventional; both compressors contain a forged crankshaft, anti-friction bearings, a forged connecting rod, and a cast crosshead.

Both the lube and non-lube compressors have the same lubrication system for these components—an oil pump, which flows oil through the crankshaft, passes through the connecting rod to lubricate the crosshead pin bushing and the crosshead. This oil is retained in the frame with a frame oil head and metallic scraper rings.

On the totally non-lubricated compressors a housing is used between the frame and the cylinder, along with a longer piston rod. This ensures that no part of the piston which is in contact with the frame-lubricating oil will pass into the cylinder.

Generally an oil deflector assembly is used on the piston rod to prevent any oil from "creeping" up the piston rod by capillary action and entering the cylinder.

The cylinder bore is honed to an 8–16 RMS finish to reduce wear of the piston rings and rider band. In addition to the normal honing to achieve this sort of finish, the non-lube cylinder is given an additional treatment of "teflon honing." In this treatment, blocks of virgin teflon are used in place of the abrasive stones on the hone sets, which impregnates the pores of the iron bore of the cylinder.

The piston rod surface finishes through the packing travel are very important. On a lubricated compressor, the piston rod is finished to 16 RMS, while on the non-lube cylinder the piston rod is micro finished to 8–10 RMS.

In addition, the piston rod will have a surface hardness in the packing travel area of 50–55 Rc.

Piston rod alignment and eccentricity must very carefully be checked in order to minimize packing wear. Excessive lateral movement or runout exceeding .003″ will prevent packing from sealing.

PISTON AND RIDER RINGS

The pistons shown in Figures 2-34 and 2-35 are of the two-piece design which is the preferred construction; however, some compressors will use pistons of the one-piece design, and some piston used in larger bores use three separate pieces, as illustrated in Figure 2-36.

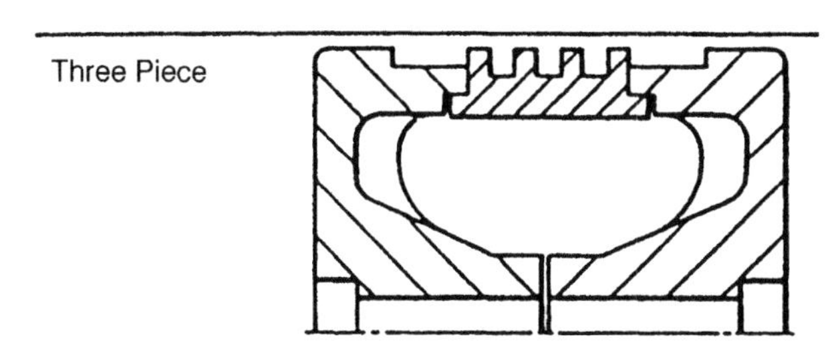

FIGURE 2-36. Three-piece piston design (*Source: Dresser-Rand, Painted Post, New York*).

A piston used in a non-lube cylinder is usually made .125″ to .250″ smaller in diameter than the cylinder bore, depending on the size.

The piston shown in Figure 2-35 has a center rider band to support the piston, while the one shown in Figure 2-34 has the rider band on the upper half of the piston. Location and number of rider bands are dictated by the design, diameter, length, and weight of the piston.

Rider bands and piston rings are made from PTFE (Teflon) with various fillers such as glass, carbon, bronze, or high performance polymers. Carbon filled Teflon is customary in low-pressure air service. It has good heat conductivity, is compatible with many cylinder materials, and is normally used with hydrocarbon gases as well as dry gases such as nitrogen and helium. Bronze may be used for pressures above 2000 psi, and graphite is used for very high temperature applications.

Rider bands are designed to support the weight of the piston and piston rod. Minimum rider band widths are given in Table 2-1.

TABLE 2-1
MINIMUM WIDTH OF TEFLON RIDER BANDS

Piston diameter, in.	Teflon band width, in.
To 7	1
7 to 11	1½
12 to 15	1¾
16 to 21	2

Piston ring widths and gaps are based on the coefficient of expansion of the materials and a 250°F temperature differential is assumed.

Example: A cast iron piston ring width for a 10″-diameter cylinder is ⅜″; a ½″ ring width is recommended for glass filled Teflon. Side clearances in the piston ring groove and the end gap are adjusted for the greater expansion of the ring material.

With the rider band installed, the diametrical clearance between the rider band and the cylinder bore should be .00125″ per inch of diameter of a piston made of cast iron, and .002″ per inch of diameter for an aluminum piston.

Rider band life can be extended by rotating a piston 120° to 180° at each overhaul.

Rider bands are available in several styles. Figure 2-37 shows various types furnished by France Compressor Products. The use of pressure relief grooves to prevent pressurizing is recommended on all split type rings.

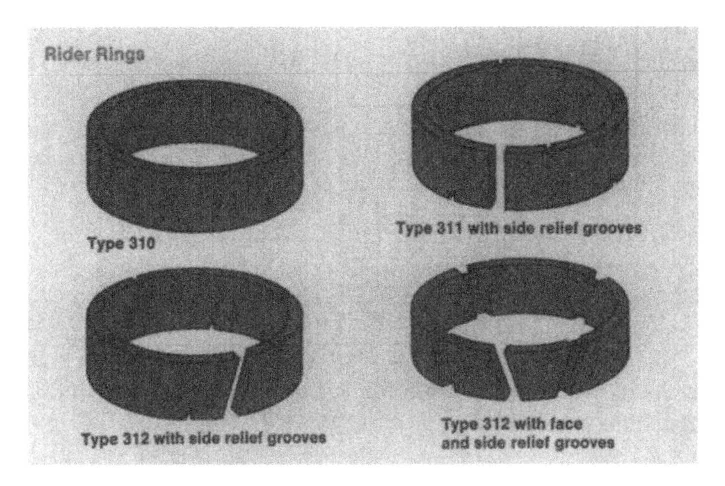

FIGURE 2-37. Rider bands for reciprocating compressor pistons (*Source: France Compressor Products, Newtown, Pennsylvania*).

INSTALLATION OF RIDER BANDS

Solid PTFE rider bands are machined with an interference fit so that, once expanded over the piston end, the rings will contract like a rubber band to provide a snug fit on the piston. Unlike piston rings that must move freely in their grooves to seal, rider rings are literally stretched in order to slide them over the end of the piston and into the ring groove.

This can be accomplished in either one of two ways, depending upon whether the rings are pre-stretched or not. Pre-stretched rider rings are supplied on a groove locked expander to fit over the piston end. Once in the ring groove, the PTFE natural elastic memory causes the ring to contract to its original dimensions. This contraction can be accelerated by the careful application of heat.

Rider rings that are not pre-stretched require a force fit over the piston end. This is accomplished by the use of a press and a special fixture as illustrated in Figure 2-38.

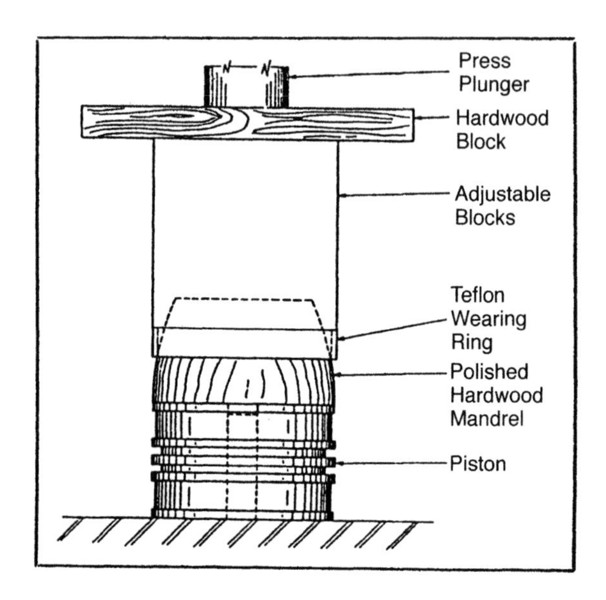

FIGURE 2-38. Assembly fixture for Teflon® wear rings (*Source: Dresser-Rand, Painted Post, New York*).

PISTON RINGS

Piston rings made of PTFE will have an end gap clearance of .020″ to .024″ per inch of piston diameter when fitted into the cylinder bore. Side clearance in the groove should be 0.010″–0.020″ per inch of width. It should be remembered that Teflon expansion rates are approximately seven times those of cast iron. Values for oil-lubricated compressors with cast iron rings are .0035″ per inch of diameter for the ring gap.

Figure 2-39 shows a piston for a non-lubricated compressor with a center rider band that has pressure relief grooves and two piston rings installed.

Figure 2-40 shows a three-piece piston used in higher pressure, non-lubricated process compressors handling synthesis gas. It consists of cast iron ends with a cast iron piston ring carrier. The rider bands are sandwiched between the end pieces and the ring carrier.

AIR FILTRATION

Another area requiring special attention for successful trouble-free operation of non-lubricated air compressors is proper filtration of the incoming air.

FIGURE 2-39. Truncated piston for conventional non-lubricated compressor pistons (*Source: Joy Manufacturing Company, Division of Gardner-Denver, Quincy, Illinois*).

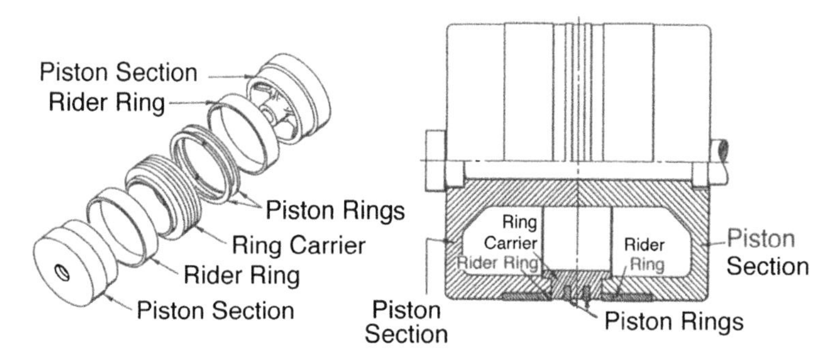

FIGURE 2-40. Three-piece piston used in non-lubricated compressor handling synthesis gas (*Source: Dresser-Rand, Painted Post, New York*).

The air filter must be of the dry type, preferably cloth or paper, and usually larger than that used for the lubricated compressor in order to provide adequate filtering area.

Air that may appear clean often contains large quantities of suspended solids such as dust, soot, and fine particles of cinder, flyash, and sand. Concentrations range from .50 grains per 1,000 cu. ft. in rural areas to 5 grains per 1,000 cu. ft. in industrial areas. Because 7,000 grains equals 1 pound, a 1,000 cfm compressor, operating 24 hours per day in an industrial atmosphere will, without filtration, ingest 1.03 pounds of dust per day. These solid contaminants accelerate abrasion, wear, and erosion.

Filter cartridge efficiency should be 99.7% for 10 micron particles, and 95% for 5 micron particles.

Suction piping from the filter to the cylinder should be clean and free of dust and scale. Steel piping should be treated with a rust inhibiting or epoxy paint. In critical applications, stainless steel or aluminum piping should be used. These materials are always preferred.

Abrasive materials can become embedded in the Teflon and will be retained, shortening ring life and accelerating cylinder wear. Thus the cost of maintenance will increase if air filtration is inadequate.

Valves and Unloaders

Valves, unloaders, and clearance pockets for non-lube units have teflon nubs or bushings for self-lubricating operation. Channel valves have teflon strips over the channels and Teflon guides at the ends. Plate valves usually have Teflon nubs between the springs and the plate. In some cases, all components are Teflon coated.

Unloaders have Teflon bushings in the sliding parts (Figure 2-41).

Piston Rod Column or Frame Loading

Each compressor is subject to a piston rod column or frame load limitation. The size of each compressor establishes a maximum allowable frame or piston rod column load. Keeping within this allowable load ensures that the frame castings and running gear parts, such as crankpins, crosshead pins, main bearings, crossheads, and connecting rods are not subject to loadings or stresses beyond their design points.

Piston rod column or frame load is the force that the pressure in a cylinder exerts on the piston and, in turn, the piston exerts on the piston rod. This load is transmitted through the piston rod back to the frame and running gear parts.

Frame load is the difference of the total loads across the piston or across one cylinder. It is the net area of the piston times the difference between the discharge and suction pressures for that cylinder.

Excessive Rod Load

Most major damage to compressors is caused by exceeding recommended rod loads. For this reason, it is important that operators and mechanics understand rod loads before they start a compressor.

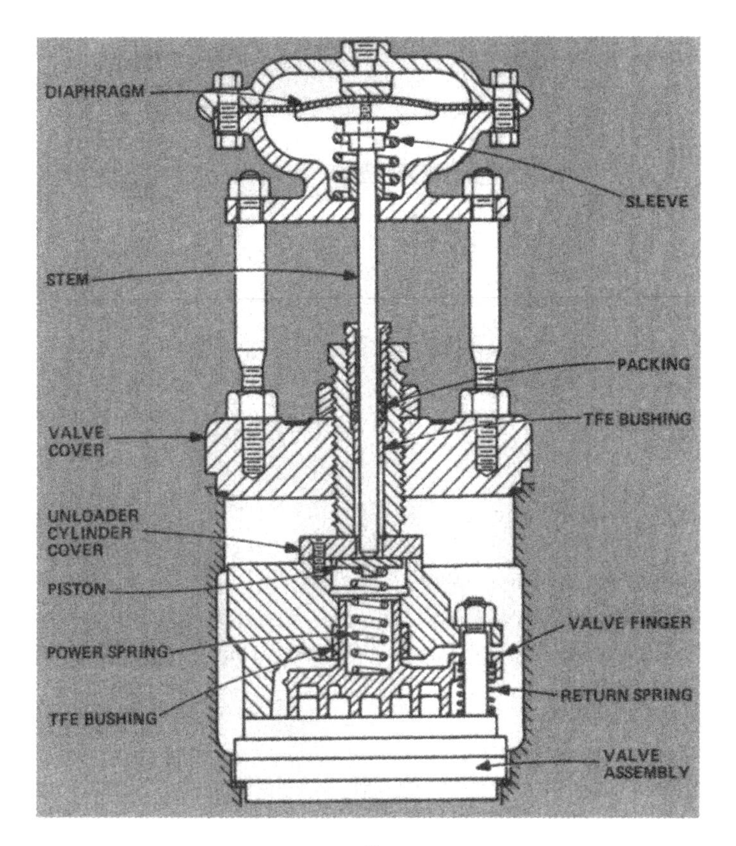

FIGURE 2-41. Unloader with Teflon® bushings are used with non-lubricated compressors (*Source: Plant Engineering, October 18, 1979*).

ROD UNDER COMPRESSION

Consider first a double acting cylinder, as shown in Figure 2-42. As the piston pumps toward the head end, the discharge pressure (P_2) on the piston tends to compress and buckle the piston rod. At the same time, gas is entering the cylinder behind the piston, exerting suction pressure force (P_1) on the back side of the piston. Two forces are opposite in direction, but since the discharge pressure is greater, the net result is a force compressing the rod. This is called *rod load compression.* It is obvious that as suction pressure is decreased, or discharge pressure is increased, the net compression on the rod increases. Therefore, if the operator lets the suction or discharge pressure deviate too far from the design conditions, the maximum permissable compressive load may be exceeded.

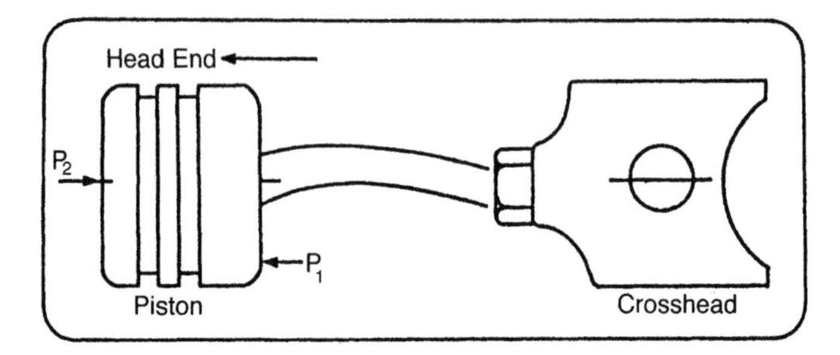

FIGURE 2-42. Piston rod deflection tendency during outward compression stroke.

ROD UNDER TENSION

As the piston discharges toward the crank end on the return stroke, as seen in Figure 2-43, the net force of the suction and discharge pressures results in a tension load on the rod. This is called *rod load tension.*

The operator can damage the compressor by decreasing the suction or increasing the discharge pressure too far above the design pressures.

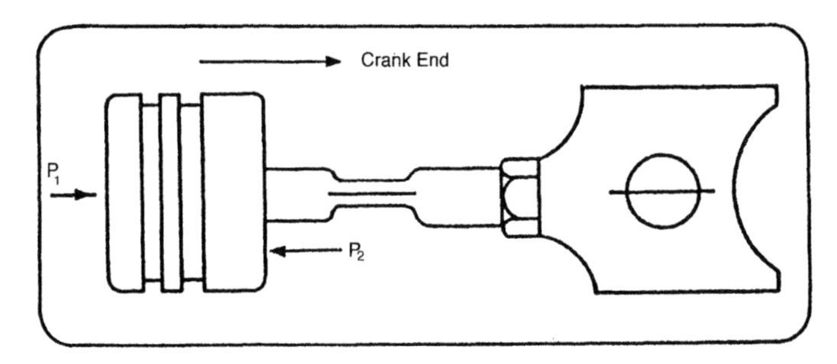

FIGURE 2-43. Piston rod stretch during compression in crank end portion of cylinder.

ROD LOAD ASSIGNMENT

The rod can easily absorb most excessive tension and compressive forces, but other parts, such as pistons, connecting rods and bolts, crossheads and shoes, bushings and bearings, are also stressed. The most highly stressed part, therefore, determines the rod load assigned by the compressor builder. This value is different for each model of compressor, but generally crosshead pin bushing pressures are limiting.

Rod loads can be calculated by simple arithmetic, but in operation, suction and discharge pressures can change so fast that the operator does not have time to calculate.

Rod Load Calculations

Force = Pressures × Area

Two methods of calculation are used, an average method and an exact method, which requires calculation of compression and tension loads.

The following formulas are used:

Average Method

$$F = P \times A = (P_d - P_s) \times \text{net cylinder area}$$

Where

Net cylinder area = cylinder area – ½ rod area

Exact or Compression and Tension Method

Piston moving toward head end or rod in compression

$$\text{(a) } Fc = (P_d \times A_{he}) - (P_s \times A_{ce})$$

Piston moving toward crank end or rod in tension

$$\text{(b) } Ft = (P_s \times A_{he}) - (P_d \times A_{ce})$$

Where

P_s = suction pressure, psia

P_d = discharge pressure, psia

A_{he} = cylinder head end area, sq. in.

A_{ce} = cylinder crank end area, sq. in. = A_{he} – rod area

The average method can be used for quick checks, but if above or very close to maximum allowable, or if the cylinder is relatively small, 10 inches diameter and below, then the exact method must be used.

Example:
Cylinder diameter = 12″
Piston rod diameter = 2″
Suction pressure = 0 psig
Discharge pressure = 100 psig
Crosshead pin = 4″ diameter × 5″ long

(A) Average Method
Cylinder area = 113.098 in.^{2}
–½ Rod area = 1.573 in.^{2}
Net cylinder area = 111.525 in.^{2}

$$F = (114.7 - 14.7) \times 111.53 = \underline{11{,}153} \text{ lbs}$$

(B) Exact Method
Cylinder area = 113.098 in.^{2}
– Rod area = 3.146 in.^{2}
Crank area = 109.952 in.^{2}

$$Fc = (114.7 \times 113.098) - (14.7 \times 109.95)$$
$$Fc = 12972.34 - 1616.29 = \underline{11{,}355} \text{ lbs}$$

$$Ft = (14.7 \times 113.098) - (114.7 \times 109.95)$$
$$Ft = 1662.54 - 12611.27 = -\ \underline{10{,}949} \text{ lbs}$$

It is important that the compression calculation be positive and the tension calculation negative. If both values come out plus or minus, this indicates that the rod is not changing from plus compression to minus tension, which means that there will be no reversal at the crosshead pin and no lubrication to this part.

Frame load considerations or piston rod column loading undergo careful analysis of the gas and inertia loads at normal operating conditions and relief valve settings for both full and part load operation. This analysis includes evaluating loading characteristic throughout the frame and running gear, perhaps graphically (see Figure 2-44). The crosshead, as illustrated in Figures 2-45 and 2-46, is an area of concern.

An examination of the combined gas and inertia forces at the crosshead pin will disclose the magnitude of the loading through 360° of rotation so that proper load reversal occurs to ensure continuous lubrica-

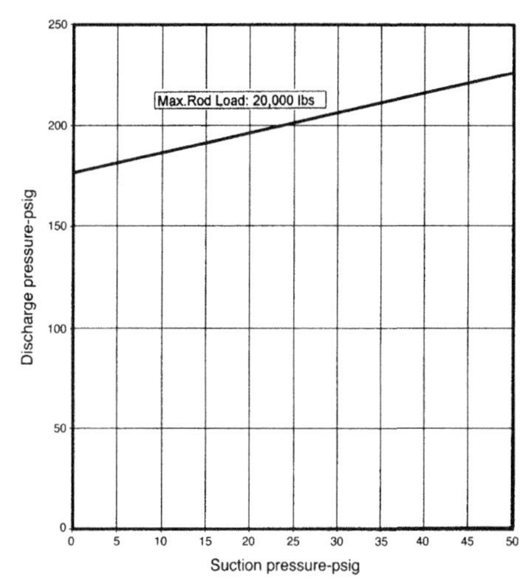

FIGURE 2-44. Example of graphic representation of rod load.

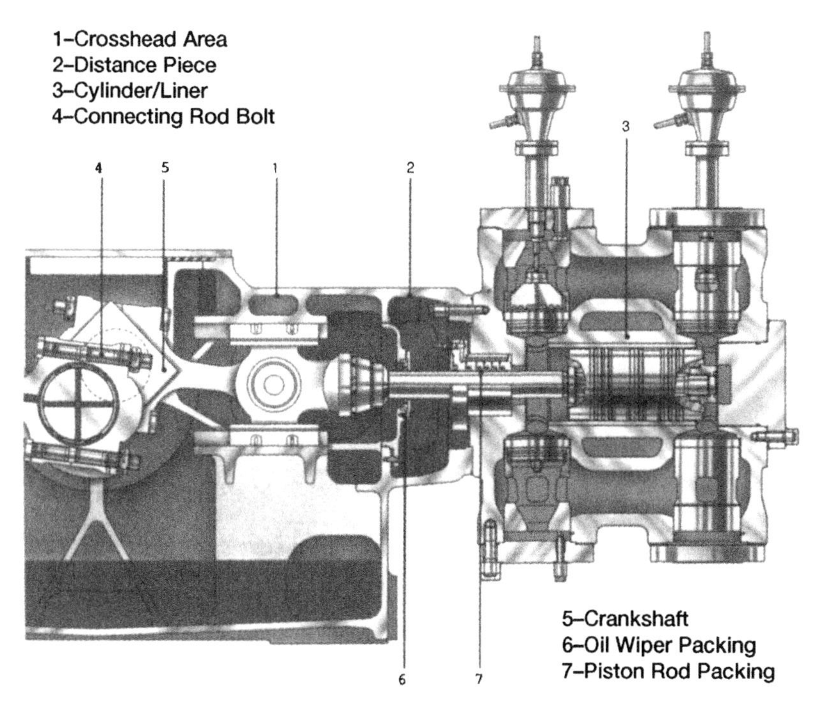

FIGURE 2-45. Crosshead area of horizontal-opposed reciprocating compressor (*Source: Babcock-Borsig, Berlin, Germany*).

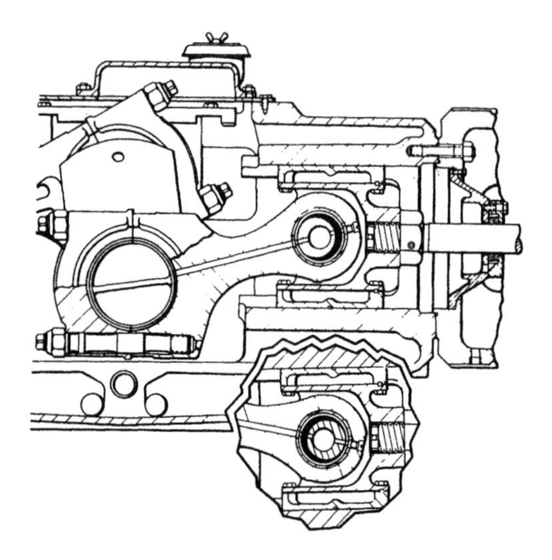

FIGURE 2-46. Typical crosshead arrangement illustrating load reversal at pin. (*Source: Dresser-Rand, Painted Post, New York*).

tion. Figure 2-46 illustrates such a reversing load at the crosshead pin with the combined gas and inertia load causing a load reversal to allow oil to fill the space, thereby achieving both lubrication and cooling.

The degree of reversal required is illustrated in Figure 2-47, which shows the magnitude and duration of the reversing load through one revolution.

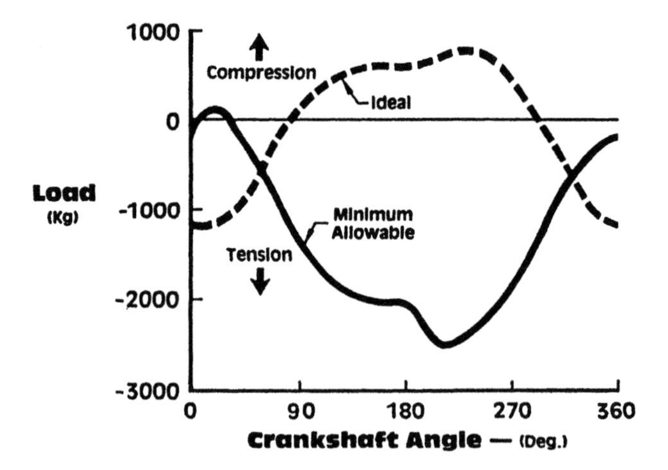

FIGURE 2-47. Typical crosshead pin loading diagram showing one full revolution.

Tail Rod Piston Design

Tail rod construction is required in those cases where reversal cannot be obtained. This design has identical pressure areas on both ends of the piston, resulting in load reversals at the crosshead pin. An additional packing assembly is, of course, required.

This particular application is highlighted in Figure 2-48 where a suction pressure of 2800 psi and discharge pressure of 3100 psi is used with a 6″ bore cylinder (28 sq. in. area) and a 3″ piston rod (7 sq. in. area).

PISTON POSITION	PRESSURE DIRECTION	NET PRESSURE AREA	PRESSURE	A x P
	←	28 - 7	3100	65,100
	→	28 - 7	2800	58,800
		← LOAD DIRECTION		6300#
	←	28 - 7	3100	65,100
	→	28 - 7	2800	58,800
		→ LOAD DIRECTION		6,300#

FIGURE 2-48. Load on crosshead pin with tail rod design for specific example conditions.

It can be seen that the frame and bearings have to withstand only 6,300 pounds; moreover, there is load reversal at the crosshead pin.

Using the same data, but omitting the tail rod, the resulting frame loading is shown in Figure 2-49.

PISTON POSITION	PRESSURE DIRECTION	NET PRESSURE AREA	PRESSURE	A x P
	←	28	3100	86,800
	→	28 - 7	2800	58,800
		← LOAD DIRECTION		28,800#
	←	28	2800	78,400
	→	28 - 7	3100	65,100
		← LOAD DIRECTION		13,300#

FIGURE 2-49. Load on crosshead pin without tail rod for the specific example conditions listed in Figure 2-48.

If the tail rod were to be omitted, the frame load would be 4½ times greater. Worse yet, the load does not reverse in direction and is, in fact, not much reduced, making it mandatory to provide a greatly increased size of crosshead pin bushing.

DISTURBING OR SHAKING FORCES

When a reciprocating compressor runs, moving parts such as pistons, piston rods, crossheads, and connecting rods are repeatedly accelerated and retarded. These velocity changes set up pulsating inertia forces. The forces and couples are of the first and second order.

The first order forces have the same frequency as the compressor shaft speed and the second order forces have a frequency equal to twice the shaft speed. It is possible, but in most cases not economical, to design the compressor in such a way that these inertia forces cancel each other. By careful design, with equal piston masses and by fitting counterweights, it is possible to reduce these inertia forces to low values.

THE EFFECT OF UNBALANCE

Compressor inertia forces may have two effects. One is a force in the direction of the piston movement, and the other is a couple or movement that is developed when there is an offset between the axis of two or more pistons on a common crankshaft. The interrelation and magnitude of these forces will depend upon such factors as number of cranks, their longitudinal and angular arrangements, cylinder arrangement, and extent of counterbalancing possibility.

Two significant vibration (movement) periods are set up:

- Primary—at rotative speed
- Secondary—at twice rotative speed

There are others that normally can be neglected.

Aside from the direct inertia forces, there are also forces set up by the torque variations on the compressor shaft.

Although the forces developed are sinusoidal, only their maximum values are considered in analysis. Figure 2-50 shows relative values of the inertia (shaking) forces for various compressor arrangements. The

Crank Arrangements	Forces		Couples	
	Primary	Secondary	Primary	Secondary
Single crank	F′ without counter wts. 0.5 F′ without counter wts.	F″	None	None
Two cranks at 180° In line cylinders	Zero	2 F″	F′D without counter wts. $\frac{F'D}{2}$ with counter wts.	None
Opposed cylinders	Zero	Zero	NIL	NIL
Two cranks at 90°	1.14 F′ without counter wts. 0.707 F′ without counter wts.	Zero	0.707 F′ without counter wts. 0.354 F′D with counter wts.	F′D
Two cylinders on one crank Cylinders at 90°	F′ without counter wts. Zero with counter wts.	1.14 F″	NIL	NIL
Two cylinders on one crank Opposed cylinders	2 F′ without counter wts. F′ with counter wts.	Zero	None	NIL
Three cranks at 120°	Zero	Zero	3.48 F″D without counter wts. 1.73 F″D with counter wts.	3.48 F″D
Four cylinders Cranks at 180°	Zero	4 F″	Zero	Zero
Cranks at 90°	Zero	Zero	1.41 F″D without counter wts. 0.707 F″D with counter wts.	4.0 F″D
	Zero	Zero	Zero	Zero

F′= primary inertia force in lbs.
$F' = .0000284\ RN^2W$
F″= secondary inertia force in lbs.
$F'' = \frac{R}{L} F'$
R= crank radius, inches
N= R.P.M.
W= reciprocating weight of one cylinder, lbs.
L= length of connecting rod, inches
D= cylinder center distance

FIGURE 2-50. Unbalanced inertia forces and couples for various reciprocating compressor arrangements (*Source: Compressed Air and Gas Data, Ingersoll-Rand Co., Phillipsburg, New Jersey*).

diagrams are plan views with the exception of the fourth arrangement with cylinders at 90°, an elevation view.

DISTURBING FORCES

Figure 2-51 shows schematically vertical and horizontal single cylinder reciprocating compressors. During one complete revolution, the piston executes an alternating motion with alternating accelerations. Gas is aspirated into the cylinder during one-half of a revolution. This gas is then compressed and discharged during the other half revolution.

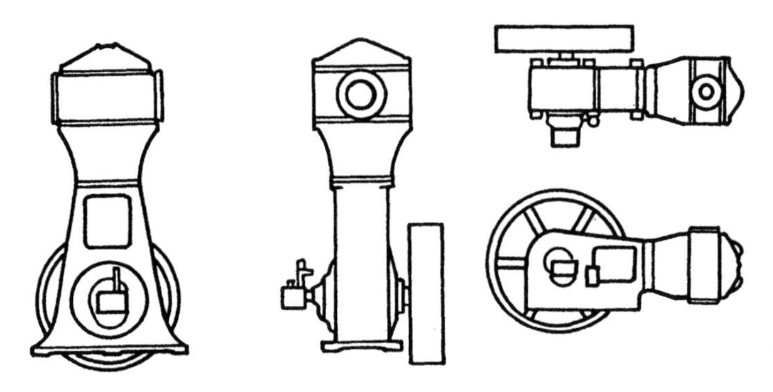

FIGURE 2-51. Schematic arrangement of vertical and horizontal single cylinder reciprocating compressors.

Gas Forces

We can see, therefore, that two types of forces are produced in an operating reciprocating machine. These instantaneous gas forces simultaneously acting on piston and head are always exactly equal and opposite in direction. It is important, therefore, to realize that *the internal forces caused by the working fluid do not produce external forces but are counteracted by stresses in the material of the cylinder and frame.*

Inertial Forces Due to the Rotating Masses

The rotating masses are the sum of weight illustrated by the unshaded crankshaft portion of Figure 2-52 and two-thirds of the weight of the connecting rod.

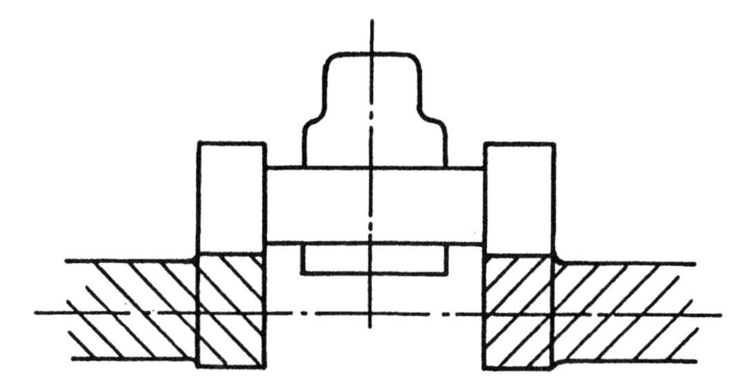

FIGURE 2-52. Rotating mass visualization on crankshaft web.

The centrifugal force created by these masses is an exciting force, having the same intensity in all directions. It is the product of mass, crank radius, and square of angular speed.

By selecting proper counterweights and placing them on the opposite side of the crank, the rotating forces can be balanced.

Therefore, *the inertial forces of rotating masses produce a centrifugal force of constant magnitude that can be completely balanced by using properly sized counterweights.*

Inertia Forces Due to the Reciprocating Masses

The inertia forces of the reciprocating masses are by far the most important. These forces are the result of the masses or one-third the weight of the connecting rod, plus the weight of the crosshead, piston rod, and piston. The weights of fasteners forming a part of the various assemblies must also be considered.

These forces, which are the result of acceleration and deceleration of the reciprocating weights, exert a variable force on the crankpin, acting along the axis of the cylinder.

As may be seen, reciprocating forces are variable (depending on crank position) and act at fundamental and even multiples of rotating speed.

The forces resulting from the rotating and reciprocating masses can be resolved into force systems consisting of two parts: primary forces and secondary forces. These are expressed in both horizontal and vertical directions, and, in the instance of multi-crank compressors, as moments or force couples.

PRIMARY FORCES

Primary forces are composed of rotating and reciprocating masses and occur at crankshaft speed.

In effect, the primary force can be likened to a piece of string weighted at one end, with the opposite end tied to a ring that rotates about a fixed steel rod (see Figure 2-53).

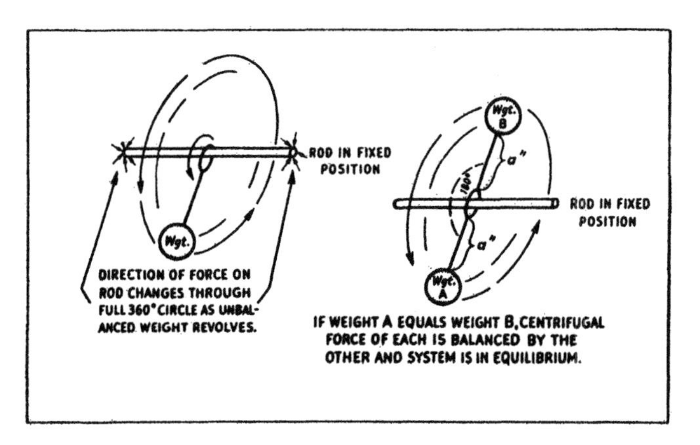

FIGURE 2-53. Unbalanced and balanced forces acting on a shaft system.

As long as the weight (mass) revolves about the rod, the string will be in tension, and the effect of centrifugal force exerted by the rotating weight will be to pull the rod in the direction of the weight. Since the weight is revolving, the direction of the force will be constantly changing through a full 360° circle. If the weight (or mass) rotates at a constant speed, the force exerted on the rod will be constant also.

If the steel rod were, instead, a crankshaft without a counterweight, the unbalanced force of the sum of the reciprocating and rotating masses would cause the crankshaft to slam violently about within its bearings. The vibrations would be transmitted to the frame and into the foundation. An excessively large foundation would be required to restrain these unbalanced forces.

But, suppose a second string were tied to the ring on the rod and an equal weight put on its end, this weight being the same distance from the center line of the shaft as the first weight. If both weights are caused to rotate in the same direction and at the same speed, with an interval of

180° between the weights, it will be seen that the ring does not pull the rod as the weights revolve about the rod. The centrifugal force of weight No. 1 is counteracted by the centrifugal force of weight No. 2, creating a situation of forces in equilibrium.

In actual practice, weight No. 2 would be a counterweight. The mass of this counterweight acts through its center of gravity at a distance sufficient to offer a force equivalent to the effective weight of the reciprocating and rotating masses acting through a point in the crankpin.

In a single cylinder compressor, the addition of counter-weighting allows a part of the primary unbalanced forces acting in the direction of the cylinder center line to be transferred into the perpendicular direction. It is important to remember that the counterweights in a single compressor do not "balance" the inertia force. They only "transfer" the force in a perpendicular direction.

In a two-cylinder "Y" compressor, if the reciprocating weights of both cylinders are equal, then the sum of the primary forces of both cylinders is a constant magnitude force rotating with compressor speed. Consequently, this force can be balanced by the counterweight. Practically, however, the same counterweights are used for several cylinder combinations and, therefore, a small unbalanced primary force is sometimes present.

We know from experience that this resultant force does not produce any difficulties in actual installations.

Secondary Forces

The secondary forces are the result of reciprocating masses only, and because of acceleration and deceleration of these masses, occur at twice crankshaft speed. They act only in the plane of the cylinder axis, and can only be balanced by counterweights rotating at twice crankshaft speed, and in opposite directions. Such an arrangement requires a separate counterweighting system, which is gear or chain-driven. Because of costs, space requirements, and design difficulties, balancing of single crank compressors has not yet been successfully employed. It may therefore be concluded that there is no practical way for reduction or elimination of secondary forces, except by suitable arrangement of cranks of multi-crank compressors.

The balanced-opposed type compressor with its multi-crank design is well suited for secondary force considerations. Assuming equal reciprocating weights in an opposed, two-cylinder compressor arrangement, all the inertia forces mutually cancel each other. Only couples are transmitted to the foundation.

In the single cylinder and "Y" type compressor, the secondary forces can neither be balanced nor transferred in a different direction. This force must be compensated for by proper design of the foundation.

COUPLES

Couples are a result of the forces acting on individual cranks in a multi-crank compressor. These forces are not in the same plane, but act on each crank in a plane at right angles to the shaft axis. The forces produce a moment (force couple) that will try to rotate the shaft about its center in a plane passing through the shaft axis.

This is the primary couple. The secondary couple, occurring at twice crankshaft speed, cancels out.

The design of a balanced type compressor permits a complete balancing of the primary as well as the secondary forces without attachment of counterweights. Because the cranks are adjacent to each other (see Figure 2-54), the arrangement permits a minimum moment arm, resulting in very small primary and secondary couples.

FOUNDATIONS FOR RECIPROCATING COMPRESSOR

The reciprocating compressor, because of the alternating movement of its pistons and various others parts, develops a shaking or inertia force that

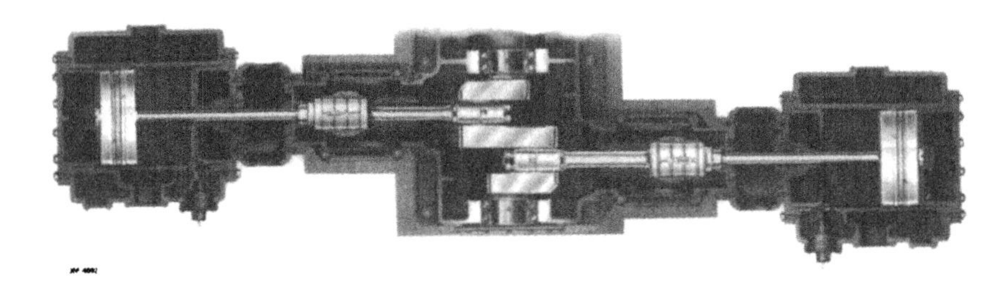

FIGURE 2-54. Balanced-opposed cylinder arrangement with closely spaced cranks minimizes moment unbalance (Source: Dresser-Rand, Painted Post, New York).

alternates in direction. The net result of this force must be damped and contained by the mounting. This is in addition to the dead weight load.

There are many compressor arrangements, and the net magnitude of the moments and forces developed varies a great deal between them. In some cases, they can be partially or completely balanced out. In others, the foundation must handle it all.

When balance is possible, this machine also can be mounted on a platform just large and rigid enough to carry the weight and maintain alignment. Some units have been mounted on a structural steel platform. Small units of the "Y" or "W" arrangement can be bolted to a substantial building floor. However, most reciprocating compressors require considerable attention to foundation design.

Foundation Movement

A reciprocating piston will tend to move the compressor and its foundation because inertia forces are set up, just as if men standing in a small boat move suddenly in one direction and then the other. Relative to the lake bottom, the boat also changes position. The heavier the load or the broader its bottom and, therefore, the greater the resistance to movement (the heavier the foundation and the broader its base), the less the movement, but movement still occurs.

Any foundation and the soil upon which it rests has a natural frequency of vibration. The soil is elastic (as a spring). The degree of elasticity depends upon its character. The foundation design, mass, and the soil characteristics must be such that this natural frequency of vibration is quite far removed from the primary and secondary vibration frequencies imposed by the compressor. Unless this is so, resonance will occur, vibration amplitude will multiply, and the foundation will have to be altered. A good foundation will have a vibratory frequency well above the operating speed.

Basic Foundation Design

A proper foundation must:

- Maintain the compressor and driver in alignment, level, and at the proper elevation.
- Minimize vibration and prevent its transmission to adjacent building structures.

There are five steps to accomplish the first objective:

1. The safe dynamic soil-bearing capacity must not be exceeded at any point on the foundation base.
2. The unit loading of the soil must be distributed over the entire area.
3. Foundation block proportions must be such that the resultant vertical load due to the mass of the machine and any unbalanced inertia force falls within the base area.
4. The foundation must have sufficient mass and bearing area to prevent its sliding on the soil because of any unbalanced forces.
5. Temperature variation in the foundation itself must be uniform to prevent warping.

In addition to the ability of the soil to carry the load the soil must be elastic. Piling sometimes may be necessary. *Since it is expensive to repair or replace a foundation once in place, it would make little sense to economize here.* Figure 2-55 illustrates typical examples of improper foundations and their correction.

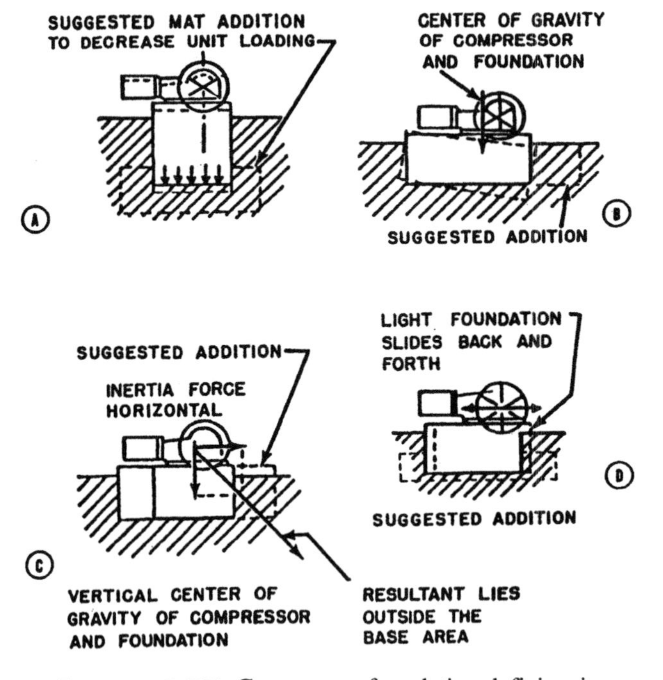

FIGURE 2-55. Compressor foundation deficiencies.

The following comments apply to the figure:

A. Too heavy loading: foundation sinks.
   Solution—extend base or footing surfaces.
B. Foundation not centered under center of gravity. Foundation tilts due to unequal soil loading.
   Solution—add more area at heavy end.
C. Foundation rocks. Resultant vertical and inertia forces fall outside foundation.
   Solution—spread foundation bearing as shown.
D. Foundation too light. Slides on soil.
   Solution—make foundation deeper and longer.

COMPRESSOR PIPING AND PULSATIONS

Pulsation is an inherent evil in reciprocating compressors. It interacts with piping to cause vibrations and performance problems.

Indiscriminately connecting piping to a compressor can be dangerous and cost money in the form of broken equipment and piping, poor performance, inaccurate metering, unwanted vibration, and sometimes noise. Piping connected to a compressor can materially affect the performance and response.

PULSATION FORM

Pulsation is caused by the periodic action of compressors propelling gas through a pipe. The piston-crank-valve mechanism generates a variable pressure which, over time, creates a pressure wave in the suction and discharge piping that resembles a composite wave as shown in Figure 2-56.

This composite wave is made up of a number of waves that are multiples of the fundamental sine wave. The multiples make up the composite wave form. Therefore, multiple frequencies of pulsation exist that can serve as exciting forces of vibration.

The dominant frequency in this wave form occurs once per revolution. In the case of a compressor with a single-acting cylinder or a double-acting cylinder with one end unloaded and operating at 300 rpm (5 rps), the dominant frequency is 5 cps.

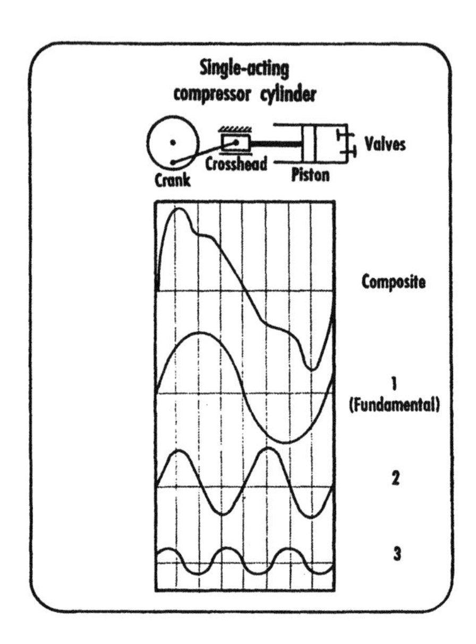

FIGURE 2-56. Pressure wave generated by single-acting cylinder.

Figure 2-57 shows a double-acting compressor cylinder with the composite wave form and the multiples that make it up. The dominant frequency in this wave form occurs twice per revolution. Therefore, the dominant frequency from a double-acting cylinder at 300 rpm would be 10 cps.

ACOUSTIC VELOCITY

The pressure waves generated by reciprocating action move through the gas with a speed referred to as the acoustic velocity. This velocity depends upon the molecular weight of the gas, temperature, and gas constants.

Depending on gas composition and temperature, it varies from a low of about 700 fps for propane, up to a maximum of about 1,500 fps for methane. This range covers most gases encountered. Hydrogen-rich gases can have acoustic velocities well over 2000 fps.

Lengths of piping elements, such as reduced diameter and expanded diameter sections, will have a pressure build-up or tend to resonate when excited by a characteristic frequency called the resonant frequency. The length corresponding to this resonant frequency is called the acoustic length.

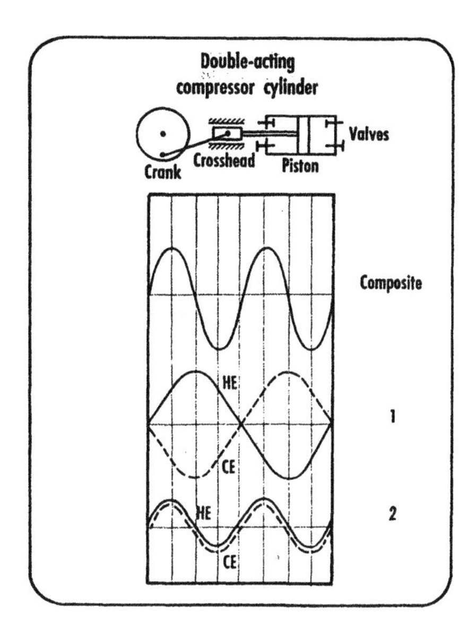

FIGURE 2-57. Pressure wave generated by double-acting cylinder.

When the physical length of a pipe or piping element is the same as the acoustic length, a condition of resonance or pressure building-up exists for a given exciting frequency.

A pipe that is in resonance has a pulsation or pressure buildup that can be several times the pulsation generated by the compressor. This pressure buildup can be substantial and can force the piping into vibration.

If the mechanical natural frequency of a pipe length should correspond to the exciting frequency of this pressure buildup, the vibration could be violent. The mechanical natural frequency is a function of pipe length, cross sectional dimensions, and material.

To control the vibration then, it is necessary to detune the system by reducing the pulsation, changing the acoustic response of the piping system and/or reducing or increasing the mechanical natural frequency of the pipe by a change in stiffeners or weight.

PIPE VIBRATION

Compressor piping vibration is generally due to pulsation. Vibration originating from unbalance can cause vibration of piping in close proximity to the compressor, but this is not usually the case. Vibration radiated through the ground from compressor unbalance can appear as a vibration of the piping, but this is unusual.

Pulsation can build up due to resonance and force piping to vibrate. Coincidence of the pulsation frequency can result in violent vibration. Therefore, it is logical and proper to detune the system by reducing the exciting pulsation or stiffening the piping to reduce the vibration or a combination of both.

Piping supports must be spaced in a way that the natural frequency of the piping span or configuration does not coincide with any exciting pulsation frequency being generated by the compressor.

Certain piping configurations should be used with precautions, or avoided if at all possible. Expansion loops, "Z"- or "L"-shaped piping overhead, cross-overs, and long sweeping bends are all susceptible to vibration from small amounts of pulsation.

Effect of Pulsation on Performance

It is well documented that the performance of a compressor can be materially affected by the piping. More or less gas can be compressed, at the expense of horsepower, because of the effects of pulsation and the interaction with the piping.

Some of the effects pulsation can have on the indicator card are shown in Figure 2-58. Good pulsation control minimizes these effects. Nozzle diameter and length between the cylinder and bottle have a great effect on the indicator card. A good rule to follow is to make the nozzle as big in diameter as possible and as short as possible.

Design Overview of Labyrinth Piston Compressors

Earlier in this chapter, we alerted the reader to a special non-lubricated reciprocating compressor configuration, the labyrinth piston, or "Laby" machine. Although available since the mid-1940s, labyrinth compressors are not as well known in the United States as they are in other parts of the world.

In labyrinth piston compressors, an extremely large number of throttling points provide the sealing effect around pistons and piston rods. No contact seals are used.

Whereas plastic sealing rings depend on permanent mechanical friction for efficient performance, the labyrinth principle embodies an extremely small clearance between sealing element and counterpart. This

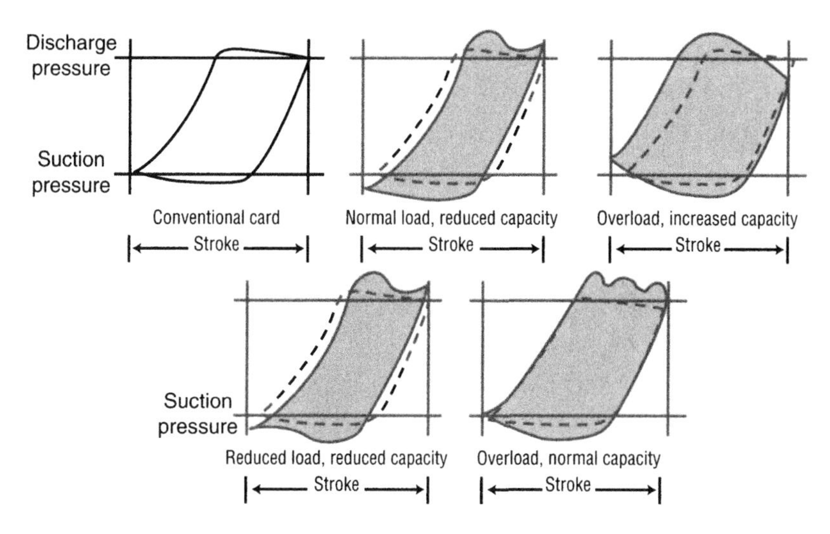

FIGURE 2-58. Effect of gas pulsations on pV diagrams ("indicator cards").

is the key to the exceptional durability, reliability, and availability of this compressor type and, therefore, to its economic operation.

Above all, the unique labyrinth sealing technique is employed for applications where no lubricants are allowed in the cylinder and where no abrasion particles are accepted in the process gas. This is particularly true for oxygen compression, where safety is the most important aspect. At the other extreme, it is also employed for applications where the process gas is heavily contaminated with impurities, such as polymerization products or other very small and hard particles. They have effectively no influence on the labyrinth seal performance, compressor reliability, wear rate, and maintenance intervals.

Piston and piston rod are guided by the crosshead and the guide bearing which are located in the oil-lubricated crankcase. Both guiding elements are made of metal and are oil-lubricated, thus assuring a precisely linear operation of the labyrinth piston as well as an extremely long life of the piston/piston rod guiding system.

The distance piece separates the gas compressing section from the oil-lubricated crankcase.

Figure 2-59 shows the essential separation into an oil-free and oil-contacted section. Figure 2-60 depicts flow within the sealing labyrinth.

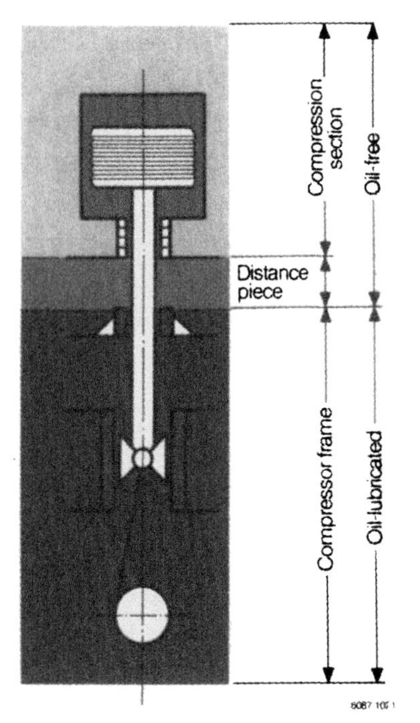

FIGURE 2-59. Oil-free and oil-lubricated sections make up the typical labyrinth piston compressor (*Source: Sulzer-Burckhardt, Winterthur, Switzerland*).

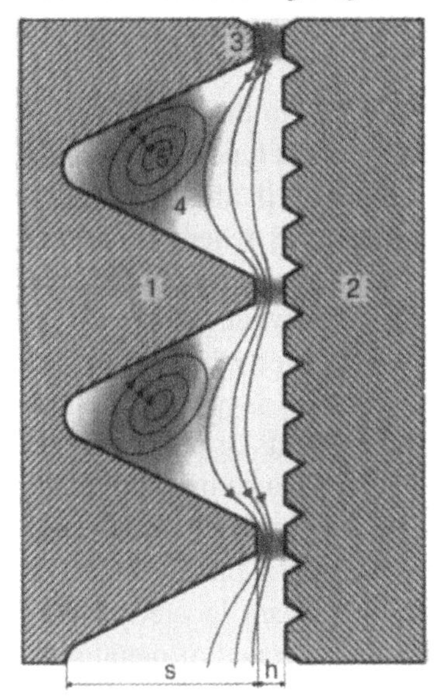

FIGURE 2-60. Gas flow within the sealing labyrinth region of a typical labyrinth piston compressor (*Source: Sulzer-Burckhardt, Winterthur, Switzerland*).

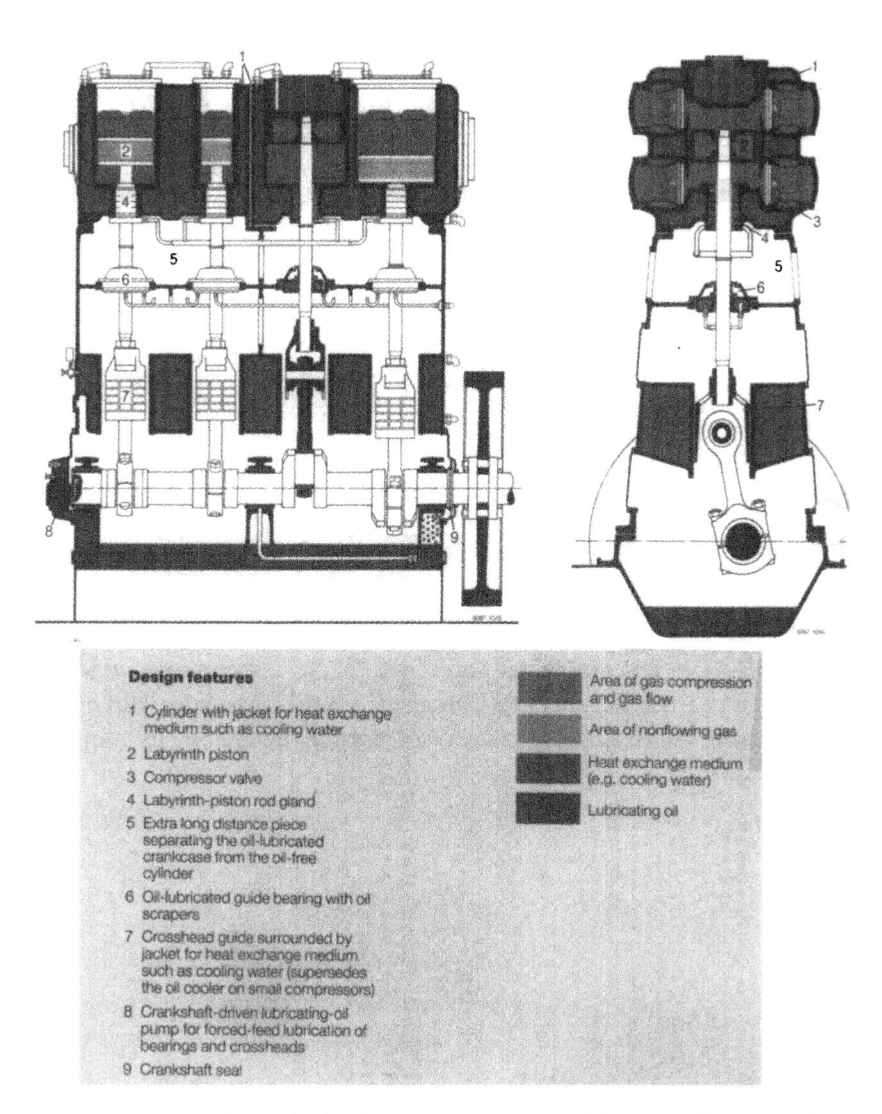

FIGURE 2-61. Labyrinth piston compressor with open distance pieces (*Source: Sulzer-Burckhardt, Winterthur, Switzerland*).

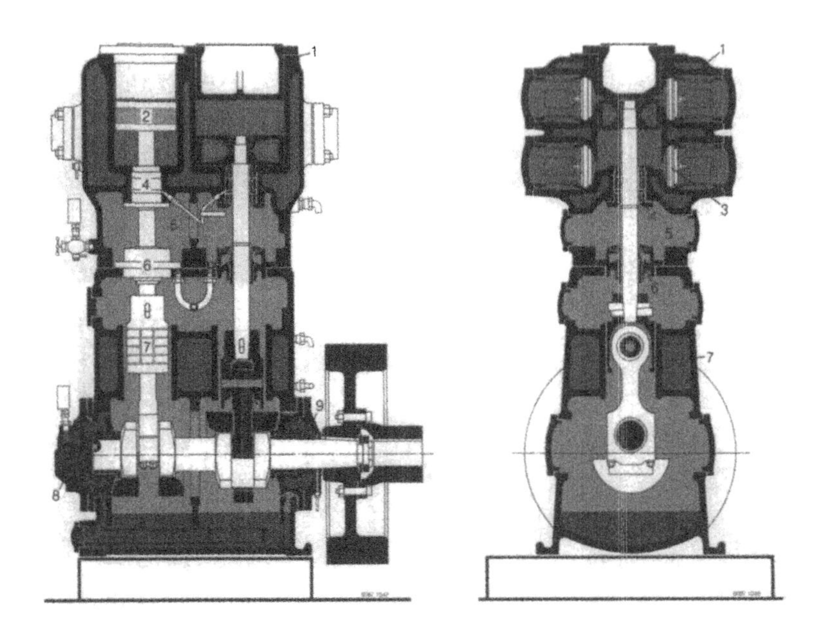

FIGURE 2-62. Labyrinth piston compressor with closed distance pieces (*Source: Sulzer-Burckhardt, Winterthur, Switzerland*).

Figure 2-61 represents Sulzer Labyrinth Piston Compressors with open distance pieces, and Figure 2-62 illustrates a similar machine with closed distance piece for certain explosive or otherwise vulnerable gas compression duties.

CHAPTER 3

Operation and Maintenance of Reciprocating Compressors

LUBRICATION OF RECIPROCATING COMPRESSORS

Perhaps the most important operational item in the field of compressing gases is proper lubrication. To provide proper lubrication involves a certain amount of intelligent care and attention.

Proper lubrication includes:

- Selection of a high quality lubricant suited for the particular service conditions
- Cleanliness in storage and dispensing
- Application of correct quantities in a manner that permits effective performance

Closely allied to the above three factors are maintenance of clean gas into and throughout the compressor and the use of regular procedures for inspecting, cleaning, and maintaining mechanical perfection, both in the compressor and its accessories. Proper lubrication will lead to:

- Reduced wear rates of sensitive and critical parts
- Avoidance of catastrophic failure and malfunction

- Increased reliability
- Reduced cost of system downtime, spare parts, new oil, and labor for repair
- Minimum power consumption

A lubricant is expected to:

- Separate rubbing parts
- Dissipate frictional heat through cooling and heat transfer
- Flush away entering dirt as well as debris
- Minimize wear
- Reduce friction loss and power required
- Reduce gas leakage
- Protect parts from corrosion
- Minimize deposits

In the lubrication of reciprocating compressors, two somewhat different groups of requirements must be considered: first, those of the bearings of the driving end, and second, those of the compressor cylinders.

In both bearings and cylinders the lubricating oil must form and maintain strong films that will minimize friction and wear. In cylinders, it must render this service with the rate of oil feed kept at a minimum and, in addition, it must protect against rusting and aid in sealing the piston, valves, and rod packing against leakage. However, bearings are supplied with large quantities of oil; this oil is used over and over for long periods.

Crankcase or Bearing Lubrication

In practically all reciprocating compressors, the oil charge for lubrication of bearings is contained in a reservoir in the base of the crankcase. Oil from the bearings, crossheads, or any cylinder open to the crankcase drains back to the reservoir by gravity. However, a variety of methods and combinations of methods are employed for delivering oil from the reservoir to the lubricated parts.

Splash Lubrication

Oil may be delivered to lubricated parts entirely by splash. In these compressors, a portion of, or projection from, one or more cranks or connecting rods dips into the oil and produces a spray that reaches all internal parts.

Where the splash method is employed, the level of oil in the reservoir should be maintained within predetermined limits in order to prevent either over or under lubrication.

Splash and Flood Lubrication

Many horizontal compressors use the flood system for bearing and crosshead lubrication. Oil is elevated from the reservoir by discs on the crankshaft and is removed from the discs by scrapers. Some oil is diverted to the main and crankpin bearings, and the remainder is led to a large pocket over the crosshead from which a stream of oil flows to the crosshead pin and crosshead bearing surfaces. In some designs, the main bearings are lubricated from this pocket and the oil flows through external tubes.

In some compressors, the oil is thrown or splashed into the pocket over the crosshead by scoops on the crank counterweights. In the compressor shown in Figure 3-1, crankshaft counterweights and an oil splasher dip into the oil reservoir and throw oil to all the bearings. Oil pockets over the crosshead and over the double row tapered roller main bearings assure an ample supply for these parts.

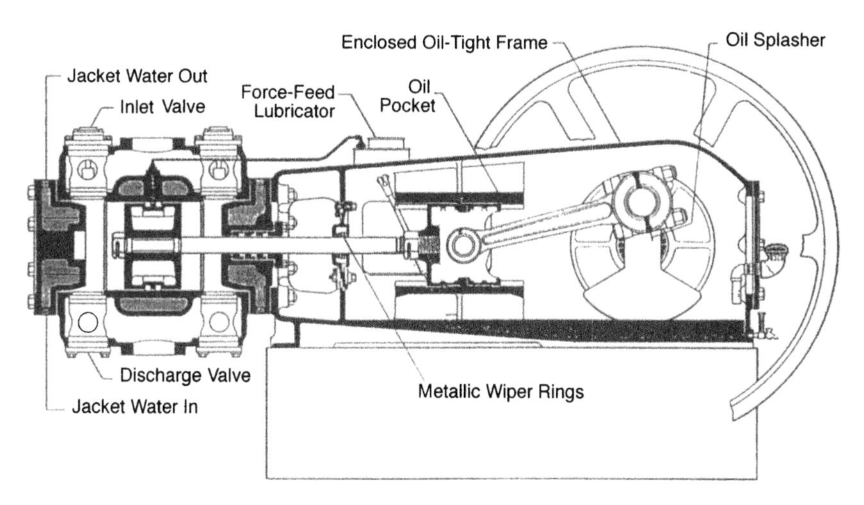

FIGURE 3-1. Splash and flood lubrication applied to a single-cylinder, double-acting compressor (*Source: Dresser-Rand, Painted Post, New York*).

Full Pressure Circulation Lubrication

In this system, as shown in Figure 3-2, a positive displacement pump draws oil from the reservoir and delivers it under pressure to the main bearings and to the crankpin bearings, and hence to the crosshead pin bearings and crossheads.

Oil is drawn up from the crankcase sump through a strainer to the oil pump, and forced through the oil filter. From the oil filter, the oil is led under pressure to the crankshaft. The main bearings are lubricated by drilled passages in the main bearing saddles. The oil is forced through drilled passages in the crankshaft and the connecting rod to lubricate the crankpin bearings, crosshead pin bushing, and the crosshead slide.

Pressure circulation systems are equipped with suction strainers and pressure relief or control valves. Oil filters, usually of the full-flow type, and pressure gauges are also provided. On larger machines where large quantities of oil are being circulated, oil coolers are provided. Oil pressure safety devices are incorporated in the system to shut the compressor down in the event of oil pressure failure.

Compressor crankcases are designed to exclude dust and other contaminants and to prevent oil leakage. All access openings have tight

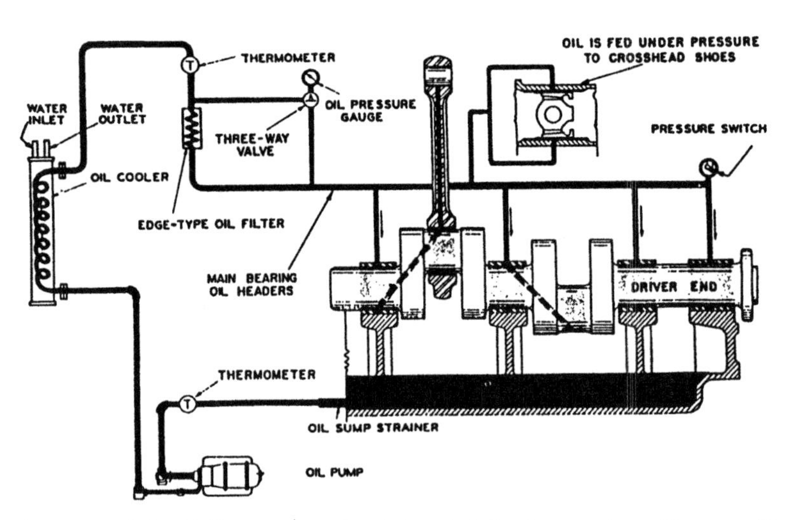

FIGURE 3-2. Full pressure circulating lubrication system (*Source: Dresser-Rand, Painted Post, New York*).

covers, and shafts are effectively sealed. Vents are provided to permit crankcase breathing, and when these are equipped with simple air filters, the entrance of contaminants is minimized.

Factors Affecting Bearing Lubrication

Fluid-Film Lubrication

In general, the factors such as load, speed, temperature, and the presence of water and other contaminants have a moderate effect on compressor bearing lubrication. All lubricated surfaces in compressor crankcases are provided with an excess of lubricating oil.

During operation, surfaces are separated by relatively thick oil films that normally prevent metal-to-metal contact and wear and reduce friction to the low value of fluid films.

Under these conditions, the most important property of an oil is its viscosity at the operating temperature. Oils of suitable viscosity will distribute rapidly and evenly to form a film that will adequately resist being squeezed from loaded areas.

Thin-Film Lubrication

During operation, fluid oil films exist between bearing surfaces. However, while stopped, adequate fluid films can no longer be maintained, and some metal-to-metal contacts may occur.

Under these conditions, and regardless of bearing type, the film strength property of oil is of increasing importance. An oil having adequate film strength will resist rupture and, thus, minimize metal-to-metal contact, friction, and wear. Diester and polyalpha-olefin (PAO)-based synthesized hydrocarbon oils are often compounded for these requirements and should be considered for modern compressor installations.

It should be emphasized that only thin films of oil exist at points of frictional contact in antifriction bearings. Hence, to minimize metal-to-metal contact and to reduce friction, a lubricating oil having adequate film strength is required.

Cooling, cleaning, and protection against rusting are also important functions of the lubricating oil, and the oil viscosity should be in a range that provides ready distribution.

The oil should be applied in such a way that only a minimum amount will remain in the path of the moving parts. Notwithstanding differences between plain and antifriction bearings, the moderate conditions in compressor crankcases are such that one weight of oil provides satisfactory service for both types of bearings.

Oxidation

Much of the circulating oil in compressor crankcases is broken up into fine spray or mist by splash or oil thrown from rotating parts. Thus, a large surface of oil is exposed to the oxidizing influence of warm air, and oxidation will occur at a rate that depends on the operating temperature and the ability of the oil to resist this chemical change.

Oil oxidation is accompanied by a gradual increase in viscosity and, eventually, by the deposition of insoluble products in the form of gum or sludge. These deposits may accumulate in oil passages and restrict the flow of oil to bearings.

Conditions that promote oxidation in crankcases are mild, however, compared to oxidizing conditions in compressor cylinders.

Water

Although water may enter compressor crankcases by condensation from the atmosphere during idle periods or possibly from leaking jackets, there is generally little water present because of the continuous venting of water vapor at crankcase temperatures.

Normally, therefore, there is little opportunity for the formation of troublesome emulsions, which could combine with dust and other contaminants to form sludges that would restrict the flow of oil to lubricated surfaces.

A good compressor crankcase oil will, nevertheless, need adequate water-separating ability to resist the formation of harmful emulsions and to permit water to collect at low points where it may be drained off.

Oils for Crankcase/Bearing Lubrication

The type of oil used for the lubrication of bearings and running gear components must comply with the compressor manufacturers' recommendations, but generally a good quality, non-detergent mineral oil should be used. These lubricants should contain rust and oxidation inhibitors and should display good anti-foaming qualities. The anti-foaming properties are particularly important in compressors using splash or flood systems.

At normal ambient temperatures, the oil should have a minimum viscosity of 200 Saybolt seconds at 130°F. This is equivalent to a SAE 30, or ISO 100 grade lubricant.

It is not desirable to use a detergent-type oil, and, if for any reason a change is made to an oil with detergent characteristics, the entire oil system must be thoroughly cleaned soon after start up. Dirt and deposits, which will be loosened by the detergent oil, must be flushed from the system.

Experience has shown that it is false economy to use an inferior, low first cost, and poor quality oil.

Frequency of Oil Changes

It is not practical to pre-define how often an oil should be changed. The oil will become contaminated with foreign material being held in suspension as well as moisture due to condensation; therefore, the time interval for oil changes is governed by the operating and local ambient conditions.

It must be remembered that the oil charge will not last indefinitely, and certainly if the compressor is shut down for an overhaul, it is poor practice not to completely clean the crankcase, change filters, and install new oil at this time. Periodic oil analyses are strongly recommended in an effort to determine optimum change frequencies. In addition to moisture content and viscosity stability, total acid number (TAN) and flash point may be worth tracking and trending. Lubricating oil analysis for component wear is an excellent maintenance technique.

Cylinder and Packing Lubrication

In contrast to the lubrication of crankcase and bearings, the lubrication system for compressor cylinders and packing must be able to reliably deliver relatively small amounts of oil at higher pressures in order to lubricate the wearing surfaces of cylinders and piston rods.

Cylinder and packing lube systems are terminating or "once-through," systems. The volume of oil delivered at each point needs to be just enough for proper lubrication. Therefore, the rate of lubrication at each point is critical, and over lubrication must be avoided.

Excessive oil volumes can cause fouling of valves, gumming of the packing, and accumulation in the downstream piping system. Because of the higher pressures and low flow rates involved, a "pressured header"-type system, similar to the frame lubrication system, is neither workable nor acceptable.

These systems are "positive displacement" systems that must be capable of accurately delivering, monitoring, and protecting the oil flow to each of the required lubrication points.

The force-feed lubrication system used for cylinder lubrication could well be the most important support system used on reciprocating compressors today. Should it fail or not work properly, the compressor units could be seriously damaged in a short time of operation.

Ironically, because of its apparent complexity, the lube system is in many cases the most misunderstood, ignored, neglected, and misused system on the compressor.

Method of Application

Oil is fed directly to the cylinder walls at one or more points by means of a mechanical force-feed lubricator or a centralized lubricator system.

The oldest and most basic lube system is the *box lubricator.* Here, an individual pump feeds each point, each with an adjustable stroke and some sight glass to view the adjusted feed rate. The suction stroke of the pump pulls oil from the reservoir and discharges it down the line.

Box lubricators are driven either from the crankshaft, another moving part of the machine, or by a separate electric motor. They contain a reservoir for oil and individual pumping elements. A camshaft operates the pumping elements.

There are two types of pumping elements used in the box type lubricator system: pumps with sight glass (Figure 3-3), and pumps with pressurized supplies (Figure 3-4).

Pumps With Sight Glass

Rotation of the lubricator cam actuates the pump rocker arm assembly to operate the pump piston. On the piston downstroke, spring pressure is exerted on the piston causing it to follow the cam. As it moves down, a pressure reduction is created between the piston and the check valve and

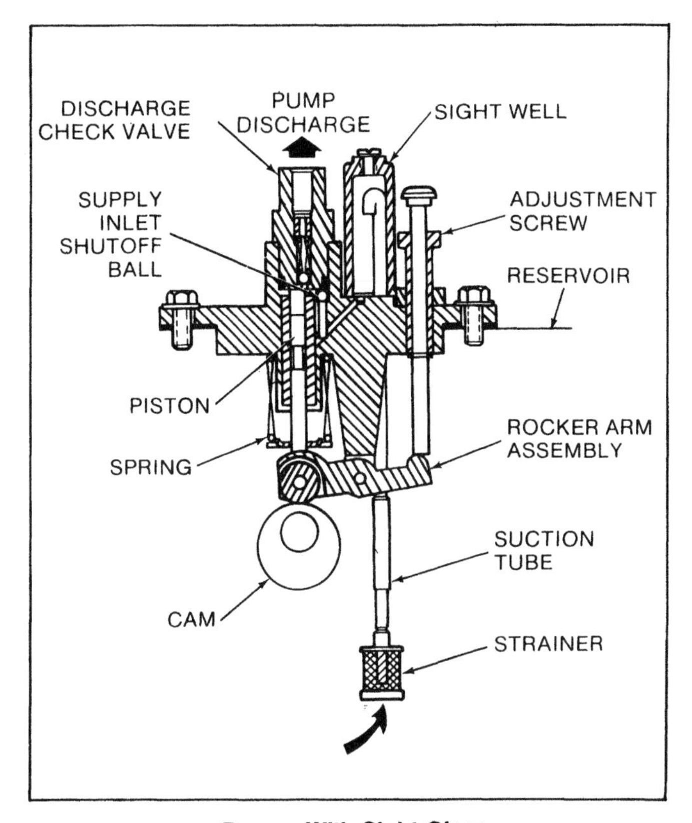

Pumps With Sight Glass

FIGURE 3-3. Force-feed box lubricator with sight glass (*Source: Lubriquip, Inc., Cleveland, Ohio*).

the valve closes. The supply inlet shutoff ball is then unseated and lubricant is drawn into the piston cylinder from the sight well. This creates a pressure reduction (vacuum) in the air-tight sight well that causes lubricant from the reservoir to be drawn into the well until pressure is equalized. On the piston upstroke, the oil in the cylinder is injected through the discharge check valve to the machine injection point. The number of drops seen falling in the sight well is the amount of oil discharged by the pump. Each pump can be adjusted by means of an external screw. This changes the length of the pump stroke, which changes the pump discharge volume.

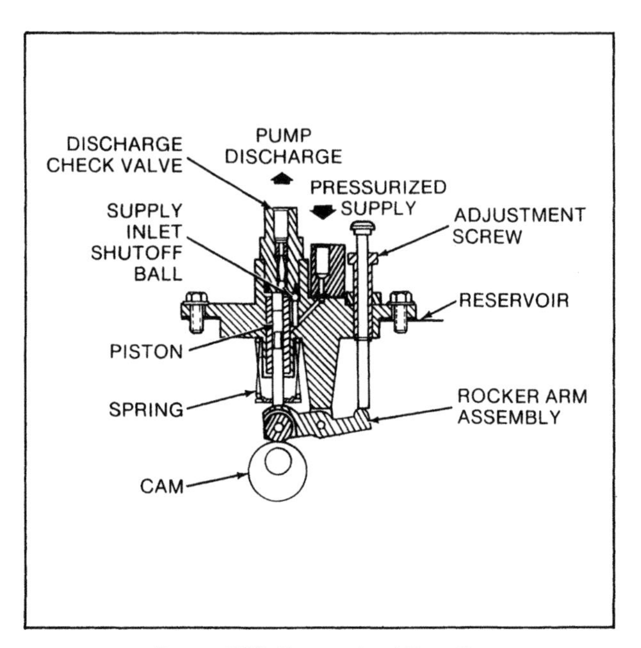

FIGURE 3-4. Force-feed box lubricator with pressurized pumps (*Source: Lubriquip, Inc., Cleveland, Ohio*).

Pumps With Pressurized Supplies

Rotation of the lubricator cam actuates the pump rocker arm assembly to operate the pump piston. On the piston downstroke, spring pressure is exerted on the piston causing it to follow the cam. As it moves down, a pressure reduction (vacuum) is created between the piston and the discharge check valve, and the valve closes. This allows the pressurized supply to unseat the supply inlet shutoff ball and pressurize the piston bore with lubricant. On the piston upstroke, the piston forces the supply inlet shutoff ball to seat and shut off the pressurized supply. Lubricant in the piston cylinder is forced out through the discharge check valve to the machine injection point. Each pump can be adjusted by means of an external screw. This changes the length of the pump stroke, which changes the pump discharge volume.

Rate of Oil Feed

The amount of oil fed to the compressor cylinders should be sufficient enough to provide lubrication and effectively seal the piston against leakage. Oil feeds above this amount are wasteful, cause oxidation, and tend to increase oil carryover to distribution lines.

Excessive Oil Feed

All oil fed to the cylinders is subjected to oxidizing conditions. Under prolonged heating, even the best quality compressor oils will oxidize to some extent. Therefore, feeding more oil than is actually needed results in increasing the amount of oxidation products formed.

Because the highest temperatures are encountered on discharge valves and in discharge passages, and most of the oil fed to the cylinders eventually leaves through the discharge valves, it is here that deposits tend to accumulate. To prevent or minimize trouble from deposits, an oil especially suitable for compressor service that permits using very low rates of oil feed should be used.

Feed rates for compressor cylinder lubrication are typically shown in drops per minute; however, the principal difficulty with the pumping elements of box type lubricators is that the measurement of drops is not reliable.

Consumption per 24 hours, not the number of drops, is the proper way to determine feed rates because drops vary in size and there also is a variance between the specific gravities of the sight glass liquid and that of the lubricating oil. Use the number of drops merely as an estimate and as a way to obtain a balance between cylinder feeds.

The quantity of oil required to provide ample lubrication for any compressor cylinder is obtained by the formula:

$$\frac{\text{Gallons/24 hours} = \text{diameter (in.)} \times \text{stroke (in.)} \times \text{RPM}}{385{,}000}$$

$$\frac{\text{Drops/minute} = \text{diameter (in.)} \times \text{stroke (in.)} \times \text{RPM}}{13{,}800}$$

The formulas are based on 40,000 drops per gallon and lubrication of 600 square feet of swept surface per drop.

The drops per minute must be divided between the number of oil feeds to the cylinder. Do not count the feed to the rod packing when making this distribution. The packing must be considered as a separate cylinder for lubrication purposes.

Rates of oil feed will generally fall between the limits shown in Table 3-1. These rates should be increased for dirty or wet gas conditions and also for the initial running of new compressors. Oil feed rates can also be obtained from Figure 3-5.

A practical and frequently used method to determine whether the proper amount of oil is being fed consists of removing the discharge valves

TABLE 3-1
RATES OF OIL FEED

	DROPS OIL PER MINUTE (Total for cylinder and packing)*					
Cylinder	**Discharge Pressure of Cylinder, psi (kg/cm²) gauge**					
Diameter, in. (cm)	**25–75 (1.8–5.3)**	**75–150 (5.3–10.6)**	**150–250 (10.6–17.6)**	**300–600 (21.1–42.2)**	**600–1500 (42.2–105.6)**	**1500–3000 (105.6–211)**
6–8 (15–20)	3–6	5–8	6–9	7–12	9–16	12–18
8–10 (20–25)	4–7	6–9	8–11	10–15	12–20	
10–12 (25–30)	5–9	7–11	9–13	12–18		
12–14 (30–36)	6–10	9–12	11–16	14–21		
14–16 (36-41)	7–12	10–15	13–18			
16–18 (41–46)	8–13	12–17	15–20			
18–20 (46–51)	9–15	13–19	17–22			
20–22 (51–56)	10–17	15–21	19–24			
22–24 (56–61)	11–18	17–23	21–26			
24–26 (61–66)	12–20	18–25	23–28			
26–28 (66–71)	13–21	20–26	25–30			
28–30 (71–76)	14–23					
30–32 (76–81)	16–24					
32–34 (81–86)	17–26					
34–36 (86–92)	18–27					
38–40 (97–102)	19–28					
36–38 (92–97)	20–30					

The figures given are for gravity and vacuum type sight-feed lubricators. For glycerine sight-feed lubricators, the figures given should be divided by three. Feeds to cylinder bores should never be less than one drop per outlet per minute under any conditions. The quantities given are based on an average of 8,000 drops per pint (16,900 drops per liter) of oil at 75°F (24°C). (Reprinted from Compressed Air and Gas Handbook, *Fifth Edition, John P. Rollins, editor. New Jersey: Prentice Hall, 1988.)*

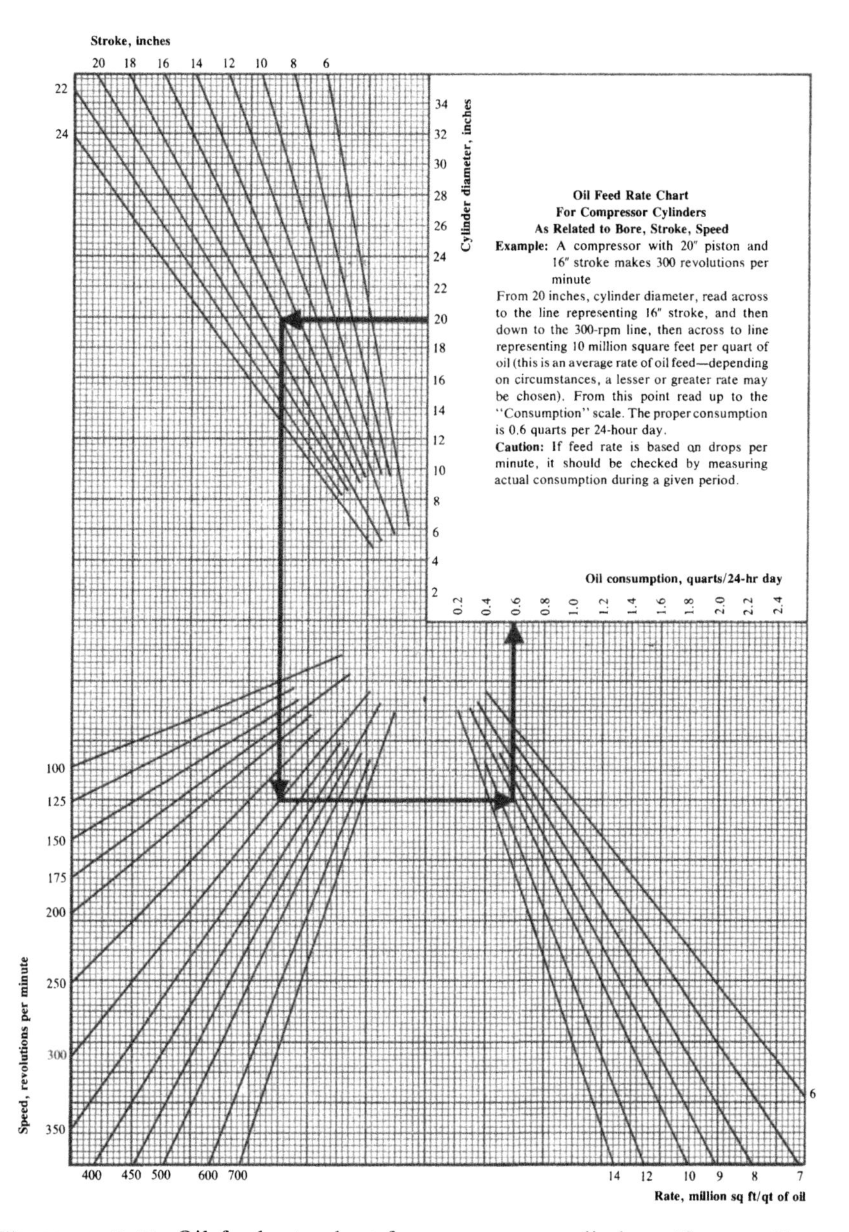

FIGURE 3-5. Oil feed rate chart for compressor cylinders. (*Source: Exxon Corporation, Marketing Technical Department, Houston, Texas*)

from time to time and examining them and all accessible parts of the cylinder, valve passages, and discharge pipe. All surfaces should have a wet appearance and should feel oily to the touch.

If they are dry or show signs of rusting, the rate of oil feed should be increased. On the other hand, if the parts have excessive oil on them, or if oil is lying in pools in the cylinders, the rate of oil feed should be reduced.

In the operation of air compressors, over lubrication of the cylinders and the subsequent carryover of the excess oil to the downstream piping system causes excessive deposits and the real possibility of explosion or fire.

The biggest problem with box lubricator pumps is that they either pump too much oil, or no oil at all. They are difficult to adjust and maintain; they often are driven too fast to deliver the required amount of oil; and they cannot provide protected delivery. Any pump can stop without being noticed, causing serious damage to compressor cylinders, packing, or other components. If very old or subject to dirty oil, the pump piston can be so badly worn in the pump body that it can no longer reliably deliver oil. If the rocker arm or cam is badly worn, the pump will not operate at its maximum stroke. If the adjusting stem is worn, missing, or not assembled or not located correctly on the rocker arm, the pump will pump at maximum stroke and its discharge cannot be cut back.

Remember—all pumps need a clean, air-free oil supply.

Divider Block or Feeder System

Recently, a great many improvements in lube system design have occurred, primarily in the areas of automatic operation, accuracy of delivery, and fail-safe protection. The more current type of lube system is the divider block system.

This system employs one central pump feeding a network of one or more divider blocks or feeders, which "divide" and "distribute" the required volumes of oil down the various lube lines. Although some earlier systems still use the "vacuum type" pumps to feed the dividers, most pumps have what is called "pressurized inlet."

In the divider block system, the drive box of the pump does not contain the oil supply. That drive box oil serves only to lubricate the drive mechanism. The system oil is supplied directly to the inlet of the pump by gravity or a pressurized system. If new oil is used, it may come from an overhead tank or through a pressurized system from a bulk tank.

The divider block takes oil from the pump and precisely meters it to as many lines as required. This is accomplished through a stack of "piston sections" or "feeder sections," each of which has a moving piston of a given displacement.

Piston displacements may vary, and the number of sections may vary, so the actual block displacement is different from one to the next. All pistons move hydraulically, actuated only by the flow into the block. Each piston moves over and back in turn, once for every block cycle.

The principle is this: each piston movement delivers a positive displacement down one line. When one piston moves, that line gets its shot. But, in addition, reduced diameters in the piston center section force the inlet oil to move the next piston in the stack. When all have moved over, a reversing passage sets up the pistons to start back down the other side.

Clearly then, each line must take its shot of oil, or that piston will not valve the flow to the next piston. This is how each line can be protected against interrupted flow.

Pumps for Divider Block Systems

There are three basic types of pumps used for these systems. All pressurized inlet pumps must have an adequate supply of pressure and volume to handle the required flow rates. The typical range of inlet pressure varies from mere inches of water column to 30 psi.

The first type of pump is the *box lubricator* converted to pressurized inlet. The sight glass and drip tube are replaced by a cap with inlet fittings for the supply lines. The suction tube is removed or plugged. Typically, two pumps are manifolded together for each lubrication zone—one active with capacity for the entire system and the other an installed spare.

The second type of pump is the *injector pump,* which employs a diesel-fuel-type piston and barrel assembly. Typically, one pump head is used for each adjustable zone of lubrication. The pumps are larger than the box lubricator type and are mounted in a heavy duty cast reservoir.

The third type is the *Gerhardt Pump,* which is a similar barrel and piston design with inlet ports and discharge check valve. The primary difference is that the flow rate is made by rotating the piston, which has a helical groove in relation to the inlet ports on the barrel.

Divider Block Components

These systems are equipped with main line flow switches and mechanical counters for monitoring. The newest type of monitoring and protection device for the lube system is the electronic monitor.

These devices give accurate control over the oil-feed rates and protection to the compressor in the event of system or oil supply failure.

Figure 3-6 shows a Trabon MV system arrangement and Figure 3-7 a Watchman Lubemation System.

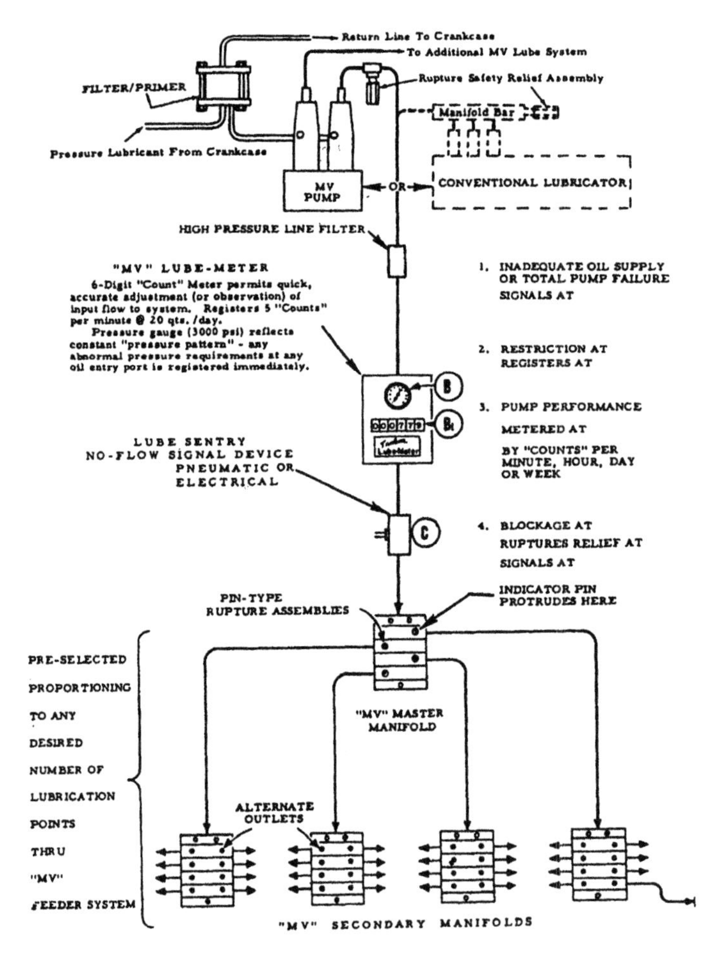

FIGURE 3-6. Divider block component system for cylinder lubrication. *(Source: Trabon/Lubriquip, a unit of INDEX Corporation, Cleveland, Ohio)*

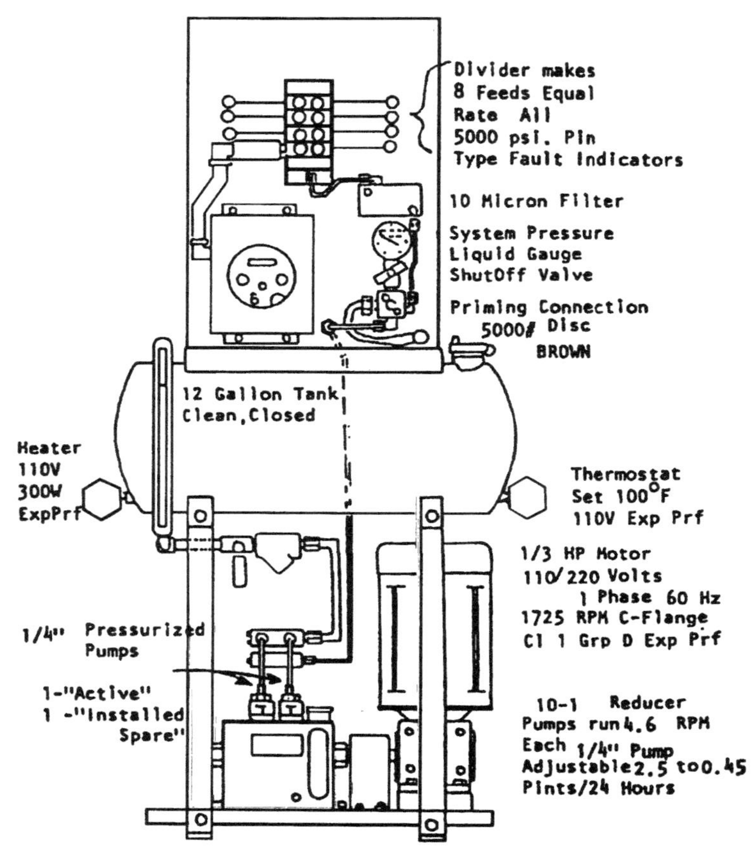

FIGURE 3-7. Watchman Lubemation System. (*Source: Sloan Brothers Company, Oakmont, Pennsylvania*)

Contamination Concerns

Dirt and air are the two primary factors involved in most lube systems failures. The lubrication system is a hydraulic system, and like any other system of this type, dirt in the oil can cause serious damage to lubrication system components. Even if it does not cause immediate failure, it can greatly reduce the reliability of the lube system and compressor. Proper filtration and clean oil are necessities.

Purging Oil Lines and Components

Although not usually the cause of damage in a lubrication system, the presence of air in lube lines and components is often the cause of lubrication failure. Finding where it lies is often difficult, and eliminating its source can also be tricky. Remember, although air cannot usually damage the components of a lube system, it certainly does not lubricate the compressor very well, so all air must be eliminated.

Purging of air in lube systems after installation, maintenance, or testing is very important. The proper purge method is to begin at the source and move progressively downstream, verifying at the inlet and outlet of each component that all air and contaminants have been eliminated.

When new tubing is run, care should be taken to avoid pumping dirt, particles of cut tubing, or other contaminants into the components. These can be very damaging.

Pump a large quantity of oil through long tubing runs before connecting them. A large displacement portable hand pump works well for this purpose.

Terminal Check Valves

The final, and certainly an important part of a lubrication system, is the terminal check valve. It must hold against back pressure, preventing backflow of gas or air in the oil line during shutdown. If it does not, gas can work back into tubing and cause air-lock problems. Therefore, a check valve must be located at the end of every oil line.

Double ball checks should be used on all compressor and packing feeds, preferably the "soft ball" or "soft seat" type, capable of positive sealing under adverse conditions. Mount the check valves horizontally or discharging up. Maintaining a wet seat aids in positive sealing.

COMPRESSOR CYLINDER LUBRICATION OIL SELECTION

Table 3-2 shows the recommended minimum viscosities for various cylinder diameters and pressures.

TABLE 3-2
RECOMMENDED LUBRICANT VISCOSITIES FOR RECIPROCATING COMPRESSORS

MINIMUM FLASH: 400°F. MAX. POUR: MIN. GAS SUCTION TEMPERATURE PLUS 10°F.					
Pressures	Minimum Viscosity @ 210°F.				
PSIG	Cyl. Diams.	up to 10"	10"-15"	15"-20"	20" & above
0-200		50 SSU	50 SSU	50 SSU	80 SSU
200-1000		60 SSU	80 SSU	80 SSU	80 SSU
1000-2500		80 SSU	80 SSU	110 SSU	
2500-4000		110 SSU	110 SSU		
4000 & up		150 SSU	150 SSU		
Hydrocarbon Gases - Use next higher viscosity from table with minimum viscosity of 80 SSU @ 210°F.					

These values are only given as a guide; different oils of the same viscosity may have different lubricating qualities. Regular examination of the cylinder bores during the first weeks of operation should be made to determine if adequate lubrication is being maintained.

The choice of a cylinder lubricant is affected by the properties of the gas to be handled. In this respect, gases usually fall into one of three classes: inert, hydrocarbon, and chemically active gases.

Inert Gases

Inert gases that do not tend to condense on the cylinder walls introduce no special problems and employ the same cylinder lubrication that is used for air cylinders operating under similar conditions. Examples of such gases are carbon dioxide, nitrogen, and helium.

Hydrocarbon Gases

Some gases, although dry at the start of compression, tend to condense on cylinder walls as pressure increases. This is because the temperature at which a gas can be liquefied by heat removal (i.e., lowering its condensing temperature) increases with pressure, and in the case of certain gases, condensing temperatures that are above cylinder wall temperatures are reached as compression proceeds.

Under such conditions, part of the gas in contact with the cylinder wall is cooled to the condensing temperature and liquefies. The condensate formed will do the following:

- Tend to wash the lubricant film from the cylinder walls
- Dissolve in the lubricating oil, reducing its viscosity.

Viscosity reduction in service due to these effects should be compensated for by the use of a heavier-bodied oil than would be otherwise selected.

In addition, higher than normal jacket water temperatures are advisable to prevent or minimize condensation, which requires heavier-bodied oil.

Chemically Active Gases

Obviously, compressor parts must be made of metals that will resist corrosion by whatever gas is handled. It is the effect of an active gas on the lubricant that is of interest.

With some gases, such as oxygen, petroleum products must not be used. Chlorine reacts with oil to form gummy sludges and deposits. Methyl chloride or ethyl chloride and sulfur dioxide also may form sludges in the presence of oil, moisture, and selective additives.

Synthetic Compressor Lubricants

Synthetic oils have become more widely used and accepted as compressor lubricants because of their higher "auto-ignition" characteristics and ability to prevent carbon buildup on valves and piston rings.

The use of synthetic oils may allow the reduction of feed rates to the cylinder by approximately one third. This reduction in feed rates will result in less oil in the downstream piping system. Reduced oil accumulation and the "fire resistant" characteristics help to prevent fires in the discharge lines. However, no oil is fireproof or explosion proof.

Types of Synthetic Lubricants

There are literally thousands of synthetic fluids and lubricants, most of which are not commercial items. There are a few fluids that are used in commercial quantities, some of which have given satisfactory service in lubrication of compressors. These fluids include:

1. Phosphate esters, such as Firequel, Pydraul, Houghto-Safe, Shell SFR, etc. These are fire-resistant lubricants that have given good service in air compressors.
2. Polyalkylene glycols (Ucon Fluids). These are available in water soluble and water insoluble formulations and when used as recommended, in the proper viscosity grade, have given good service in gas compressors. They are not fire resistant.
3. Fluorocarbons (Halo-carbon). These are mentioned because they are completely fireproof. They are suitable for oxygen compressor lubrication; however, they are very expensive.
4. Diesters and polyalpha-olefins (PAO). These are used most frequently as compressor synthetic lubricants. They include, Synesstic (Exxon), Androl (Tenneco), MOCOA, IR SSR coolant (Tenneco), Sullair 24KT (silicon based).

If a change is made to synthetic oil in a compressor that has been operating with petroleum oil, it will require extensive cleaning of the piping system as well as the compressor cylinder and its components. The synthetic will loosen and possibly wash off deposits left by the petroleum oils. The unit should be shut down after a short run with synthetic oil for this extensive cleaning.

Regular Inspection and Maintenance

Compressor installations should be inspected regularly at intervals that will have to be determined by the severity of operating conditions. Necessary maintenance work such as replacing worn or broken parts; installing new packing; changing crankcase oil; cleaning the crankcase, the force-feed lubricator and the air filter; and removing deposits from valves, discharge passages, coolers, and water jackets should be carried out promptly.

Following an overhaul, the entire system should be thoroughly inspected. Typical tasks should involve the following:

- The lubrication system:
  1. Clean filter and fill sump to proper level.
  2. Build up pressure. Units not equipped with a motor driven oil pump usually have a hand priming pump.
  3. Check for leaks, proper pressure, and operation of alarm and shutdown devices.
- Compressor cylinders:
  1. Disconnect each force-feed lubricator line at the compressor cylinder.
  2. Fill lubricator sump with proper grade oil.
  3. Operate lubricator and check flow of oil. Units not equipped with motor driven lubricator can be operated by hand cranking or hand pumping each pump feed.

Caution: Failing to prime and purge each line before starting the compressor can cause serious damage due to inadequate lubrication. Running a motor-driven lubricator too long (with oil lines connected) before starting must be avoided.

Deposit Removal

Excessive deposits on valves and in discharge areas indicate dirty or chemically contaminated suction air, excessive oil feeds, or the use of an unsuitable oil. Deposits should be controlled by regular removal and by correcting conditions responsible for deposit accumulation.

Changing Crankcase Oil

The frequency of changing crankcase oils depends largely on the cleanliness of the atmosphere surrounding the compressor. The used oil should be drained while still warm, and the crankcase should be wiped clean with lint-free rags before introducing new oil. Lint-producing rags should not be used for wiping. Solvent cleaners are not usually required and in no case should a flammable solvent be used.

Force-Feed Lubricators

Maintenance of force-feed lubricators ordinarily involves cleaning the sight-feed glasses, pumps, and reservoirs. Where liquid filled sight-feed glasses are used, the liquid may become cloudy or may gradually be displaced by oil. In either case, the glasses should be drained, cleaned and refilled with clean liquid. Contaminants such as dust and dirt gradually accumulate in the reservoirs, and the reservoirs and pumps should be cleaned at intervals of not over one year depending on the cleanliness of the surroundings. Lint-free wiping cloths should be used for this purpose.

Oil Containers

Only clean containers should be used for storing and dispensing compressor oils, and the containers should be well covered when not in use. Dirty containers obviously constitute a source of oil contamination.

Galvanized (that is, zinc-coated) iron containers should not be used for lubricating oils. Zinc may react with certain constituents of oil, compounding materials, or oxidation products to form viscous metallic soaps that thicken oil and act as catalysts promoting oil oxidation.

Cleaning Air Filters

Intervals between the cleaning of air filters should be determined by local conditions and will depend on the type and capacity of the filter, volume of air handled, and the amount of dust in the air. Clean filters impose very little restriction on the flow of air, and contaminants should not be allowed to accumulate to the extent that this restriction becomes excessive.

Oil Analysis

Regularly scheduled lubricating oil analysis is essential for trouble-free operation and must be part of any maintenance program. If this can not be handled with in-house facilities, there are several companies that specialize in this service. One-month sampling and analysis intervals are typically recommended.

Records should be kept of these monthly analyses and trending of wear particles performed in order to determine which components may be wearing and are in danger of failure.

Extended Shutdown Periods

Before stopping a compressor prior to an extended shutdown, the oil feeds to the cylinders should be increased or the force-feed lubricator should be operated by hand to provide an extra supply of oil. Condensation of moisture will occur during the shutdown period, and the extra supply of oil will provide additional protection against rusting.

A practice often adopted where large compressors are used consists of opening the cylinders and swabbing them with special oil to protect against rusting during long shutdowns.

Before starting a compressor after an extended shutdown, the oil lines to the cylinders should be disconnected at the cylinders, and lubricators should be pumped by hand until oil flows at the disconnected end. The lines should then be reconnected, and the lubricator should again be operated by hand to provide an initial supply of oil to the cylinders. Furthermore, about twice the normal rate of feed should be maintained for the first hour of operation.

An alternative method often used to protect compressors during long periods consists of running them unloaded for a few minutes each day or blanketing them with an oil-mist environment.

Conventional, long-term storage protection would include at least the following 13 steps:

1. Remove all cylinder valves, unloaders, and rod packing. Coat and store in warehouse.
2. Blind opening at rod packing box.
3. Blind compressor gas inlet and discharge casing connections, after filling cylinder and gas passages completely with an approved preservative oil. (Allow some space for thermal expansion.)
4. Do not clean or coat cylinder cooling passage. Flush only with water, drain and air dry. Plug inlet cooling passage. Leave low-point drain cracked open and wire valve in this position (this permits self-draining of any atmospheric condensation).
5. Blind opening at wiper ring adjacent to crosshead compartment.
6. Fill crankcase and crosshead compartment completely with a suitable preservative oil. Install a valved vent (at high point, if possible) to permit addition of rust preventative later if required.
7. Blank off pulsation bottles and knock out drums after coating with an approved preservative (fill and then drain vessels).
8. Coat intercooler/aftercooler gas passages (not water passages) with suitable, approved preservative by filling and draining. Blank or

plug all connections on these components. Keep low-point drain valve on water side slightly open and wire valve in this position (water side need only be flushed and air dried).

9. Fill all lubricators, lube pumps, and filter on compressor frame with an approved preservative oil. Plug all connections and vents.
10. Coat flywheel with protective grease and cover with guard.
11. Coat all exposed shafts and linkages with protective grease. Cover crankshaft opening from compressor with tape.
12. Remove all gear type couplings (hubs, spacers, keys) and disc-pack of non-lube type couplings. Coat with an approved protective medium and store in warehouse.
13. Protect all instrumentation and control panels according to appropriate procedures.

Operational Problems and Maintenance of Compressor Valves

In a reciprocating compressor cylinder, the valves, without a doubt, have the greatest effect on the operating performance of the machine, both from an efficiency standpoint and from a mechanical reliability standpoint.

The valves are installed in the compressor cylinder directly in the air or gas stream, and are subjected to considerable abuse. It is surprising that we generally have as little valve trouble as we do.

Stop and consider for a moment just what is expected of a compressor valve. It must open and close with every stroke of the piston, many times each minute. The number of times each valve operates in eight hours, a day, a week, and a year at different compressor speeds is shown in Table 3-3.

TABLE 3-3

VALVE ACTUATIONS PER TIME UNIT WITH VARIABLE COMPRESSOR SPEED

	Compressor speed, rpm		
Time	300	600	1000
8 HOURS	144,000	288,000	480,000
24 HOURS	432,000	864,000	1,440,000
1 WEEK	3,024,000	6,048,000	10,080,000
1 MONTH (30 DAYS)	12,960,000	25,920,000	43,200,000
1 YEAR (365 DAYS)	157,680,000	315,360,000	525,600,000

A valve must do this directly in the path of the gas stream and is subject to any entrained liquids, foreign particles, corrosive gases or materials, and occasionally must operate practically covered with sticky or "gunky" material. It is subjected to all types of destructive forces such as tension, compression, impact, twisting, bending, abrasion, erosion, and extreme temperature variations.

If the valve fails to function properly, the compressor cannot supply compressed air or gas and must be shut down for repair. So, anything that can be done to improve the operation of a compressor valve also improves the overall operation of the compressor itself.

Successful operation of a compressor valve begins with sound basic design to meet the specific requirements of a given installation. The valve must be adequately maintained and used in a normal manner.

VALVE THEORY AND DESIGN

What Is a Compressor Valve?

Compressor valves are devices placed in the cylinder to permit one-way flow of gas either into or out of the cylinder. There must be one or more valves for inlet and discharge in each compression chamber (cylinder end).

Basic Requirements of a Compressor Valve

Basically, an automatic compressor valve requires only three components to do the job it is required to do:

1. Valve seat
2. Sealing element
3. A stop to contain the travel of the sealing element

However, a valve comprising only the above components installed in a modern compressor would not fulfill the expected life and efficiency requirements. Due to the high demands on today's reciprocating compressors, the valves require a much more elaborate design than outlined above.

Modern compressor valves either incorporate or exhibit the following:

1. Large passage area and good flow dynamics for low throttling effect (pressure drop)
2. Low mass of the moving parts for low impact energy
3. Quick response to low differential pressure
4. Small outside dimensions to allow for low clearance volume
5. Low noise level
6. High reliability factor and long life
7. Ease of maintaining and servicing

Basic Function of a Compressor Valve

A compressor valve regulates the cycle of operation in a compressor cylinder. Automatic compressor valves are pressure activated, and their normal movement is controlled by the compression cycle.

The valves are opened solely by the difference in pressure across the valve; no mechanical device is used. The best illustration of a compressor valve cycle is obtained by correlating the piston movement to the pressure volume diagram.

To visualize the sequence of events, we align a schematic drawing of a horizontal single-acting reciprocating air compressor (top Figure 3-8) directly above its piston-velocity graph (center Figure 3-8) and its cylinder-pressure graph (bottom Figure 3-8). P_1 represents inlet pressure and P_2 represents discharge pressure.

The piston is shown at its top dead center, momentarily motionless at the end of its compression stroke, Point D in the pV-diagram. At this moment, the discharge valve has just closed and the suction valve has not yet opened.

Gas Intake

When the piston starts moving to the right on the suction stroke, the small amount of gas remaining in the cylinder, its clearance volume, is expanded from P_2 to P_1 and lower. The resulting slight underpressure permits the suction pressure, P_1, to push open the suction valve, and gas from the suction plenum is drawn into the cylinder (point E in the pV-diagram).

As the piston nears the end of its suction stroke, its deceleration reduces the gas speed through the open valve, and in a properly designed valve, the spring load closes the valve at the moment the piston reaches its bottom dead center (point A in the pV-diagram.)

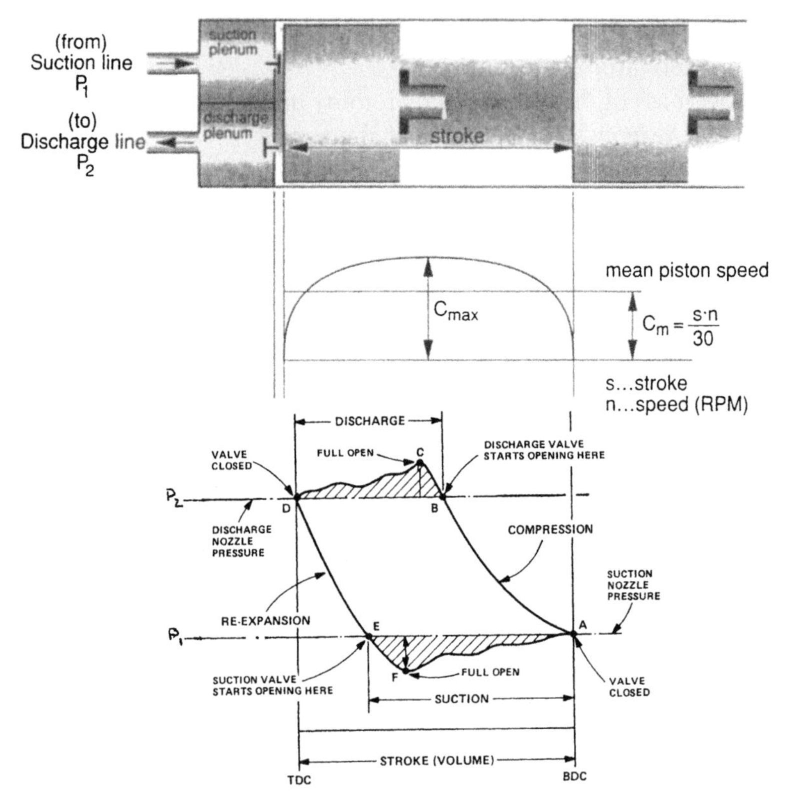

FIGURE 3-8. Piston stroke position superimposed on pV-diagram (*Source: Hoerbiger Corporation of America, Ft. Lauderdale, Florida*).

Compression

With both suction and discharge valve closed, the piston's return stroke to the left compresses the gas in the cylinder and reduces its volume while it increases its temperature and pressure, until the pressure exceeds the desired delivery pressure, P, by the amount sufficient to open the discharge valve (point B in the PV-diagram).

This excess pressure is necessary to overcome the equalization static pressure on the valve plate and to lift the valve plate against the spring load.

Gas Discharge

When the discharge valve opens, the excess pressure decreases in diminishing waves to P_2. Just before the piston reaches the end of its leftward compression stroke (point D in the PV-diagram), the discharge is automatically closed by its springs.

Compression Work

The area enclosed by the curve E-A-B-D-E in the PV-diagram represents the total work performed during compression. Areas D-B-C-D and E-A-F-E indicate energy expended activating the valves and in overcoming flow resistance in and out of the cylinder.

Schematic of Suction and Discharge Valves

Figure 3-9 shows the sequence of events of a suction and discharge valve per compression chamber.

Inertia must be overcome to break the valve open

When the valve is closed, part of the valve plate or valve ring is firmly set against the seat lands.

The sealing element, strip, plate, ring, or channel initially lifts off the seat land very slowly, but then accelerates rapidly for the following reasons:

1. It is evident from Figure 3-9 that the cylinder pressure is initially only applied to a reduced area of the valve plate, that is, valve plate area minus area covered by seat lands.

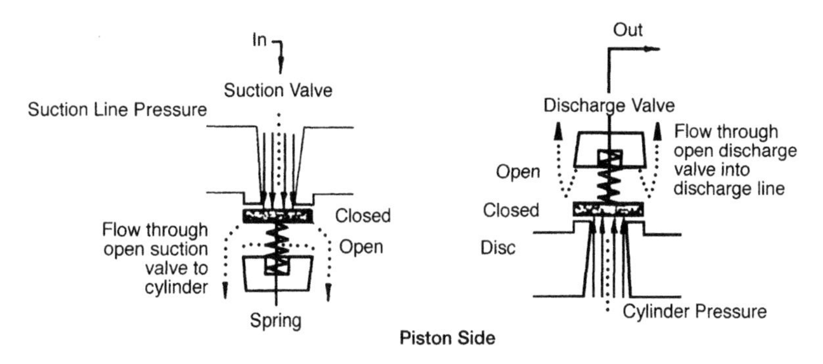

FIGURE 3-9. Valve action schematic for reciprocating compressors.

2. Lubrication and condensate further delay separation of the two contact surfaces between valve plate and seat land. This results in a sticking effect.
3. The pressure that eventually will force the valve plate to lift has an approximate linear increase over time (Figure 3-10). As it increases sufficiently to overcome the spring load, the valve plate will begin to accelerate.

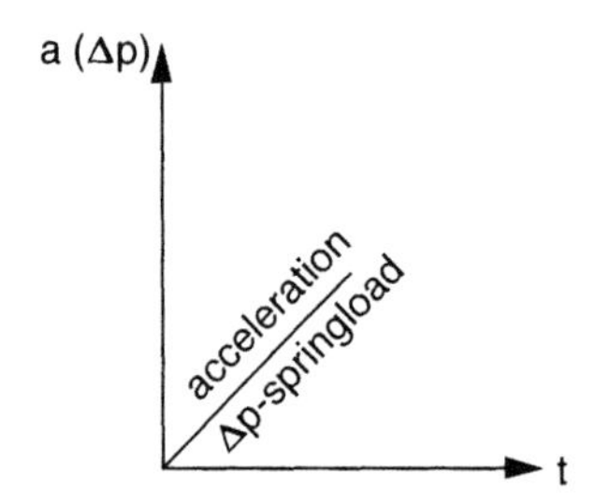

FIGURE 3-10. Valve acceleration and pressure differences versus time.

The velocity increases gradually

According to the laws of physics, the velocity will begin at zero and increase gradually (parabolic). Valve lift, or the distance that the valve plate travels, lags behind valve velocity (Figure 3-11).

The combination of the above three factors creates the effect of the high initial differential pressure between the cylinder and plenum, which can be seen on all pV-diagrams. A pressure differential is necessary and, on

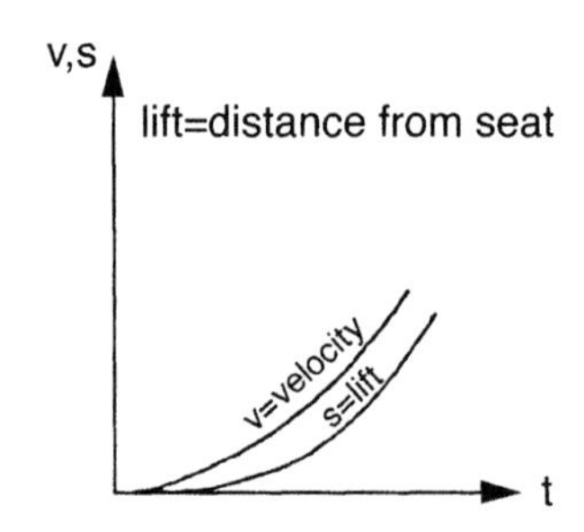

FIGURE 3-11. Valve velocity and lift versus time.

suction valves, the pressure has to be reversed. These factors explain why a pressure differential is necessary inside the cylinder versus outside the cylinder to lift the valve plate off the seat. The difference in area of a sealing element is normally 15 percent to sometimes as high as 30 percent between exposure underneath, seat side, and exposure on top, guard side.

Because there is always some leakage through the closed valve plate along the seat lands (Figure 3-12), there is a certain amount of pressure build-up in this area. Therefore, the actual pressure differential needed to break the valve open is only 5 percent to 15 percent over the line pressure and not higher, as would theoretically result from the above mentioned differential in area.

As the valve plate lifts off the seat lands, it accelerates the valve plate rapidly against the spring load toward the guard. The valve plate or sealing element impacts against the guard causing the so-called opening impact and, at this stage, the valve is considered fully open.

The flow of the gas

The flow of gas through the seat keeps the sealing element open. As the flow diminishes due to decreasing piston speed, the springs or other cushioning elements found in most valves force the sealing element to return to the seat lands and close the valve. Preferably, the valve is completely closed when the piston is near dead center.

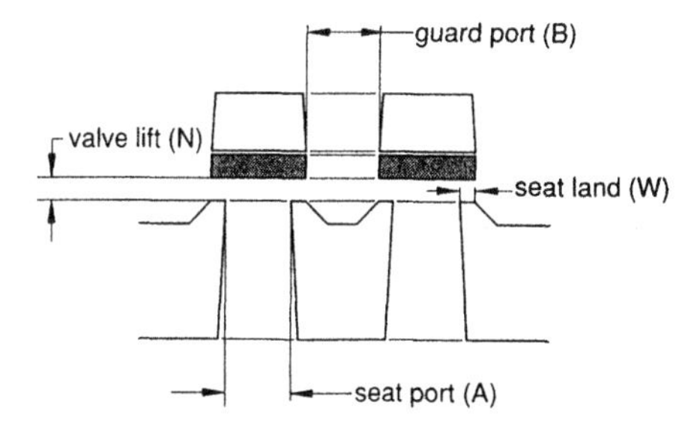

FIGURE 3-12. Valve seat land and lift representation.

The difference between suction and discharge events

The basic cycle repeats itself over and over again and is representative for every valve of every design and configuration.

The differences between suction and discharge events are the piston velocities at which the suction versus the discharge valves open and close, the time the total cycle consumes, and the pressure and temperature at which the gas is being processed.

These criteria are important to remember because they lead to the design differences between suction and discharge valves.

VALVE TYPES USED IN RECIPROCATING COMPRESSORS

Prior to the development of automatic valves, compressors used mechanically operated suction and heavy poppet-type discharge valves. These valve types were efficient and satisfactory at relatively low rotative speeds, but limited the speeds at which the compressors could operate. The lightweight, fully automatic valve allowed for higher rotative speeds, greater output, and increased efficiency from the compressor.

There are many different valve designs available for modern reciprocating compressors. They all function as check valves, opening and closing as a function of differential pressures as well as keeping the fluid in the compressor cylinder during the compression process.

However, these designs can be classified into relatively few types, defined by the shape of moving elements and the nature of damping used to control the motion. Table 3-4 summarizes these valve types and the following paragraphs give a brief description of each type.

TABLE 3-4
TYPES OF VALVES AND DAMPING METHODS

		Damping	
Plane View Shape	**None**	**Mechanical**	**Fluid**
Rectangular	X		X
Concentric Ring	X		X
Ported Plate	X	X	X
Disc (Poppet)	X		X

Rectangular Element Valves

Typical valve designs that use rectangular elements are feather valves, channel valves and reed valves. These valves generally make excellent use of available valve port areas to provide flow area at moderate valve lifts (.100″–.200″).

The rectangular shape of the elements eliminates the use of a cross-ribbing system to provide structural integrity of the valve seat and guard. This enables the use of long uninterrupted slots in the seat and guard and provides good flow area.

Feather Valves

Feather valves (Figure 3-13) use a long and narrow metallic strip as a sealing element. These strips open as a simply supported, uniformly

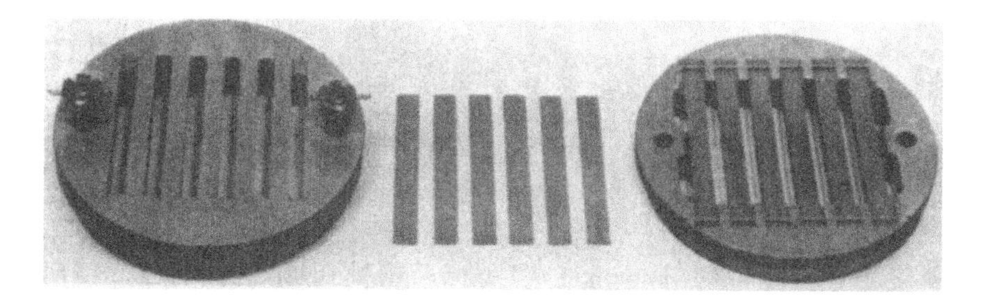

FIGURE 3-13. Feather valve incorporating long, narrow metallic strips as sealing elements (*Source: Dresser-Rand, Painted Post, New York*).

loaded beam. The guard provides a large radius stop to limit the lift. Common applications are low-to-medium-speed machines and medium pressures. These valves have not proven successful in higher speed applications or for higher pressures.

Channel Valves

Channel valves (Figure 3-14) use a long and narrow metallic channel-shaped sealing element. Springing is provided by a pre-curved long and narrow strip similar to a single element leaf spring. This spring deflects against a flat stop that causes the spring rate to increase as the channel

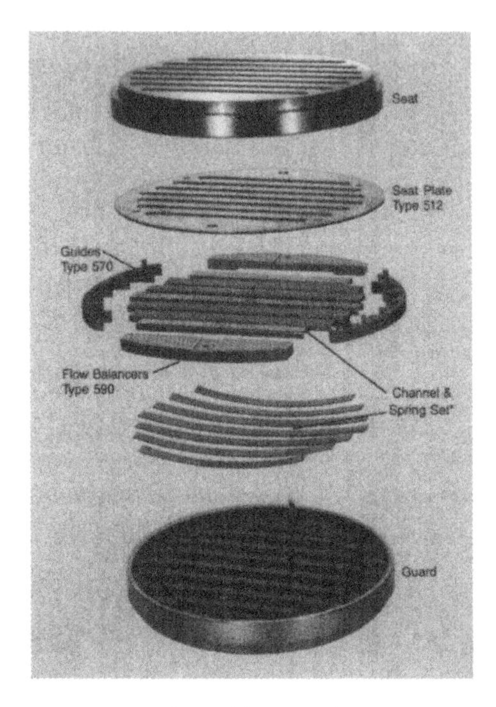

FIGURE 3-14. Channel valves (*Source: France Compressor Products, Newton, Pennsylvania*).

nears a maximum lift. The spring fits into a channel with a close clearance that can provide some gas damping. The application is similar to the feather valve, but can operate with higher pressure differentials and under more severe conditions of moisture and foreign materials.

Rectangular valves, both feather and channel valves, have straight elements and are inherently efficient. Due to their geometry, the strips or channels must incorporate relatively high lifts. Because they use metal elements, they can resist only moderate impact velocities and require clean, non-corrosive conditions. They can, however, run with narrow sealing surfaces.

Rectangular valves are the valve of choice when the conditions of service allow low-impact velocities at high lift and when the gas is clean and non-corrosive. Under these conditions, they provide reliable service and the highest efficiency obtainable.

Reed Valves

Reed valves are similar to feather valves in that the sealing element is the spring. One end of the reed is fixed so that it deflects as a cantilever beam. The stop is either contoured or a bumper that limits the deflection of the free end of the reed. These valves are used in small horsepower, intermittent-duty compressors.

Concentric Ring Valves

These valves use one or more relatively narrow metallic or nonmetallic rings arranged concentrically about the center line of the valve. Spring is provided by either wave springs or, more commonly, coil springs. There are many variations of springing configuration, but the most popular in use today features multiple small coil springs on each ring.

In many cases of oil-free applications, a nonmetallic "button" maintains a separation between the spring and ring.

Lift is controlled by a flat stop that limits the ring motion. In some cases, the ring grooves merge into a close clearance groove that provides fluid damping. (See Figure 3-15).

Concentric ring valves have a wide application range and are used in low- to high-speed applications. Lifts are as high as .160″. Experience has shown that the top end of application speed is in the area of 1200 rpm. Lifts are approximately .100″. The flow efficiency of these valves is not as good as straight element valves, but it is good for the pressure ratios generally used.

These valves can easily be used with plug-type unloaders by eliminating one or more of the inner rings.

The rings have either a rectangular cross section as shown, or a more aerodynamically streamlined shape (Figure 3-16).

Valves that use a rectangular ring are easier to manufacture and maintain. This, plus dynamic considerations, favors the rectangular configuration in most applications.

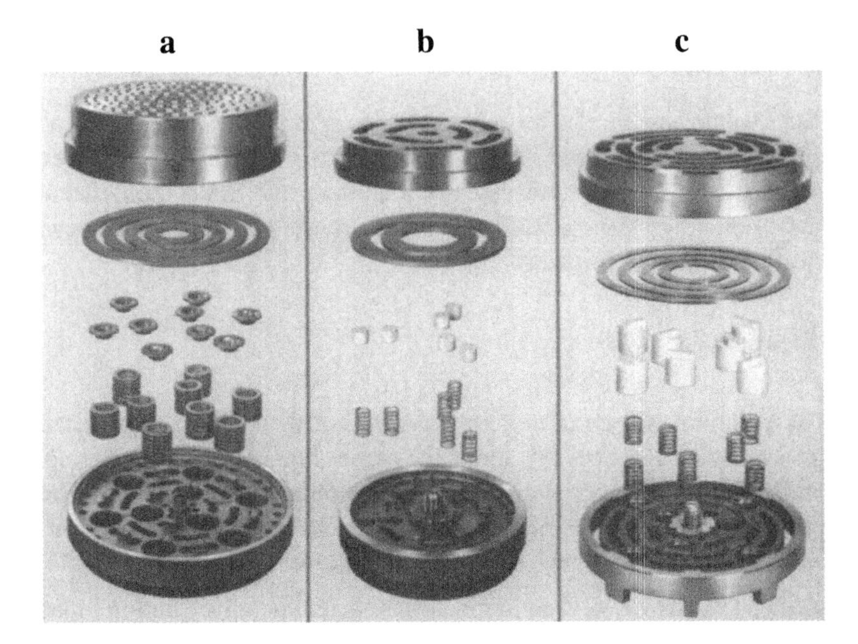

FIGURE 3-15. Concentric ring valves: (a) Typical lubricated, ring-type reversible valve of dual spring design. Each spring activates two rings. (b) Typical lubricated, ring type reversible valve of series spring design. Each ring is activated by three or more individual springs. (c) Typical non-lubricated, ring-type reversible valve employing spring loaded dynamic guide bearings. (*Source: France Compressor Products, Division of Colt Industries, Newtown, Pennsylvania*)

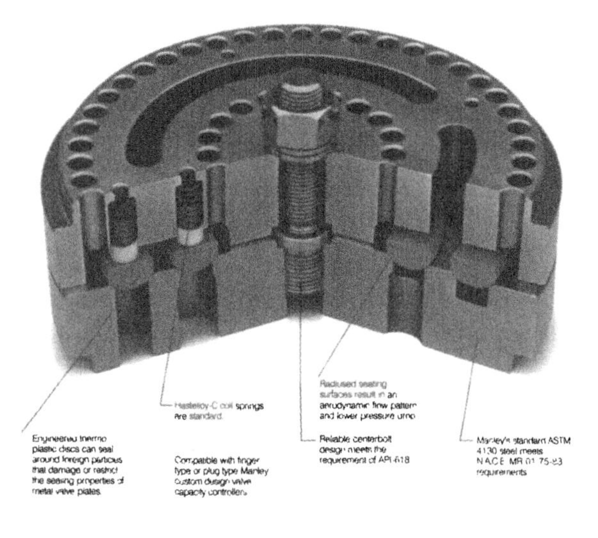

FIGURE 3-16. Aerodynamically shaped concentric ring valves (*Source: Cook-Manley, Houston, Texas*).

Ported Plate Valves

Ported plate valves are similar to the concentric ring valves, except the rings are joined together into a single element (Figure 3-17).

There are two advantages to this when compared to the concentric plates:

- The number of plate edges available for impact is reduced.
- Mechanical damping by using damping plates to absorb opening impact is easily introduced.

Spring is generally provided by coil springs that directly contact the plates, eliminating the buttons used in concentric ring valves.

The flow efficiency of these valves is roughly equivalent to the concentric ring valves, but the geometry of the sealing element is not as flexible as the concentric ring valves. For this reason, the ported plate valve

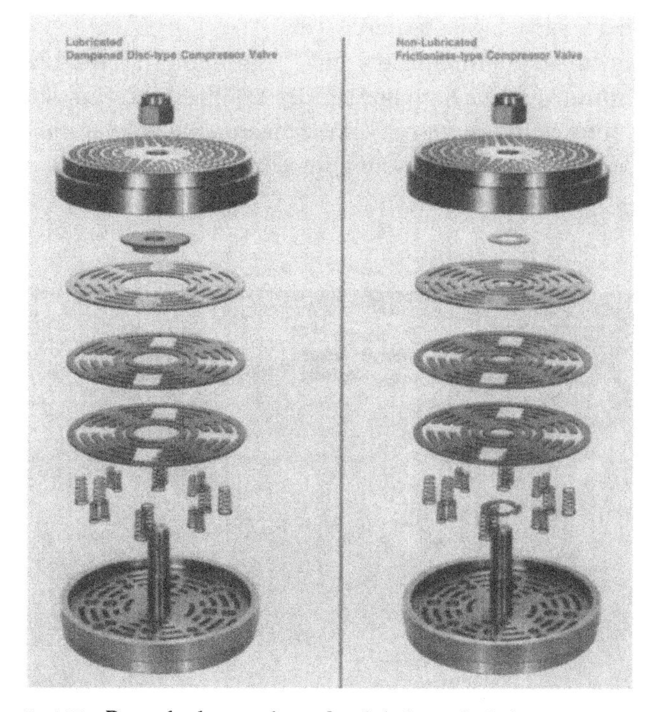

FIGURE 3-17. Ported plate valves for lubricated (left) and non-lubricated compressors (right) (*Source: France Compressor Products, Newton, Pennsylvania*).

is considered a high-speed valve although it can be applied equally well to low-speed machines.

Plate valves, concentric ring and ported types, are not the most efficient nor the least expensive valves, but they are more widely used than other types for one reason: flexibility. They can use metal or plastic plates, can use mass damping, and can be made with many narrow plates or a few wide ones, and thus can be effective over a wide range of lifts.

Disc Valves

Disc valves find many applications. The geometry varies from a single small disc in small valves to multiple discs in large valves. They can be found in applications ranging from a few hundred psi, as in poppet valves for gas transmission compressors, to over 50,000 psi in poppet valves for hyper compressors.

The design characteristics of poppet valves for gas transmission service are multiple nonmetallic and streamlined sealing elements in single- or double-deck valves. (See Figure 3-18).

Maximum lift is approximately .300″. Springing is provided by individual coil springs for each poppet. Early designs used fluid damping via unvented spring wells. However, experience indicated that this was not a good idea, and the spring wells are now vented.

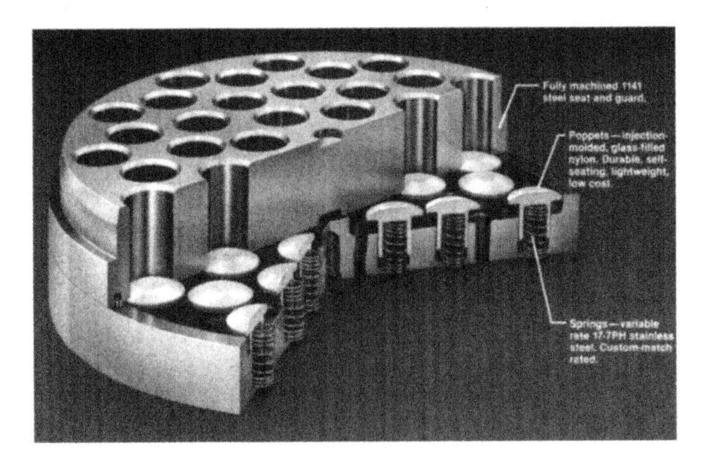

FIGURE 3-18. Disc valve with molded nylon-glass poppets (*Source: Cooper Energy Services, Mount Vernon, Ohio*).

Flow efficiency is excellent because of the high lift and streamlined poppet head. This makes an ideal valve for low-pressure ratio, high-gas density applications where valve losses are very important and sufficient pressure drop can be generated to drive the valve open.

The poppet valve can be applied in other than gas transmission service, but care must be taken to limit both valve lift and operating speed. Poppet valves at speeds above 600–700 rpm have proven somewhat unreliable.

Poppet valves require relatively large sealing surfaces because the poppets are nonmetallic and because the seat is angled. In addition, the poppets have to be relatively large. Thus poppet valves also work best at high lifts; values of .250″ or higher are common.

With plastic poppets, poppet valves can run with corrosive gas and with relatively high impact velocities, but with the very high lifts required for optimum performance they can only be used under conditions of service that create relatively low-impact velocities; typically low-speed, high-valve area and low-to-moderate gas density.

Because of the high lift and streamlined poppet shape, poppet valves are very prone to flutter. Therefore, to be efficient and reliable, they require relatively dense, medium-to-high pressure gas. Poppet valves also tend to have high-clearance volumes that preclude their use for high-pressure ratio applications.

Thus poppet valves are ideally suited to the low-speed, medium-pressure, low-ratio efficiency service typical of gas transmission applications.

Table 3-5 gives an overview of the suitability of valve types for various gas conditions.

TABLE 3-5
SUITABILITY OF VALVE TYPES FOR VARIOUS GAS CONDITIONS

	Gas Conditions				
Valve Type	**Clean**	**Dry**	**Wet**	**Dirty**	**Corrosives**
Feather (Strip)	•	•	•		
Channel (Strip)	•	•	•		
Plate	•	•	•	•	•
Ring	•	•	•	•	•
Disc (Poppet)	•	•	•		•

VALVE MATERIALS

Valve components are made of various materials depending on the operating pressures and type of gas to be handled.

Materials used for valve seats and guards include:

- Cast iron—most common for low and medium pressures
- Ductile iron—for medium pressures
- Cast steel—for high pressures
- Steel (bar stock)—for high pressures

Depending on the pressures, combinations of the above are commonly used, that is, cast iron guards with the seat made of ductile iron or steel.

For highly corrosive gases, various alloy materials are frequently used. Among these are stainless steel (316, 410, 420), 17-4 PH stainless, high nickel and cast iron.

Valve seats and guards machined from steel bar stock typically conform to AISI 4140 or 1141. Some valve suppliers use 410 stainless bar stock for seats and guards.

Sealing Elements

Materials used for sealing elements fall into two classes: high strength steels and plastic composites.

Feather valve strips, valve channels, and springs are made of steel, usually 410 stainless that is stress relieved, hardened, and tempered for maximum wear resistance. For more corrosive applications, channel may be made from 316 stainless. Feather valve strips may be made of Inconel 17-4 PH materials for greater corrosion resistance.

As indicated in Table 3-6, valve discs or plates may be metallic or plastic composite. The advantage of metal plates is their ability to withstand high temperatures and high differential pressures.

Their disadvantages are:

- Even stainless steels are subject to corrosion attack by substances frequently found in the gas compressed (for example, sulfides and chlorides).
- They fail by impact at relatively low-impact velocities (typically 20 to 25 ft/sec.).

TABLE 3-6
VALVE PLATE MATERIAL SELECTION
(WOOLLATT, DRESSER-RAND CO.)

CONDITION	COMMENTS	GLASS NYLON	410 SS	OTHER METALLIC	CUSHION
Corrosive elements in gas	H_2S, chlorides, CO_2 and H_2O	E	P	G	G
High discharge temperature (> 300°F or 180°C)	High ratio gas with high k-value	X	E	E	G
Dirty gas	Dirt will embed in glass nylon	G	P	P	P
High differential pressure (> 1000 psi)		X	E	E	E
Unloading by plate depressors	Plug unloaders preferred	X	G	G	G
Liquid slugs		P	P	P	P
High impact	Can occur with high MW, at high pressure or at high speed	E	G	P	E
Flutter	Occurs if springs too heavy or lift too high, especially at low speed with low MW gas	P	X	X	X

E = EXCELLENT G = GOOD P = POOR X = CANNOT USE

- They cause seat wear. Serious damage can result if portions of failed elements get into the cylinder.
- They are easily damaged by dirt and liquids.

For most applications, 410 stainless steel is used but 302, 316, 420, 17-4 PH stainless steel, monel, Inconel, or even titanium are used for special applications.

Plastic Composites

Plastic plates, if correctly formulated and manufactured, have lower differential pressure and temperature limits, about 1500 psi and 400°F for PEEK (polyether-ether-ketone) based compounds and somewhat lower for the nylon based materials.

They are, however, far superior to any metal element in resistance to impact fatigue, corrosion resistant, and the ability to resist damage from dirt and liquids.

Plastic elements cause very little seat wear and cannot cause damage to the cylinder if they fail.

For optimum valve plate service at temperatures not exceeding 300°F, asbestos-bakelite is a good choice. For higher temperatures, glass melamine or glass epoxy is used, and, of course, the more recent PEEK formulated materials should be given prime consideration.

Springs

The springs used on the sealing elements of valves include AISI 410 and 420 for feather and channel valves; for coil springs 17-4 PH, Inconel, and cadmium-plated chrome-vanadium steels are used, depending on the manufacturer and service conditions.

For all valve elements not only is the proper material important, but rigid specifications and quality, such as flatness, dimensional tolerances, and surface finishes are key considerations.

FUNCTIONS OF COMPRESSOR VALVE COMPONENTS

The compressor valve internal components are designed to survive the simultaneous impact stresses imposed by a large number of cycles and environmental stresses such as high temperatures, corrosive gases, entrained liquids, gas pulsations, etc.

The compressor designer is responsible for choosing a valve assembly to suit the application; however, it is the responsibility of the operator and maintenance personnel to see that the valve assembly is properly maintained.

- Valve seats must resist the differential gas pressure and wear on the surfaces in contact with the valve sealing element (discs, strips, or channels).

- The guard or stop plate, frequently the most misunderstood part of a valve assembly, has a threefold purpose:
  1. Provides a guide for motion of the valve sealing element
  2. Controls the lift of the valve
  3. Retains the valve return spring

  Because guards are not exposed to differential pressures, they need only to resist the impact forces of the sealing elements making contact.
- Valve sealing elements function by alternately moving to open the valve when it is exposed to differential pressures (allowing gas to pass through) and then moving in the reverse direction (blocking gas flow).
- Valve springs close the valve at the end of the piston stroke when the differential pressure across the valve approaches zero. Each spring is designed for the particular environment it will encounter. The substitution of springs with different characteristics may introduce serious reliability and safety risks.

VALVE LIFT

Valve lift is defined as the distance the valve sealing elements move from closed to fully open. The life of any valve depends upon the amount of lift, which varies according to compressor speed, valve diameter, pressures, and molecular weight of the gas.

One authority stated that the first question to be answered when designing or selecting a valve for a given set of operating conditions is "What is the proper valve lift?" In any valve application to a compressor, lift affects both efficiency and durability.

Too high a lift will cause premature valve failure due to impact fatigue. Too low a lift will result in excessive gas velocity through the valve, high losses, and consequently inefficiency. These facts must be reconciled in a sound commercial design.

Rotative speed is the most important factor influencing valve lift. This, of course, involves the speed with which the valve must open and close. Experience has shown that a good valve design has good durability with a 0.200-inch lift at 300 rpm; whereas, the lift must be reduced to 0.040 inch when running at 1800 rpm. In any event, maintaining valve lift is critically important and proper maintenance and repair techniques are essential. For example, worn seats or guards can often be refinished, but the original lift may be altered if refinishing is incorrectly done.

Lift limitation is obtained in different ways in different designs and, therefore, cannot be readily specified. Manufacturer's instruction books are usually quite detailed in this regard and should be carefully followed.

Above all, do not alter the original lift without consulting the manufacturer.

VALVE FAILURE ANALYSIS

When valve failure analysis is undertaken, component appearance must first be investigated to determine causes of failure.

Valve failures can be classified as resulting from three general causes:

1. Wear and fatigue
2. Foreign materials
3. Abnormal mechanical action

Wear and Fatigue

Wear, as such, cannot be completely eliminated. It can be minimized by proper lubrication, design, and selection of materials.

Most valves require some means of guiding. Wear at the guides, if severe, can result in sloppy action, cocking, poor seating, and failure. Wear must never be permitted to develop to this degree.

Fatigue is the result of repeated cyclic stress. The stress level, as well as the stress range, must be considered in any design. However, barring abnormal action, well-designed valves have a good record of withstanding fatigue.

Both wear and fatigue are adversely affected by abnormal mechanical action.

Foreign Materials

Foreign materials may be listed in the following categories:

1. Liquid carryover
2. Dirty gas
3. Carbon formation
4. Corrosive elements

- Liquid carryover from a process or from interstage coolers is apt to cause premature failures, particularly on intake valves. A slug of liquid is particularly hard on valves and may even break a seat. Liquid also destroys lubrication, thus, accelerating wear.

  Carryover can result from poor separation of condensate ahead of the compressor, or improperly designed piping where low spots permit liquid accumulation. A sudden flow change can then cause a carryover.

  It is important that interstage separators be drained regularly. Manual drainage at specified intervals is preferred to automatic traps, particularly at higher pressure levels. Automatic traps, if used, should have a bypass piped for visual observation and a check on their operation.

  Liquid slugs can also be formed when saturated gas comes in contact with the compressor cylinder walls. This can be prevented by maintaining the compressor cylinder jacket water temperatures 10 to 15 degrees above the temperature of the incoming gas stream.

  Examination of the valve sealing element will sometimes indicate liquid carryover or momentary high pressure damage, but liquid in a cylinder is generally discernible by other means.
- Dirty gas causes all sorts of problems. It accelerates wear very rapidly at all guiding points since it acts as a grinding compound. Foreign matter between the coils of springs is a frequent cause of spring failure and subsequent failure of other valve parts.
- Carbon or sludge resulting from an unfortunate combination of a particular oil and the gas being compressed may hinder proper valve action. The additives in an oil in combination with a certain gas may sometimes lead to problems. This is usually unpredictable, but in some cases, simply trying another oil has been successful.

  Too much oil can be as undesirable as too little oil. The rate of feed must be determined from experience and should be no more than is necessary to properly lubricate a cylinder. Refer to Figure 3-5.

  Too much lubrication of the discharge valves often causes carbon build-up on the valve surfaces and becomes a flake-shaped impurity. This affects the action of the valve and flakes breaking loose may cause leakage in the subsequent stages.
- Corrosion can cause high localized stress and subsequent failure. Springs are apt to fail first. The solution to corrosion problems is not always simple, and is rarely inexpensive. Valve materials can be changed. In some cases, properly sized scrubbers or chemical washers can eliminate or reduce contaminants before they enter the com-

pressor. Low velocity of the gas is mandatory for effective liquid separation in suction vessels. In a few cases simply increasing jacket water temperature or preheating the gas has eliminated condensing out of corrosive elements.

Corrosion may not cause valve failure for several months. When it does, it is best to change out and repair all the remaining valves, because they are probably close to failure. When there is a corrosive condition that cannot be eliminated, it is best to have a complete set of spare valves. This will result in less downtime and less expense.

Abnormal Mechanical Action

There are four causes of abnormal operation, although there is some interrelation between them:

1. Slamming
2. Fluttering
3. Resonance or pulsation
4. Flow pattern

- Slamming can occur when a valve opens or when it closes. Valves normally have little tendency to slam except possibly when the discharge valve is opening.

  Valve opening can be cushioned either mechanically, by springs or by a gas cushion, or by both.

  If a valve closes late, whatever the reason, backwash gas flow rather than the springs will close the valve and slamming is certain to occur. Slamming is suspected if the sealing element has a hammered or mottled appearance where it contacts the seat. Listening will often confirm slamming.
- Fluttering is a result of insufficient pressure drop through the valve. The pressure drop is a function of velocity, density, and flow coefficient. If springs are too stiff, a valve may be unable to open fully and, instead of being held securely against its stop, it remains somewhere between zero and full lift and oscillates. Springs usually suffer under these conditions.

  In severe cases, a valve may strike the seat or guard several times during one stroke of the piston. The inertia of the valve may cause it to fully open against its stop, but pressure drop is insufficient to hold it there. Consequently, it starts to close again.

Inertia may close it too far, even back to the seat before flow forces cause it to open again. By the end of the piston stroke, it may have oscillated several times.

Furthermore, at the end of the stroke, inertia may carry the valve open instead of back toward the seat, and when backwash finally closes the valve, it is with a slam.

Examination of the valve and seat may confirm fluttering. If the stop plate shows no markings, it may indicate that the valve has not fully opened. Normally the back of the valve plate itself will have some sort of pattern from the springs. In a circular or plate valve, if there is no definite pattern, and the plate appears to spin, then fluttering is likely to cause failure.

The solution to fluttering is to lower the lift and/or to use lighter springs.

- Resonance or pulsations can upset normal valve action. The amplitude of pulsations is of considerably less consequence than the phase relation of the pressure wave and crank angle. A pulsation can cause late closing, thus slamming, as previously mentioned.

  A series of high peak-to-peak pressure fluctuations surround the valve, usually as the result of an improperly designed piping system. As the sealing element tries to follow this pressure change, it flutters erratically.

  In some cases, the sealing elements will pound against the seat and guard many times during a single piston stroke, causing breakage from impact fatigue.

  Pulsations of relatively high frequency, for example, between 50 and 100 Hz, are detrimental to compressor valve life.

  Figure 3-19 depicts on a time basis what happens. The valve is not only open at dead center, but the rate of pressure change is high. The valve is slammed shut by back flow after dead center.

  The problem can be solved by eliminating pulsation through piping changes. This is a case where the problem is outside the valve designer's control and valve changes will not correct the problem.

- Flow pattern failures are infrequent, but they have happened. There may be a disturbance within the cylinder passage leading to a valve that causes some type of erratic valve action.

  The cure is to change the lift and/or springing (depending on the type of failure), or even to change the valve type.

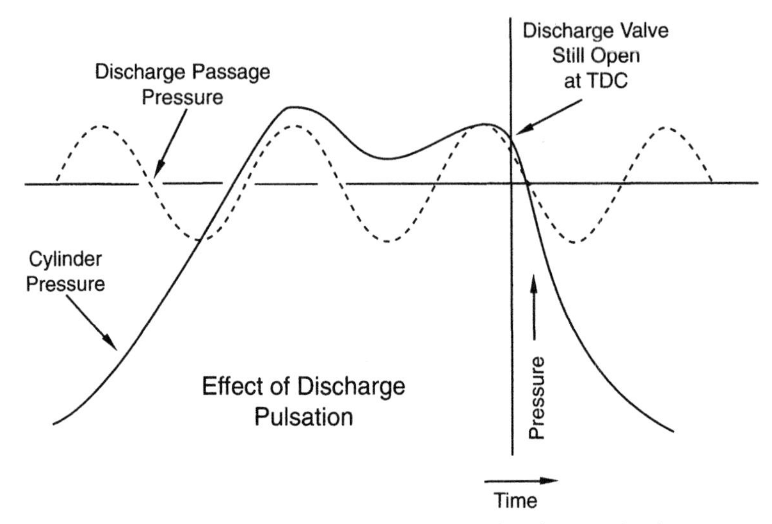

FIGURE 3-19. Time-pressure diagram illustrating how pulsations can cause abnormal valve closing.

MAINTENANCE AND REPAIR OF COMPRESSOR VALVES

The efficiency and reliability of a compressor depend to a great degree on proper valve maintenance. Valves should be inspected periodically and worn components reconditioned or replaced.

A guideline for the recommended frequency of valve inspection cannot be given, because this depends too much on the type of compressor, the operating conditions, as well as the type of gas compressed.

In new compressor installations, compressor valves should be inspected and cleaned within the first 1,000 hours. At this time, the wear on valve plates and seats should already give some indication of the necessary frequency of maintenance.

Valve Failure Groupings

When compressor valve failure analysis is undertaken, effects are investigated to determine causes. The effects are usually excessive wear, fatigue, fracture, or a combination of these. They fall into two groups:

1. Environmental effects
2. Abnormal mechanical action

Environmental Effects

Listed here are combined influences on valve life and performance derived from the gas itself. These are:

1. Corrosive Elements

   Substantial amount of corrosive contaminants in a gas will usually show up on the valve. Even small amounts, though they will not rust away the valve, can cause stress corrosion and lead to such damage as breakage of the sealing element (seat and plate). Certain compounds become corrosive only if moisture is present in the system or develops after shutdown. This moisture, in combination with the contaminants in the gas, can corrode valves.

   Coping with corrosiveness may require metallurgical upgrading of valve plate, strip, or channel, or, in severe cases, the materials of all valve components. Since this involves a major expenditure, be sure to investigate which measures are called for.

   Although hydrogen is not corrosive by itself, it can, under certain conditions, cause embrittlement due to molecular penetration of the metal (hydrogen embrittlement) of at least the top layer. The ensuing cracks lead to deterioration and subsequent fracture. Hydrogen embrittlement can be prevented by changing materials.
2. Foreign Materials and Impurities

   Despite proper filters or scrubbers, foreign particles can get wedged in the valve and prevent its proper operation, causing damage. Examine the seat lands and impact surfaces of the valve plates for traces of debris. Minor indentations and imprints of particles may show up between the plate and seat. Make certain filters, separators, knockout pots, and drains are working properly and are sized to handle any impurities from upstream.
3. Liquid Carryover

   Liquid slugs can have a devastating effect on valves. The plate or strips will be subjected to extreme destructive forces and will crack. Slugs occur when entrapments get carried through. They are formed when saturated gas contacts the cylinder walls. Prevention involves raising the cooling water temperature 10 to 15° above the incoming gas temperature. If liquids are coming from the upstream feed line, separators can be checked and much of those liquids eliminated by proper construction and sizing (for low gas velocities) of separator internals.

4. Improper Lubrication

   Valve life is shortened by excessive lubrication. It can cause sticking of plates or strips, which delays reseating of the sealing element. A delayed closing normally results in excessive slamming forces. Excessive lubrication acts as liquid carryover and can cause the same damage as condensate slugs or water contamination. Too much lubrication of the discharge valves, especially if they are exposed to high temperatures, often causes "coking." A carbon buildup will form on the valve surfaces and become a flake-shaped impurity.

Abnormal Mechanical Action

This term covers the effects that substantially alter the normal opening and closing motion of the valve. Included are:

- valve flutter
- slamming from delayed closing or other pulsations
- multiple impacting from excessive pulsations

Well-designed valves, with the proper spring load for a given application, will do a good job. Their opening and closing motion will be such that no harmful pulsations will occur. However, many valves are standardized for average conditions of operation and, when applied outside this range, may malfunction. The right spring load of a valve depends on, among other factors, its operating pressure, the gas velocity, and the specific gravity of the gas.

It is difficult to detect abnormal mechanical action, but careful analysis is warranted of the surface where the sealing elements impact against either the seat, upon closing, or guard, upon opening. If these surfaces show wear related to impact (a hammered finish), assume that flutter or multiple impact is the problem.

Abnormal mechanical action can also result from pulsations in the gas stream. These may be caused by improper pipe selection or manifold sizing, but with today's design technology available to the major compressor builders, this is rarely the case.

The flow of gas to and from the valve is channeled around cages, through cylinder openings, and cavities under and above the valve. The

TROUBLESHOOTING VALVE PROBLEMS

Failure	Reason or Action to Be Taken
Suction valves only	1. Liquid carryover 2. Intake pulsations 3. Make sure correct parts are used
Discharge valves only	1. Discharge pulsations 2. Over lubrication 3. Make sure correct parts are used
Both suction and discharge valves	1. Discharge failures due to broken pieces from suction valves 2. Rust, scale, or foreign matter 3. Insufficient or improper lubrication 4. Pulsations on both suction and discharge 5. Make sure correct parts are used
Failure following reconditioning	1. Incorrect reconditioning procedure 2. Make sure correct parts are used 3. Reusing worn parts
Rapid wear of suction valves only	1. Incorrect or insufficient oil 2. Cylinder jacket water temperature is lower than incoming gas temperature 3. Gas may be wet and washing inlet lubrication from valves
Seasonal failure of valves	1. If failures occur in winter months only, liquid is the probable cause and piping should be insulated or steam traced and/or knockout pots placed close to the cylinder
Valve seat gasket failures	1. Gasket improperly installed and pinched 2. Valve loose due to improper tightening or being knocked loose due to liquid slugs
Valve seat fractured	1. Valve loose by improper tightening 2. Valve tightened against broken pieces of gasket 3. Improper tightening of valve jam screw or cover nuts 4. Valve loose due to liquid slugging

gas flow does not impact the closed valve evenly around its perimeter; thus, the wobbling opening and closing motion results.

Uneven impact can also cause multi-ring valves to open unsynchronized, one ring opening first and taking the most severe impact, and the others following the lesser impacts against the stop plate. Such a pattern will cause one ring of the set to fracture more frequently than the others.

VALVE PROBLEMS

Experience has shown that after the "shakedown" or initial commissioning of a new compressor installation, very few valve problems occur and the compressor operates trouble-free.

Valve problems start to become more frequent after a period of six or seven years of operation and are due to normal wear. If improper reconditioning and rebuilding is then performed, these problems become frequent and cause continuing unscheduled shutdowns.

Invariably the valves are condemned and valves of a different type or manufacture are purchased and installed. The valve troubles now disappear, not because of the different style valve or manufacturer, but simply because new valves have been installed.

The same results might have been obtained at considerably lower cost if the original valve had been properly maintained and properly reconditioned.

Again, keep in mind that improper maintenance is the major cause of valve failures. Valves are often rebuilt with parts that have not been reworked and restored to the original manufacturer's tolerances or have been machined with no regard to flatness, finish, or hardness effects.

Record Keeping

Experience has also shown that a systematic study of valve problems will provide the best and often the quickest solutions to valve problems. Record keeping is necessary to successfully evaluate and solve valve problems. It is obvious that records should be sufficiently detailed to show which parts are involved and where these parts were located in the compressor. A picture of some sort is helpful in spotting a pattern of valve failure. A typical valve location chart is shown in Figure 3-20.

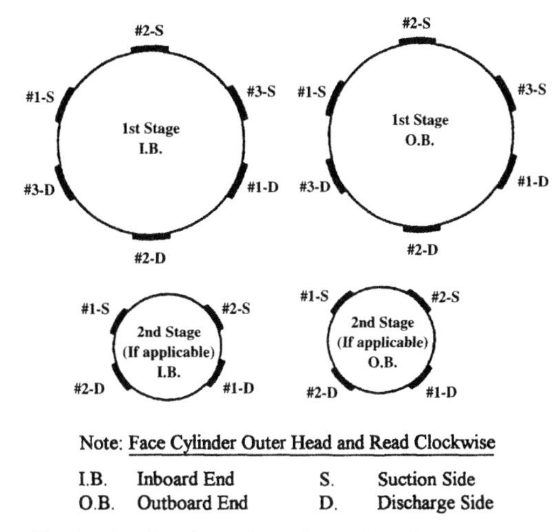

FIGURE 3-20. Typical valve location chart showing a common valve numbering system (*Source: Dresser-Rand, Painted Post, New York*).

When a valve has to be changed, it should be removed and tagged with identifying marks, as shown in Figure 3-20. For example, 1st stage, IB, 2D refers to first stage frame end, center discharge. A properly reconditioned replacement valve should now be installed. The valve that was removed should then be examined and a record made, identifying the failed part and its location. Figure 3-21 shows the type of valve record or report which should be maintained.

DISASSEMBLY, ASSEMBLY, AND INSPECTION PROCEDURES

Disassembly

Refer to Figure 3-22A for a typical valve assembly with conventional restraint.

1. Remove valve screw cap and back off valve jam screw a minimum of four complete turns.
2. Remove valve cover nuts and remove valve cover.
3. Loosen set screws holding valve yoke and remove yoke.
4. Remove cover and valve seat gaskets and discard.
5. Carefully clean gasket seating surfaces on valve covers and valve seats.
6. Visually inspect valve seats for any signs of damage such as burring, wire drawing, and breakage.

VALVE FAILURE RECORD

Date of valve problem	Compressor number	Replaced by (initial)	Location of valve problem	How long has valve been in service (hours)	What failed?	Is lubrication on cyl. bore and valve adequate? (lubricated units only)	Suspected reason for failure

FIGURE 3-21. Typical valve failure record (*Source: Dresser-Rand, Painted Post, New York*).

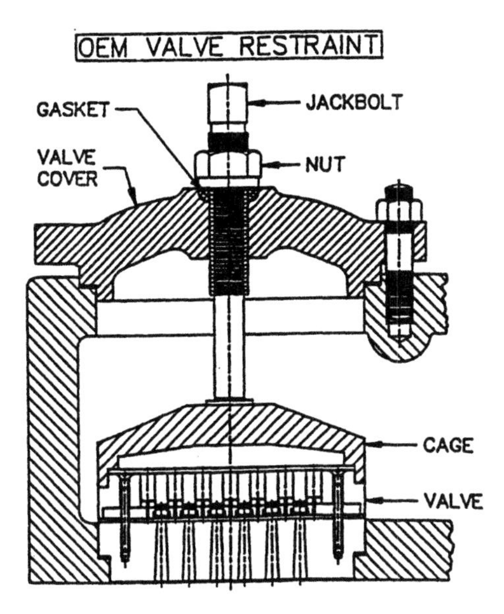

FIGURE 3-22A. Conventional valve restraint (*Source: Cook Manley, Houston, Texas*).

Assembly

1. Install new valve seat gaskets of the proper diameter, thickness, and material.
2. Place valve assembly on seat, taking care that the gasket stays in place and is not pinched.
3. Place yoke against valve and tighten set screws to hold assembly in position (lower valves). Extreme care must be taken so as not to "cock" valve assembly off the seat or over tighten set screws, as this can cause cocking of the valve assembly. Improper tightening may cause cocking of valve, loose valve, and leaking valve.
4. Install new valve cover gaskets.
5. Install valve cover, making sure the jam screw is backed off and not in contact with yoke. Tighten valve cover nut evenly in crisscross fashion and torque to proper value.
6. Tighten valve jam screw and torque to proper value.
7. Install valve jam screw cap, using new gasket.
8. After the compressor has been brought up to pressure and temperature stabilizes, check to be sure there are no leaks or signs of abnor-

TABLE 3-8
TORQUE VALUES—WRENCH READING

Thread Size	Valve Covers Torque ft-lbs	Jam Screws Torque ft-lbs
½″–10	24–28	
⅝″–11	47–56	
¾″–10	85–102	60–70
⅞″–9	138–166	110
1″–8	207–248	150–165
1⅛″–8	307–370	225–250
1¼″–8	435–522	315–350
1⅜″–8	595–714	430–475
1½″–8	790–948	570–630
1⅝″–8	1025–1230	740–820
1¾″–8	1305–1566	950–1050
1⅞″–8	1619–1944	1170–1300
2″–8	2189–2356	1450–1600
2¼″–8		2100–2300
2½″–8		2900–3200

mal valve operation. The valve cover nuts should be retorqued to the proper value (see Table 3-8).

9. Check and record valve cover temperatures using an infrared temperature measuring device.

A problem documented with some compressors is the single-bolt valve restraining system. Under high torque, the gasket in the valve cover can distort, allowing gas leakage and reduced loading on the cage and valve. Continued operation with loose jack bolts can result in broken bolts, cages, and valves. Debris in the cylinder and massive gas release can also occur.

A multi-bolt valve restraint assembly is available for retrofit application to some compressors with weak single-bolt restraints (see Figure 3-22B).

RECONDITIONING AND REBUILDING

The valves removed from the compressor should be sent to the valve manufacturer's shop for cleaning, reconditioning, installation of all new components (sealing elements, springs, etc.), and testing.

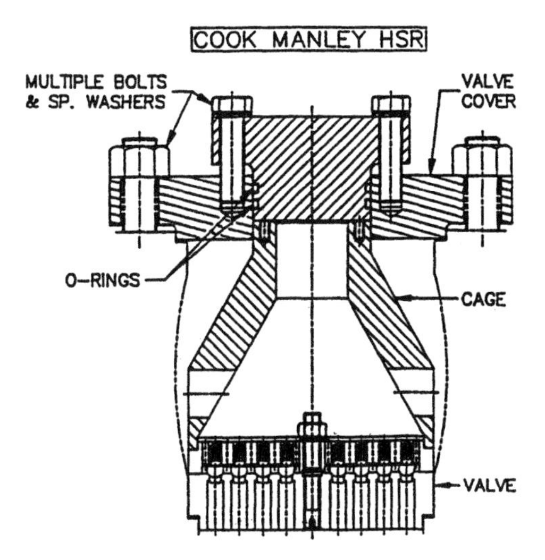

FIGURE 3-22B. High-security valve restraint (*Source: Cook Manley, Houston, Texas*).

It is generally not recommended that valves be rebuilt by other than the manufacturer of those valves or a proven and qualified rebuilder of compressor valves.

Disassembly of Valves

When compressor valves are disassembled, always work from a clean work bench to prevent dirt getting into the components.

Valves should never be clamped in a vise to disassemble or to loosen the locknut from the center bolt. Do not hammer on the wrench when loosening or tightening the nut. To properly hold a valve in place for disassembly or assembly, a device similar to the one shown in Figure 3-23 should be used.

If a Hoerbiger-type holding fixture is not available, a simple device as shown in Figure 3-24 should be made and used for holding the valves during disassembly and assembly. Spacing and size of the pins are made to suit the valves being serviced.

With the lock nuts removed, separate the seat from the guard. Keep the seat and sealing element the way they came apart. If a valve plate can be reused, it must be matched to the same seat that it was used with. Due to tolerances, the seating pattern of every valve sealing element is different. Not matching these parts can result in valve leakage.

FIGURE 3-23. Hoerbiger valve holding fixture (*Source: Hoerbiger-Mitchell, Houston, Texas*).

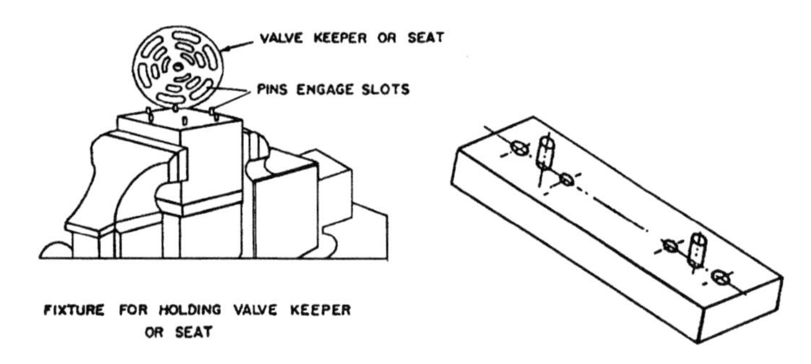

FIGURE 3-24. Elementary valve holding fixture.

1. Clean all parts. Use a cleaning fluid and a soft brush, taking particular care to free the ports of the seat and the guard from all foreign matter. This will ensure full seat area in operation. Never use wire brushes or tools with sharp edges to clean the seats and plates.
2. Check the condition of all parts:
   - seats for damage to seat lands
   - damping plates for cracking
   - valve guard for damage
3. Repair or replace worn parts:
   - Valve plates. When a valve plate or damper plate shows signs of wear, it must be replaced even if no breakage has occurred. Valve plates should never be reground or inverted.

- Springs should be replaced. Do not mix new springs and previously used springs. All springs in the valve should be replaced.

  In the case of conical springs, care should be taken to insert them into the spring nest with the large diameter resting at the bottom of the spring nest.
- Seats. To ensure high valve efficiency, it is important that the seat face is flat and free of any traces of wear, thus preventing valve leakage. If the seat face is damaged in any way, remachine and lap the areas.

  Do not remachine valve seats below the specified minimum dimension; otherwise, there is danger of breakage. Remachining has to be extended over the entire seat face, including the entire center part. When a valve seat surface is remachined, new valve plates or rings should be used to assure proper sealing.

Assembly of Valves

When assembling the valves, cleanliness is extremely important. Be sure to have all parts in the proper sequence. Make sure all parts are properly aligned. Before tightening nuts, press the seat and guard together by hand to ensure that the parts are properly aligned.

The lock nuts should be tightened with a torque wrench to the following values:

Thread	Torque (ft-lbs)
⅜″–24 UNF	15–18
½″–20	26–32
⅝″–18	65–80
¾″–16	129–156
⅞″–14	178–214
1″–12	221–268

After the valve has been assembled, check the sealing element for free movement. This should be done with a small brass rod that will not damage the element.

Testing

Even if valve maintenance has been given a high priority and all parts have been reconditioned as carefully as possible, one more check should still be made on every reconditioned valve—the so called *kerosene test.* Actually, kerosene is rarely used; a commercial solvent substitute is more common. Here, the objective is to set an assembled valve over a pail with the seat in the top position. The valve is then filled with solvent and left for a few minutes (see Figure 3-25).

No valve is free of leakage of the solvent, and some will leak more than others. Some reconditioned valves will not retain solvent at all and will leak through immediately. In some cases, the valve should be disassembled and examined for flaws.

A valve that does not retain solvent for at least one minute may leak soon after start up.

A word of caution: the solvent test reveals only whether the sealing elements are sealing against the seat. It does not reveal leakage between the seat and the stop plate (guard) or around the center bolts. It is important that these other leakage paths also be checked.

FIGURE 3-25. "Kerosene test" for valve leakage (*Source:* Plant Engineering Magazine, *January 7, 1980*).

Testing reconditioned valves with air pressure is preferred even though a special fixture will be required. Figure 3-26 shows a test fixture that can be made relatively simply.

Air pressure of 90–100 psi is applied, and the leak is determined. All valves will leak somewhat over time, but a properly reconditioned valve will leak down the pressure relatively slowly. Again, a one-minute test may be appropriate.

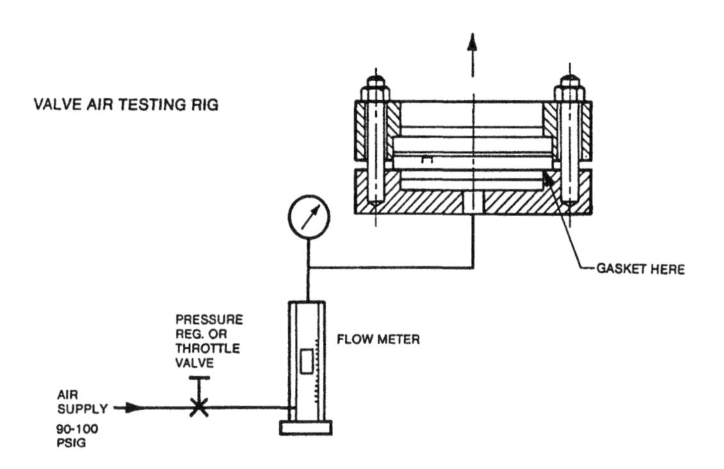

FIGURE 3-26. Simple fixture for air testing of compressor valves.

Figure 3-27 shows another test stand, available from France Compressor Products.

COMPRESSOR PISTON ROD PACKING

Piston rod packing is almost always the "full floating" type. It consists of several annular cups, segmental rings, and a flange-like gland held in the cylinder head stuffing box as a complete assembly by properly proportioned studs and nuts that secure it against a sealing gasket.

The segmental packing (sealing rings) is contained in the cups and is held together as an assembly by garter springs that hold the rings firmly on the piston rod. These rings are free to "float" in the cups.

This packing is essentially a precision mechanical seal with ground or lapped surfaces. Although of rugged construction, it must be carefully handled and examined to see that it is clean and free of nicks, burrs, and

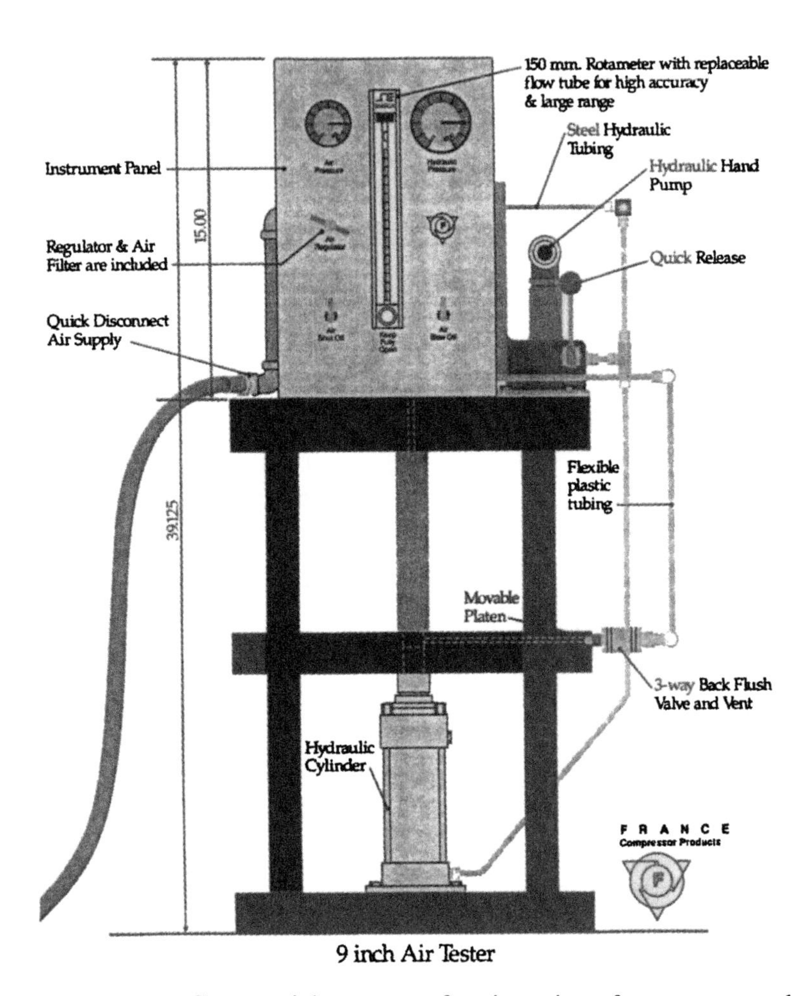

FIGURE 3-27. Commercial apparatus for air testing of compressor valves. (*Source: France Compressor Products, Newtown, Pennsylvania*).

scratches on the sealing surfaces. Packing must be properly installed according to instructions and carefully checked for wear.

Compressor packing can be fully lubricated, mini-lubricated, micro-lubricated or fully non-lubricated. It can also be cooled by oil, water, or thermo-syphon cooling.

A typical piston rod packing arrangement is shown in Figure 3-28.

FIGURE 3-28. Typical piston rod packing arrangement. (*Source: C. Lee Cook, Louisville, Kentucky*).

PACKING TYPES

There are several types of packing available. These include:

1. Fully lubricated. Oil is supplied to the packing from the mechanical lubricator. The cases are drilled so that oil will be carried to the oil cups and then to the piston rod through a drilled hole in the oil cups.
2. Mini-lubricated. Same as fully lubricated, except that oil to the packing is decreased approximately 30 percent of normal flow.
3. Micro-flow. There is no oil supplied directly to the packing. The only oil that reaches the packing is carried over on the piston rod from the crankcase.
4. Fully non-lubricated. No oil is supplied to the packing and no oil carryover from the crankcase is allowed to reach the packing. The crankcase oil is prevented from reaching the packing by means of an oil deflector collar attached to the piston rod between the oil wiper rings and the packing assembly.

Cooled Packing Type

Cooled packing includes the following:

1. Oil cooled. The packing cups contain internal and external passages for circulation of coolant through and around the cups in the stuffing box. An O-ring is contained in the circumference of the outer cup to confine the coolant within the stuffing box. The source of the coolant may either be the frame cooling system or a separate system.
2. Water cooled. This type of packing provides passageways through the cups between the packing rings for circulating cooling water. The coolant completely encircles each cup and first travels to the hottest point of the case, then travels progressively to the cooler points.

   The possibility of leakage is eliminated with precision cup manufacture and O-ring seals between each cup. The coolant must be free of any foreign material and deposit-prone chemicals. The required coolant flow through the packing is small and should be regulated by a needle valve. The flow must be adjusted to provide an outlet temperature only slightly greater than the inlet coolant temperature. Excessive flow is neither necessary nor helpful. A rate of one to three gallons per minute is adequate.
3. Thermosyphon. This type of packing is similar to water-cooled packing with passageways through the cups also. Circulation of the coolant through the packing is by convection. Heat is dissipated by ambient air passing over externally mounted finned tubes. Coolant level must be maintained within limits indicated on a sight flow gauge.

Packing Rings

Packing rings are the heart of the packing assembly, sealing along the piston rod and between themselves and the packing case (see Figure 3-29).

The tangent ring is cut into three segments so that each cut lies on the side of an equilateral triangle. The cuts of this ring maintain sealing contact regardless of variations of the ring's inside diameter. As wear occurs, the ring segments will close radially to compensate while still maintaining sealing contact at the tangential joints, Figure 3-30 and Figure 3-31.

The clearance provided at the end of each tangent joint to allow for such compensation allows a direct leakage path. To seal these joints, a ring cut radially into three segments is paired with and pinned to the tangent rings so that its segments form a seal along the rod at the edge of the

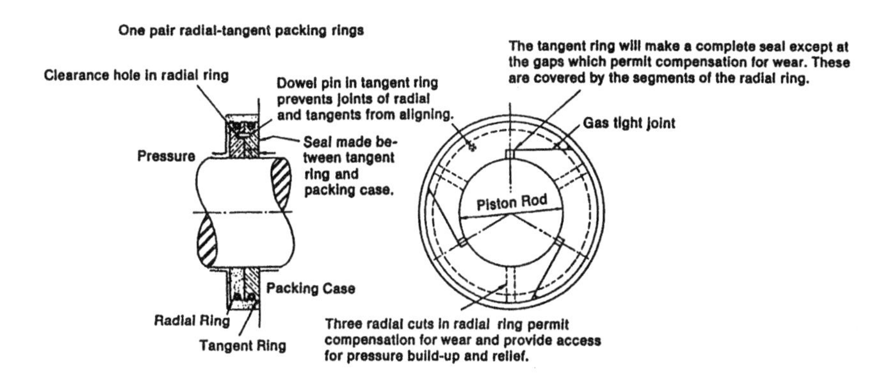

FIGURE 3-29. Tangent-cut sealing rings. (*Source: France Compressor Products, Newton, Pennsylvania*).

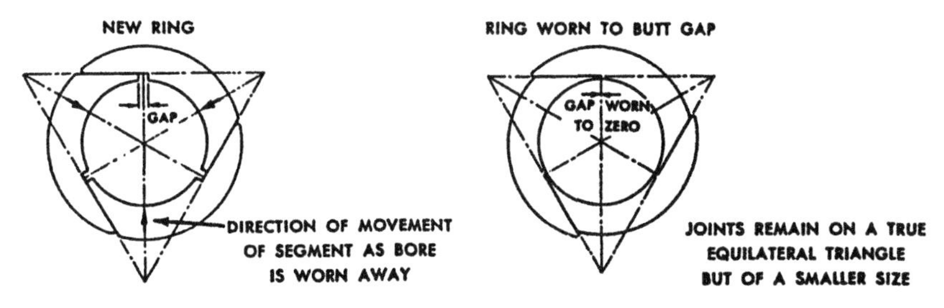

FIGURE 3-30. Sealing ring condition as wear takes place. (*Source: France Compressor Products, Newton, Pennsylvania*).

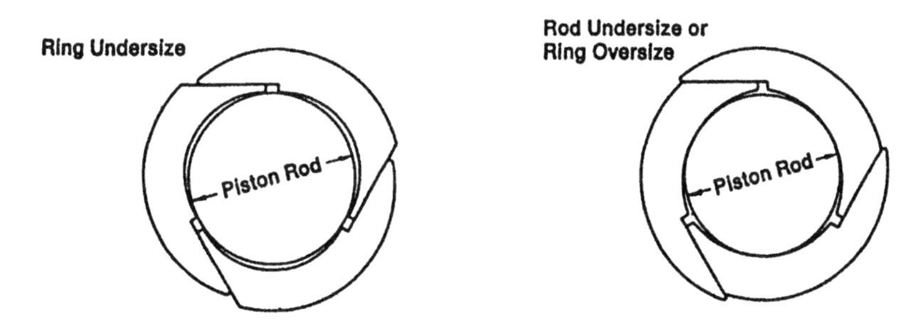

FIGURE 3-31. Sealing ring alignment with variations in bore fit. (*Source: France Compressor Products, Newton, Pennsylvania*).

tangent ring. The gaps in the radial ring provide access for pressure buildup and relief over the ring's outside diameter.

How Compressor Packing Works

As previously described, a packing set consists of a series of sealing units. Compressor packing is not bottle-tight, but the amount of leakage is extremely small, a fraction of one percent of the capacity of the machine and usually within tolerable limits. In case of toxicity of the gas or danger of explosion or corrosion, leakage may be vented to a safe place.

Because the individual rings will leak slightly, it is through the series of rings that the pressure is broken down from discharge to atmosphere (see Figure 3-32). The higher the pressure, the more seal rings will be required.

The volume of leakage will increase with an increase in the differential pressure across the rings. It will also increase with the time that the differential pressure exists. In other words, the higher the differential pressure and the longer the differential exists, the greater will be the volume of leakage.

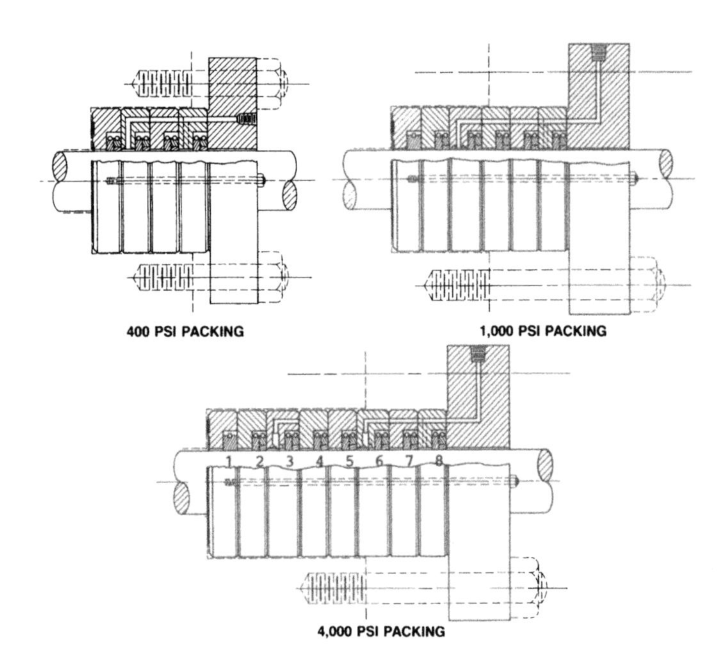

FIGURE 3-32. Conventional packing cases for different pressure levels (*Source: C. Lee Cook, Louisville, Kentucky*).

In addition to the normal vent to the atmosphere, packing cases may be arranged to vent higher pressure or introduce pressure between any of the individual rings. This is done generally to control leakage or provide means for recovering leaked gas. For cases vented as shown in Figure 3-33, normal uses include preventing air leakage into the cylinder when suction pressure is below atmospheric pressures or maintaining zero leakage of a dangerous or expensive gas.

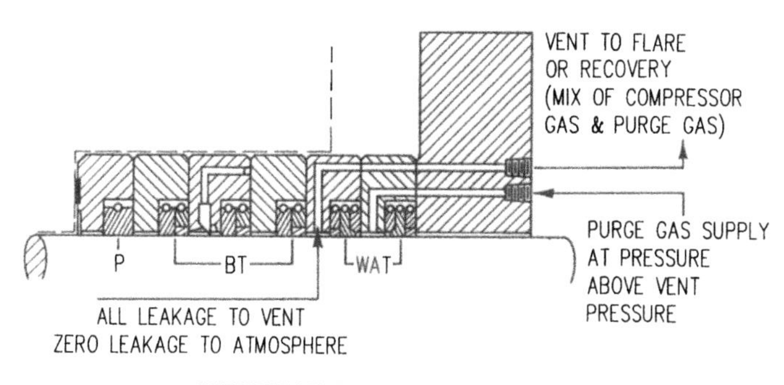

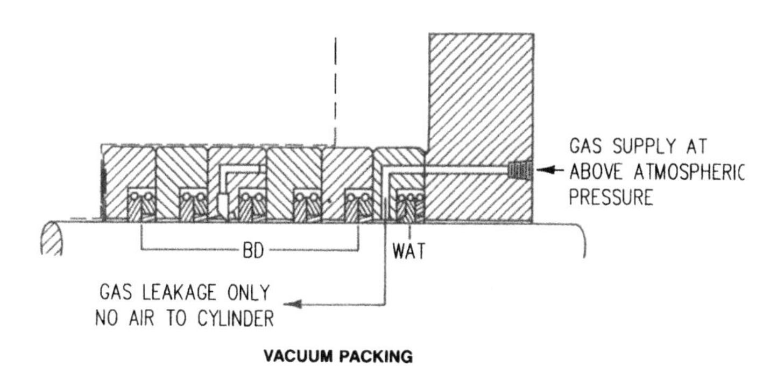

FIGURE 3-33. Special venting provisions on packing cases. (*Source: C. Lee Cook, Louisville, Kentucky*).

Vented packing cases can also be fitted with an effective standstill seal "Static-Pac"™ (see Figure 3-34).

In normal compressor operation, the minimum pressure that will exist in the cylinder, and consequently, to which the packing will be subjected, discounting variations due to valve function, will be equal to the suction pressure. The differential pressure across the packing will fluctuate

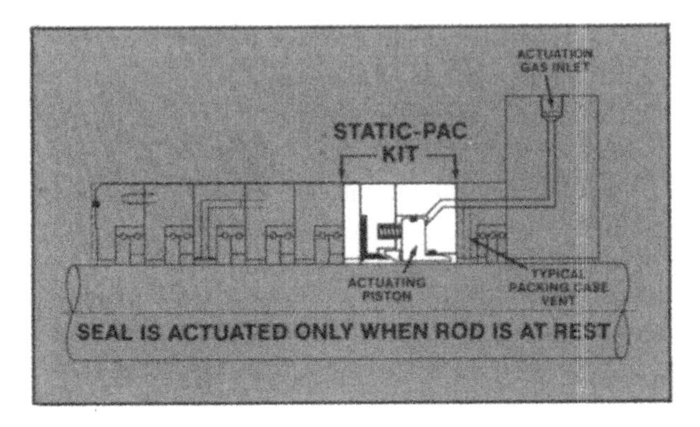

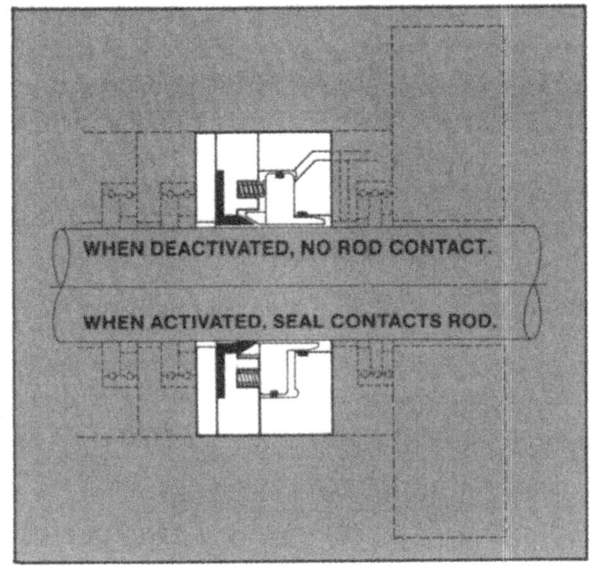

FIGURE 3-34. Static-Pac™ standstill seal for compressor packing cases (*Source: C. Lee Cook, Louisville, Kentucky*).

between suction or discharge pressure and atmospheric pressure on each compression stroke, with the full discharge pressure existing only for a short period on each stroke. The problem then is one of sealing a minimum of the differential between suction pressure and atmospheric pressure for a short period of each stroke. These conditions of time and pressure breakdown across a set of packing are illustrated in the following.

Condition 1. Both suction and discharge pressures are sealed by the first pair of rings.

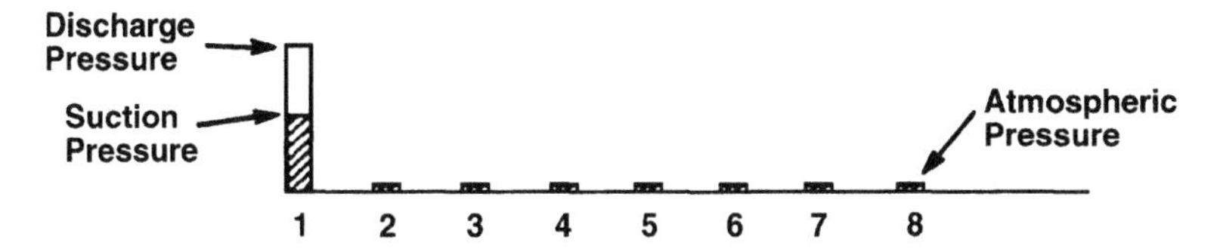

FIGURE 3-35. Condition 1. First pair of rings seals completely.

Condition 2. The suction pressure is sealed by the last pair of rings and the discharge by the first. Such a condition is believed to be transitory between Condition 1 and Condition 3.

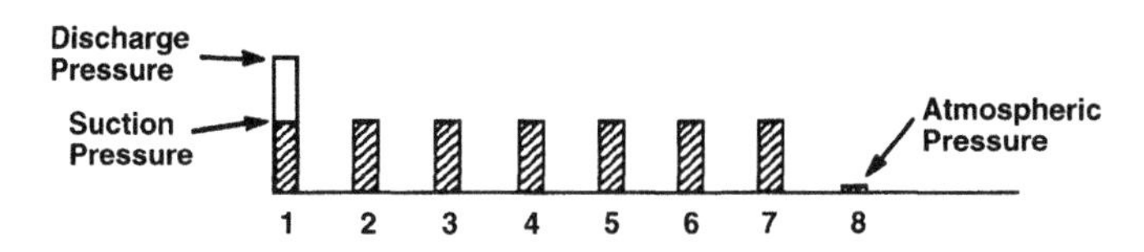

FIGURE 3-36. Condition 2. First pair of rings seals discharge. Last pair seals suction pressure.

Condition 3. This is a breakdown of suction pressure in increments across the rings, with each increment reducing the differential pressure and consequently the amount of leakage by any pair of rings. The discharge pressure is generally sealed by the first pair of rings. Should pressures higher than suction exist beyond the first pair of rings, it would leak back to the cylinder on the next suction stroke because the rings will seal only in one direction—with the radial ring facing the pressure.

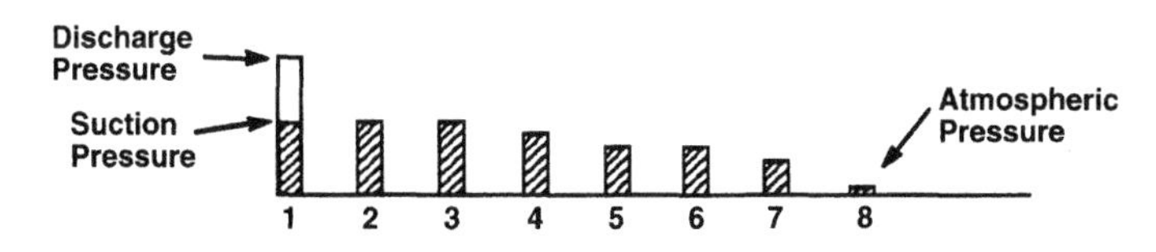

FIGURE 3-37. Condition 3. First pair of rings seals discharge. Suction pressure breaks down through packing.

Each of these conditions or patterns is considered to be "normal" and will exist alternately during normal operation.

PACKING RING TYPES

There are several types of packing rings available.

1. Radial and Tangent Set

   This is a true sealing ring, made up of a pair of rings, the first of which is radially cut with considerable clearance at each end. The second ring is tangentially cut forming an overlapping seal joint that prevents gas passage from the outside periphery toward the rod bore. The inner radial cuts of this tangent ring are arranged so that they occur approximately between the radial cuts of the first ring in the pair. In this manner, gas passage along the rod is blocked so that there is no through escape.

   The rings are doweled to maintain this staggered cut arrangement. When the rings are together as shown in Figure 3-38, no gas passage from the outside is possible because of the tangent cut. As indicated, this type of ring is single-acting or directional in that it seals pressure from one side only.

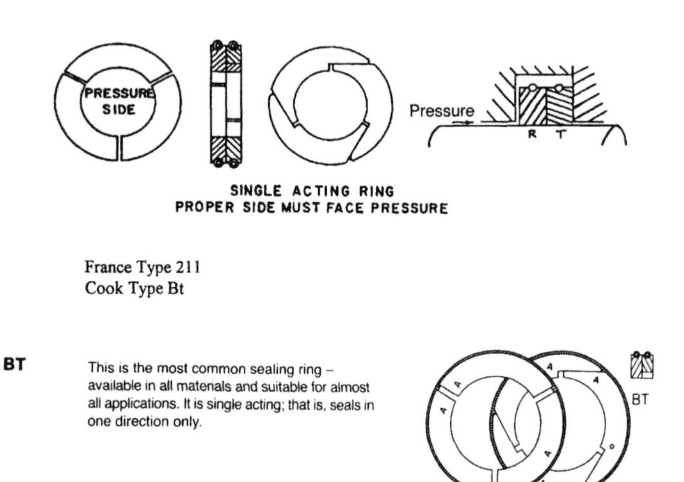

FIGURE 3-38. Single-acting sealing rings. Radically cut side must face pressure (*Source: Worthington Compressor Co., Service Manual*).

Be sure rings are installed in proper order with identification marks on ring segments facing the pressure.

2. Double Tangent Ring Set

   This type consists of two tangent rings doweled together. This assembly seals in either direction, making it double-acting. It is used in vacuum applications, behind the vent, or for any other reverse pressure purpose. This ring set normally requires a differential pressure of 15 to 20 psi to obtain a good seal.

3. Pressure Breaker Type

   This is the simplest form of packing ring and, because it does not have overlapping or seal type joints, it acts to break or slow down

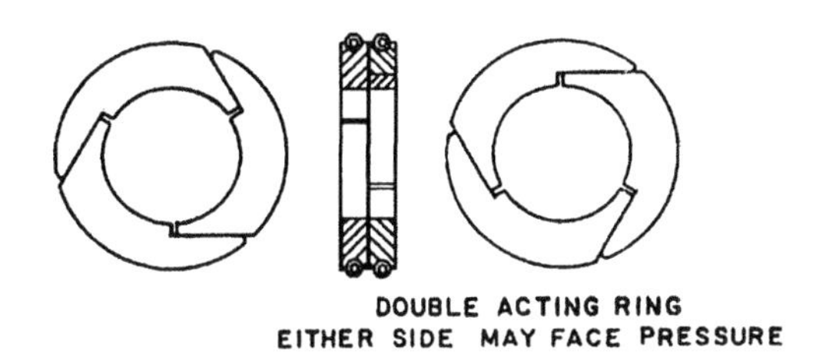

France Type 212
Cook Type BD

BD This is a double-acting sealing ring in that it effects a seal regardless of the groove side it contacts. It is intended to be used when pressure is insufficient to continually hold the ring against one face.

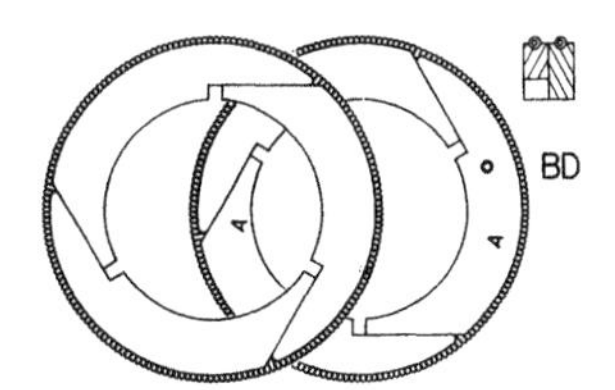

FIGURE 3-39. Double-acting sealing rings are bi-directional (*Source: Worthington Compressor Co., Service Manual*).

gas passage only without actually sealing it completely. It is, therefore, in effect a form of labyrinth.

Note from Figure 3-40 that this ring may be installed with either face toward the pressure and that it has a limited clearance at the cut only. Some clearance at this point must always be maintained, although it should not be greater than .010″ per inch of bore diameter. Obviously, this ring should not be allowed to operate without end clearance because its life would be very short. Any wear that takes place at the rod contact would tend simply to enlarge the bore

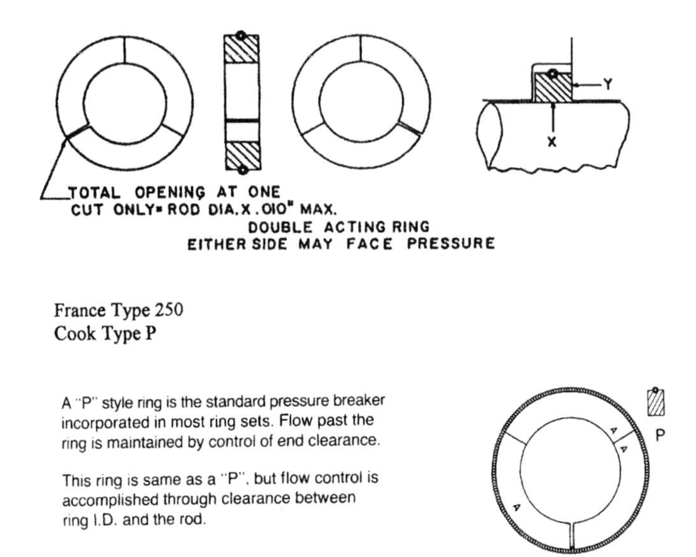

FIGURE 3-40. Standard pressure breaker ring (*Worthington Compressor Co., Service Manual*).

of the ring, and it would dangle. The most important function of breakers in compressor service is to retard backflow from the packing case toward the cylinder on the intake stroke.

4. Radial, Tangent, and Back-Up Set

   This is the basic sealing unit used in higher pressure lubricated and non-lubricated compressors, where Teflon (PTFE) ring material is used and is a combination of three rings in one cup (see Figure 3-41).

   It is made up of one radial and one tangential packing ring, with one radial-cut back-up ring completing the combination. The radial packing ring is installed in the groove on the pressure side toward the cylinder. The tangential-cut ring is installed between the radially cut packing ring and the back-up ring.

   The back-up ring may be metallic or Teflon and is bored .003″ per inch of outside diameter larger than the piston rod. It has butted ends with zero gap clearance and does not grip the rod under pressure loading. Its function is to prevent extrusion of the Teflon member of the packing rings and to conduct heat away from the rod surface by light contact with the rod.

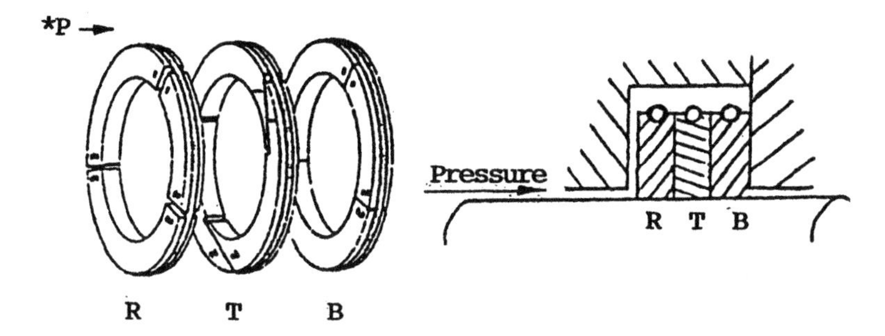

France Type 223
Cook Type BTR

FIGURE 3-41. Radial and tangential rings followed by backup ring. (*Source: Worthington Compressor Co., Service Manual*).

5. Tangent and Back-up Ring Set

This design consists of a Teflon tangential cut seal ring with a metallic radial-cut back-up ring (see Figure 3-42). It is ideal for narrower ring grooves, and it is used for higher pressure applications.

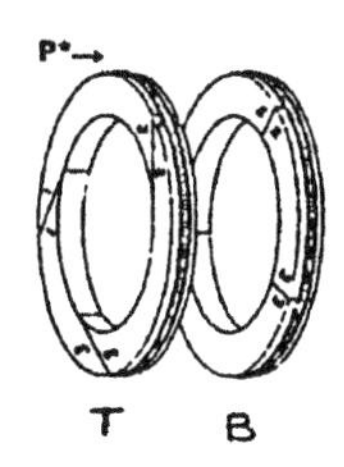

France Type 236
Cook Type TR

TR The "TR" is a sealing ring designed to be made of relatively comformable material. The tangent-cut ring is made of plastic while the radial-cut ring, which serves as an anti-extrusion ring, is of metal or rigid plastic. Radial notches insure the ring seals in one direction only.

TU This ring is like the "TR", but the anti-extrusion ring is uncut.

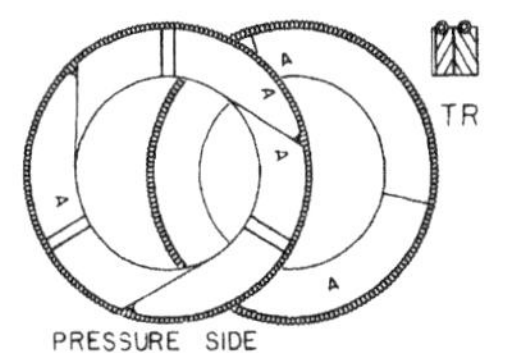

FIGURE 3-42. Elastomeric tangent-cut sealing ring backed up by an anti-extrusion ring. (*Source: Worthington Compressor Co., Service Manual*).

The metal back-up ring prevents extrusion and aids in removing friction heat from the rod surface. The ring is bored a few thousandths larger than the rod diameter, with zero clearance at the joints to assure a tight gas seal.

6. Pressure Balance Set

   The pressure balance ring set consists of a radial-cut packing ring and a tangential seal ring (see Figure 3-43). The seal ring is fastened directly to the packing ring with two screws in each segment. The radial packing ring must always face the pressure side. These rings are replaced only as a complete set.

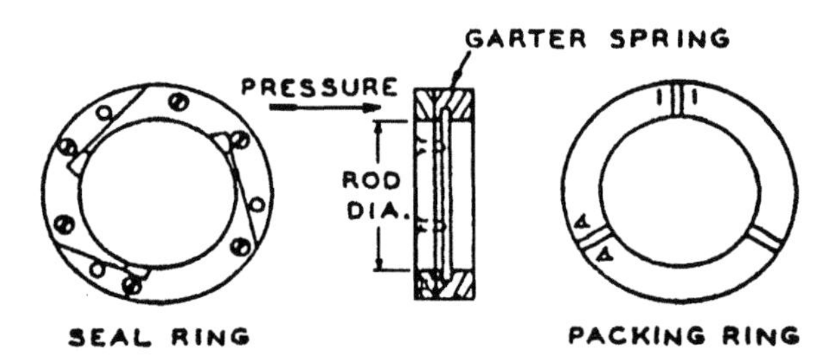

FIGURE 3-43. Pressure balance set with seal ring fastened directly to the packing ring (*Source: C. Lee Cook, Louisville, Kentucky*).

PACKING RING SIDE CLEARANCE

To ensure that the packing rings are full-floating under operating temperatures, they must have proper side clearances. Side clearance is the axial opening, or depth of the cup less the thickness of the rings that are installed in the cup.

Table 3-9 shows the required side clearances for each cup depth and ring material.

MAJOR SOURCES OF PACKING PROBLEMS

There are several indicators of "troubles with metallic packing" that are frequently misunderstood and in reality are a manifestation of trouble originating elsewhere.

Excessive or rapid wear is an example of this. Barring improper selection of mating materials or of materials and gas, excessive wear can be caused by several factors. To fully understand and diagnose *abnormal* wear, *normal* wear should be understood.

TABLE 3-9
PACKING RING SIDE CLEARANCE

Groove Depth (Nom.)	Side Clearance, in.		
	Metallic & Carbon	PTFE	Laminated Thermosetting Resin
.615	↑	.015	↑
.572	.005	.025	.009
.447	.011	.012 .025	.021
.383		.013 .025	
		.011	
.375	↓	.025	↓

Normal Wear

Wear of sliding parts appears to be inevitable, and it is desirable to some extent. Normal wear can be classified as *wearing in* and *wearing out.* We generally consider "wearing in" to be a burnishing of the rubbing surfaces.

With packing, the problem is somewhat different. Packing rings must first undergo a period of wearing out, followed by a period of wearing in, then followed by the normal wearing out. The first wearing out period is caused by the fact that the rings must have a gas-tight fit with the piston rod and must obtain that fit by wearing out to a point of making that contact.

Temperatures may vary substantially from front to back of packing and between rod and rings. In addition to the temperature difference, the coefficients of expansion of rod and rings may differ substantially (see Figure 3-44). This means that while the fit between rod and rings is perfect at room temperature, it may be mismatched at operating temperature. To accommodate the temperature gradient through the packing and the difference in expansion rates, each pair of rings must wear to a running fit to make a satisfactory seal. Having obtained this fit we should now expect the surfaces to burnish and the original high wear rate to be replaced by a much lower normal wear pattern.

What then will change this pattern of normal wear, accelerating it to unacceptable limits?

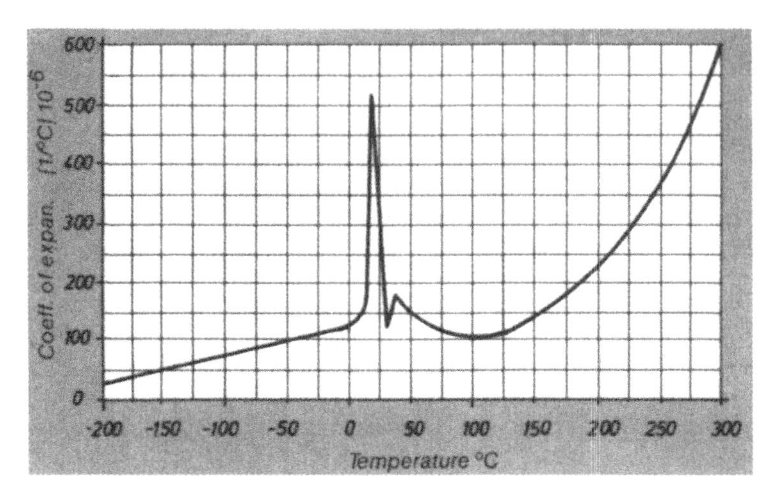

FIGURE 3-44. Unfavorable increase in coefficient of thermal expansion of PTFE with temperature (*Source: Burgmann Seals America, Houston, Texas*).

Causes of Abnormal Wear

Causes of abnormal wear include the following:

- Temperature. Changes in temperature can put packing back to the high wear rate of the initial wear out/wear in cycle. Radical changes in cylinder or packing temperature can be caused by failure of the cooling system, excessive cylinder temperature, and sticking of the rings, that is, excessive friction.
- Lubrication. If lubrication is thinned beyond the point of effectiveness by the gas or by excessive temperatures, metal contact can occur, resulting in high friction temperatures and excessive wear. Buildup of gummy or sticky carbon deposits due to poor lubrication or reaction with the gas can cause restriction of free movement of the rings, further contributing to excessive wear.
- Blow-by. Caused by improper break-in or erosion due to leakage past the bore of the rings after a sudden temperature change, blow-by can lead to excessive wear.
- Moisture. Causing pitting, erosion, or corrosion to take place on the rubbing surfaces during shutdown, moisture can lead to unusual abrasive conditions.

- Contamination. Gas contaminated by foreign material, such as welding beads, sand, pipe scale, rust, and catalyst will also cause rapid wear.
- Improper break-in. Break-in can be a source of rapid wear if the rings are improperly worn-in and a burnished rubbing surface is not obtained. If this condition exists, the rings may continue to wear until a burnished condition is conceivably achieved at some later date.

Causes of Abnormal Wear of Teflon Packing Rings

Because Teflon is widely used as a packing ring material, certain peculiarities in its operating characteristics should be understood. While the same conditions that cause abnormal wear in metal packing rings apply to Teflon, there are certain additional phenomena that should be understood.

It must be understood that because of the self-lubricating properties of Teflon materials, they will run completely without external lubrication or they will run as any lubricated material requiring a bare minimum of lubricant.

When operating without lubrication, the rings will lay down a coating on the mating surface of the piston rod and, once established, will permit the packing rings to run on this film at very low rates of wear and friction. When operating with lubrication, this film is not established unless the lubricant is completely withdrawn. If oil is introduced after this film has been established, it will displace the Teflon on the piston rod and the rings will operate as lubricated ones. If the lubricant is then withdrawn, the rings again establish a Teflon film on the mating piston rod surface. During this coating period, high wear rates of the packing rings will be experienced. It can be seen that conditions which cause a succession of lubrication events alternating with periods of operation without lube oil will severely curtail ring life.

Leakage

It is apparent that wear itself is not the root cause, but rather, a manifestation of one or more problems. Leakage, on the other hand, can be both a by-product and a problem in itself.

Leakage can be caused by one or more of the following:

- Wear can be the most obvious cause of leakage. Complete wearing out of the rings and failure to compensate for additional wear would, of course, permit clearance and leakage. Wear on the sealing face of the cup can prevent sealing of the rings against the cup, causing leakage. Wear or damage of the cup may be caused by inadequate or poor lubrication, corrosion, buildup of carbon, or abrasive material in the gas.

- Damage to the piston in the form of scuffing or scoring can be a source of leakage. If excessive wear has occurred on the rod, and rings of the original diameter are used, leakage will occur.
- Improper assembly is a readily detectable source of leakage. Rings installed improperly, that is, with the radial ring facing away from the pressure, will allow the packing to leak. If the set has been assembled in such a way that the packing cups are not perpendicular to the rod, the packing rings will not be perpendicular to the rod and leakage may occur.
- In these instances, leakage is actually a symptom rather than a root cause; however, leakage can in itself become a problem. During break-in, where a discrepancy may exist between the rings and the rod due to expansion, blow-by can destroy the oil film and cause excessive packing temperatures, causing further expansion and blow-by. In this case, leakage is indeed the problem.

Packing Problems

Some common problems associated with packing repairs in the field are as follows:

1. Segment reversal. This term refers to the mismatching of the individual ring segments. Rod packing segments are match-marked at the factory in some manner to assure proper assembly. Improper assembly of ring segments can result in excessive leakage and premature packing failure.
2. Mislocated packing. This term refers to having single-acting packing where double-acting packing should be installed.
3. Wrong side of packing facing pressure. A good rule of thumb to follow is that the match marks on the individual packing segments always face the pressure.
4. Break-in time. With engineered plastic packing, such as Teflon, break-in time is about 20 to 30 minutes. However, with metallic

packing, break-in time could be between six and eight hours with no load. Frequently, a load is applied to the machine as soon as it is warmed up. This is extremely harmful to the new packing and could result in catastrophic or premature failure.

5. Prelube. The packing box should be properly lubricated prior to starting. Refer to the instructions for maintenance and servicing of packing.
6. Packing box lubrication. Lubrication is extremely important for some types of packing. Metallic packing and Micarta packing require lubrication. However, engineered plastics, such as Teflon and Ryton have natural lubricity. Depending on the packing, the rate of oil flow must be adjusted accordingly. Adjustment procedures vary depending on the lubricator. Again, too much oil can be as bad as not enough oil.
7. Particulate matter. Trapped matter may be carried with the gas into the cylinder. It could settle on the piston rod and be carried back into the packing rings where it could mix with the oil to form an abrasive slurry.

Tapered Piston Rods

Because lubricating oil films may tend to fill up some small passages, packing can usually function in the presence of slight piston rod taper. Generally, tapered conditions are found at either end of the stroke, while the in-between portion of the rod may remain unaffected. Excessive amounts of taper will destroy the seal in the following manner.

When the rings are riding on the tapered part (see Figure 3-45), one edge of the ring bore is in contact, while the other edge is away from the

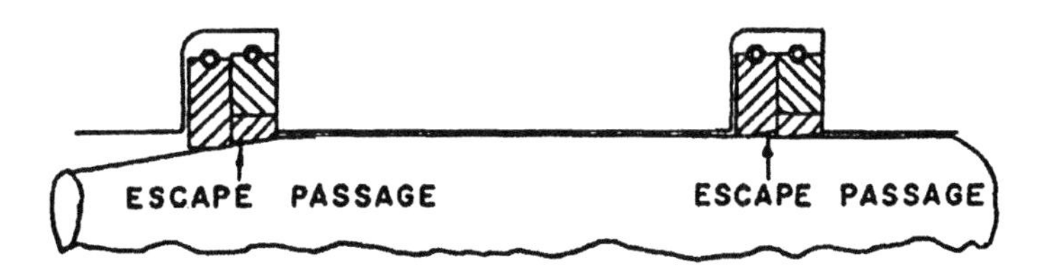

FIGURE 3-45. Effect of tapered rod contours on packing ring sealing function (*Source: C. Lee Cook, Louisville, Kentucky*).

rod. As the rod progresses into or out of the packing, the other edge comes into contact. This abnormal condition causes two separate contact surfaces to exist in the ring bore. In either position, there is a gas passage along the bore from one radial cut to the other and gas escape is made possible.

Undersized Piston Rods

If a piston rod is undersized but concentric and there is no taper, the packing can form an effective seal within reasonable limits because bore contact will be in the center of each segment and the cut in each ring is overlapped by its mate. Nominally sized rings can be used if the piston rod undersize dimension does not exceed .002″. Whenever certain undersize conditions exist, wear-in time is required before the packing will properly seal. If the piston rod is greatly undersized, leakage will occur at the tangent cut of the packing rings.

Oversized Piston Rods

Where the packing rings have a somewhat smaller bore than the piston rod diameter, the segments will ride at their respective ends. Since the center part will now be away from the rod at the exact point of the cut in the other member, gas leakage will result along this line.

This condition will possibly be corrected by wear-in, although in the case of either undersized or oversized rods, lubrication is obviously being blown away by the gas passage. Hence, dangerous overheating and dry contact may occur to the point of destruction. If the condition is severe, however, a break between the tangential cuts of the packing rings is likely to occur.

Figure 3-46 shows packing ring breaks that can be caused by undersized and oversized piston rods.

Garter Spring and/or Ring Breakage

As already mentioned, all packings operate somewhat on the labyrinth principle so that there is penetration of gas into the assembly.

As will be noted from Figure 3-47, considerable time exposure exists on the compression line. It is during this period that the gas enters the

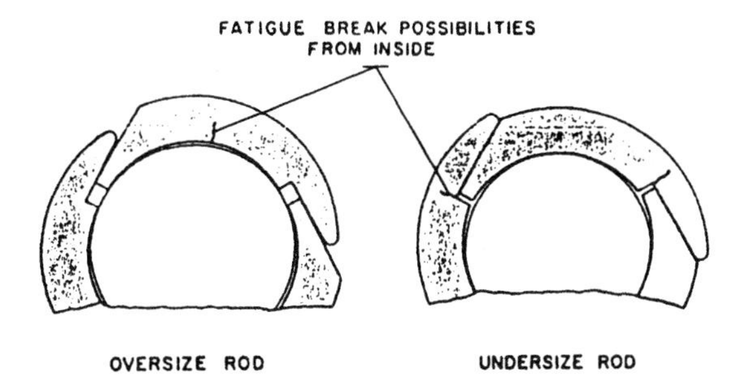

FIGURE 3-46. Packing ring breaks caused by oversized and undersized piston rods.

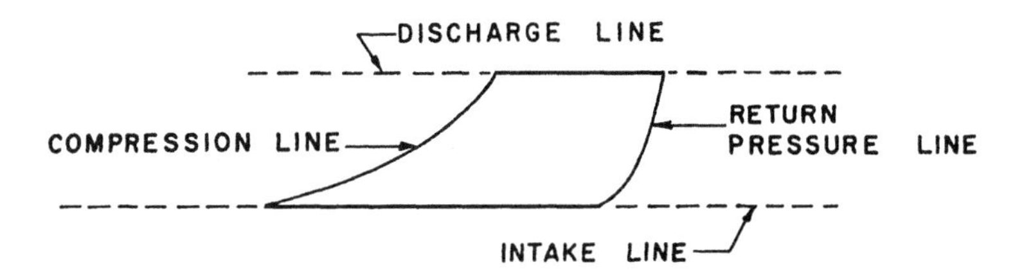

FIGURE 3-47. Typical compressor indicator diagram.

packing assembly, although it may not penetrate far enough to escape at the outside.

On the reverse, or intake stroke, the pressure return line has a considerably shorter time exposure, and any gas trapped in the case tends to reverse itself and travel again to the cylinder where the pressure is rapidly dropping to intake value. This causes an exploding action that leaves tangent-type rings particularly vulnerable. These rings may now strike the back of the groove diameter with considerable force.

Garter spring damage and ultimate breakage is a distinct possibility. It is also possible that the tangent lip will break at the point indicated in Figure 3-48. Another cause of garter spring breakage is damage in handling, particularly at the tangent ends. This is also shown in Figure 3-48.

Garter spring breakage and ring damage due to back pressure explosion is usually found in the rings nearest the pressure side of the packing assembly. Proper maintenance or the provision of breaker rings is an effective preventive measure. The labyrinth principle of the breakers slows down or minimizes the explosive action.

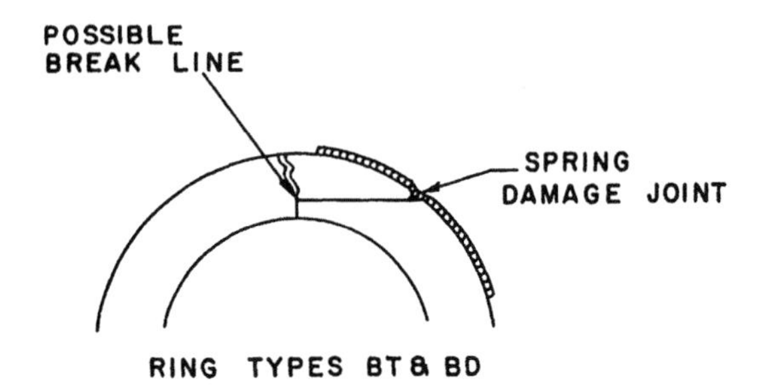

FIGURE 3-48. Packing ring damage caused by incorrect handling.

RATE OF LEAKAGE

Leakage amounts depend upon the amount of misfit between the rings and the piston rod. Leakage can also be the result of disturbances along the faces that must be in contact to form proper seals, as described for the various ring styles. Under the best of circumstances, packings will weep or bleed in small amounts that cannot be detected by ordinary means. Some of this escape is due to separation at the various points of contact as might be caused by vibration or temporary lubrication loss.

Also, some of this weeping is due to gas being forced into the rod pores or into the lubrication film so that it escapes when the rod comes out of the high pressure areas of its stroke into low pressure or atmospheric conditions. These latter amounts are negligible and more serious blow-by or gas loss will depend on the amount of disturbance at the packing and sealing contours and faces.

It is possible, with additional cost beyond ordinary practice, to provide packings of extreme accuracy, such as pre-lapped surfaces. In order for this provision to be effective, it is also necessary to achieve perfect alignment and perfect case maintenance, particularly as to true and smooth contacts. Moreover, the rod surfaces must be perfectly true and smooth.

If packings are allowed to leak excessively over long periods of time, the destructive forces will progress to the point where they are difficult to correct. New rings cannot be expected to perform satisfactorily if the rods are tapered or scratched, and if the cases have also become damaged in like manner. This would be similar to putting a new valve in an automobile engine without doing anything to its damaged seat.

Even after routine maintenance and replacement, adequate time must be allowed for all of the sealing surfaces to mate or wear in. The packing will then become virtually tight and remain so until normal wear or some foreign influence disturbs its condition.

MAINTENANCE OF PISTON ROD PACKING

Figure 3-49 shows an exploded view of typical cup-type packing with the various parts in proper order and each part named to highlight both functions and terms used.

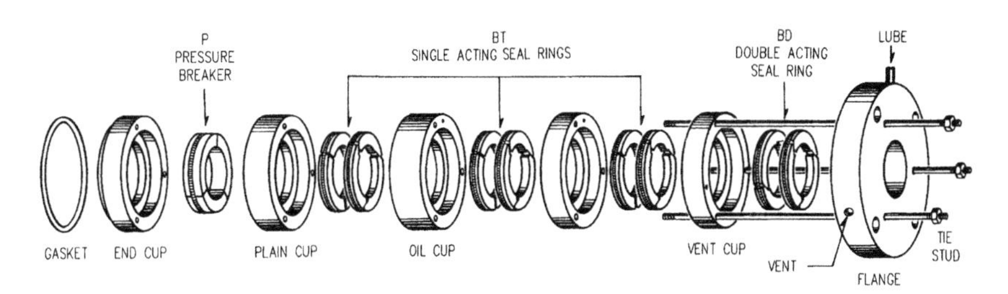

FIGURE 3-49. Exploded view of typical cup-type packing (*Source: C. Lee Cook, Louisville, Kentucky*).

Figure 3-50 shows various styles of rings in common use and illustrates proper assembly and position with respect to the pressure side.

It is strongly recommended that new packing rings and springs be installed when required, and reconditioned packing cases be used.

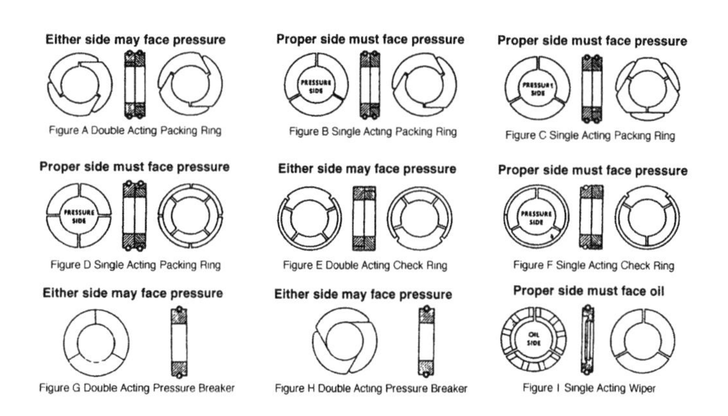

FIGURE 3-50. Different styles of packing and required assembly position (*Source: Dresser-Rand, Painted Post, New York*).

Reconditioning should be done by a qualified service shop, preferably one of the packing manufacturers that specializes in packing case rebuilding.

The best general practice suggests having available a spare set of packing that has been reconditioned by the packing manufacturer. The set that is removed at the next overhaul should be reconditioned by a competent repair facility prior to being returned to stores.

Whenever the packing is opened for inspection, the end clearance of the rings should be checked and the original clearances restored, if necessary. This adjustment is made by filing an equal amount off the end of each segment. Do not remove all the material from one segment.

Table 3-10 gives typical end clearances for various styles of rings and different rod diameters.

TABLE 3-10
PACKING RING END CLEARANCE AT EACH CUT
(REPAIRS ONLY)

Ring Type (See Figure 3-50)	Rod Diameter, Inches 2	3	4	6
a, b, c	1/16″	3/32″	1/8″	5/32″
d, e, f			1/16″	3/32″
g	.004″	.006″	.008″	.012″
h	No End Clearance			
j	1/32″	1/16″	3/32″	1/8″

It is a far better maintenance practice to install new rings and springs than to rework the old rings. The cost of these parts is small in comparison to compressor shutdowns caused by packing failures.

SERVICING PRESSURE PACKING

When cleaning packing rings, wire edges may be found around the bore of the rings if considerable wear has occurred. These wire edges must be removed with a file. However, care should be taken to not break the corners where any two surfaces of a packing ring match.

If packings are to be paired with old rods, check the rod for size and trueness. If scores or shoulders are present, the rod should be refinished or replaced. All scores and abrasions should be removed by stoning.

Packing cups may adhere to cases due to the contact of lapped surfaces and oil. Sharp tools such as chisels should not be used to separate the cups because sealing surfaces can, of course, be damaged by nicks and scratches, etc.

Packing is a precision component and should be treated as such.

When servicing and installing piston rod packing, the following points must be closely observed to prevent packing failures:

1. The cups, rings, and piston rod must be clean.
2. The ring seating surfaces must be parallel with the face of the cup, and cups must be parallel to each other.
3. The side clearance of the rings must be within the tolerances given by the manufacturer. Sufficient end clearances must exist to permit rings to seat properly against the piston rod.
4. The faces of the packing rings must be parallel. Tangent cut rings must contact each other squarely to prevent leakage.
5. The bore of the packing rings should be no more than .002″ larger than the diameter of the rod. To prevent blow-by on starting, the rings are frequently lapped to fit the rod.
6. Garter springs should be of the correct length and be free in their grooves to give the packing the required pressure on the rod. If they are worn or have lost their tension, replace them.
7. When assembling the packing on the rod, cleanliness is essential. Dirt on the ring seating surface will stick to the rings. Dirt on the faces of the cups will throw the packing case out of line and also permit leakage between the cups.
8. Care must be taken to assemble the packing in the correct sequence. If this is not done, the packing will not seal the pressure. All ring segment ends are lettered or match-marked and must be assembled accordingly.
9. The packing rings and the wearing faces of the packing case cup should be coated with a small amount of clean lubricant at assembly, except on non-lubricated packings.
10. The face of the pocket in which the packing case gasket rests must be clean and free from burrs or scars to provide a flat seat for the packing.
11. If the packing case gasket is solid copper or a triangle gasket used previously, it should be annealed before using it a second time. It is always preferable to use new gaskets.

12. When installing the packing case and gasket, care should be taken in tightening nuts on the packing case studs. These should be tightened evenly to keep the bore of the packing case centered on the piston rod. This can be checked at various times when tightening the packing case stud nuts by means of feelers between the bore of the packing case and the rod. After the case has been installed, the clearance between the rod and the bore of the packing case should be checked with the compressor piston at both ends and in the center of its stroke.
13. After installing the packing case, connect the lube oil and vent lines and fill the oil line by venting and operating the lubricator.

PISTON ROD OIL WIPER PACKING

The stuffer in the frame head of the cylinder housing or crosshead guide contains a set of piston rod oil scraper rings and a gland (see Figure 3-51) that prevents crankcase oil from being carried out of the crankcase along the rod.

A typical gland is designed to allow no more than .005″ total side clearance for the scraper rings, generally .002″ to .003″. If the clearance is insufficient for the rings to float on the rod, they will not function properly. Too much clearance will cause the rings to act as a pump instead of wiping oil from the rod.

Care must be taken to maintain a smooth rod surface. Nicks or dents in the rings or scores on the rod will prevent a tight seal.

If, after considerable service, the rings should start leaking oil, examine them. If the ends butt together, the rings should be replaced with new ones.

Renewal oil scraper rings are furnished in sets of two. These segments are lettered, and adjacent segments should match.

COMPRESSOR CONTROL SYSTEMS

Installation and start-up procedures, troubleshooting, and maintenance are important to a successful compressor installation. Of equal importance is the subject of *capacity control.*

Compressor capacity control is the method by which the output of a compressor is regulated to provide, as nearly as possible, only that amount of air or gas required at a particular instant. The ideal governing

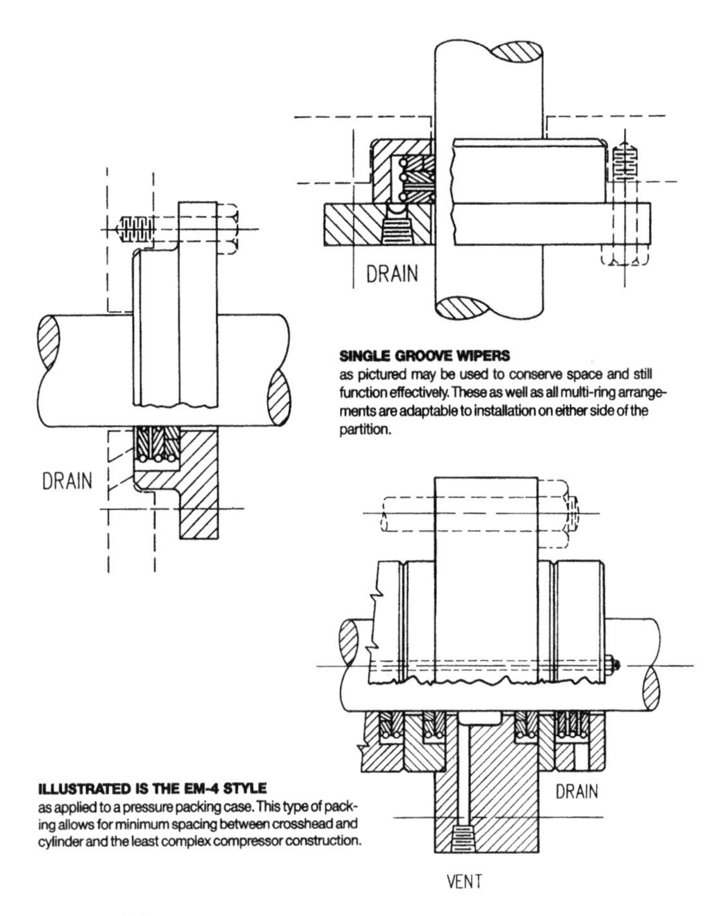

FIGURE 3-51. Piston rod oil wiper packing. (*Source: C. Lee Cook, Louisville, Kentucky*).

arrangement would be to obtain such infinite control that the compressor capacity would be exactly equal to the demand at all times.

For an installation to be successful, the controls on a compressor must function properly. It is, therefore, extremely important that operators and maintenance personnel be thoroughly familiar with the compressor controls. This familiarity must include an understanding of the reasons the controls are put on the compressor in the first place, as well as knowledge of their operation.

A steam- or internal-combustion-engine-driven compressor approximates this condition through its ability to change speed and, therefore, to change capacity. Within the limits of its maximum and minimum speed, the capacity of such a compressor is infinitely variable.

The majority of compressors sold, however, operate at fixed speeds and use synchronous or induction motor drives. Because of the constant speed nature of these prime movers, the capacity of the compressor must be altered by some means other than speed.

Compressors are controlled to limit either capacity or horsepower. The need to limit horsepower is probably the lesser known reason for compressor controls. Figure 3-52 illustrates an application of a compressor that is controlled so as not to exceed available horsepower.

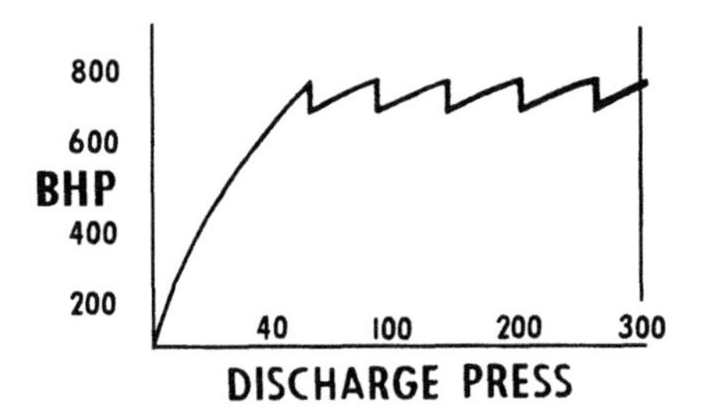

FIGURE 3-52. Capacity unloading to limit power draw.

In this case, the motor had to be loaded between 90% and 100% over a wide range of discharge pressures. When the motor becomes fully loaded, as the discharge pressure is increased, the capacity is reduced slightly before the discharge pressure is increased further.

CAPACITY CONTROL

Controlling a compressor for reasons of capacity accounts for the largest number of control applications.

Compressor regulation falls into two categories—*manual* and *automatic*—and involves elements needed to accomplish three basic functions:

1. Sense pressure changes
2. Relay pressure changes
3. Unloading mechanisms

Manual Control

With manual control, the operator is responsible for adjusting compressor capacity either up or down in accordance with pressure variations. He may control the compressor by starting or stopping the driver, throttling the driver, or manually adjusting the capacity. Other control modes are occasionally employed in special applications.

In order to manually adjust the capacity, the compressor cylinders can be equipped with suction valve unloaders, suction valve lifters, and fixed or variable clearance pockets.

Automatic Control

In automatic control schemes, the basic functions of sensing, relaying, and activating are performed by a combination of standard devices. These devices usually employ mechanical and electrical operations and principles. Most of these devices are extremely simple and have been in common use since the dawn of the machine age.

The elements found in most automatic control schemes make use of five basic mechanisms:

1. Pistons and cylinders to perform work. Inside unloaders use pistons and cylinders for operation.
2. Diaphragms. The outside-type unloaders and some pressure switches are diaphragm-operated.
3. Solenoids. These devices use the basic principles of electromagnetism to perform work and are used widely for relaying. Solenoid-operated relays control the flow of water and air and are used extensively in the circuits of modern control schemes.
4. Valves of all types. Both manual and automatic valves are used throughout automatic control schemes to control and regulate the flow of air, water, and oil.
5. Electronic and electric switches and relays. These perform the simple task of opening and closing contacts. They are applied in most electronic and electrical circuits found on modern capacity controls.

Bypass Control

The simplest, and therefore, the most widely used agent of regulation is suction valve unloading of which the bypass control is typical.

The term bypass control, when used in a discussion of compressor capacity control is somewhat misleading. The implication of the term is that the gas flow is diverted around the compressor so that it "bypasses" the compressor cylinder. This is not true, of course, for the actual process involves drawing the gas into the cylinder on the discharge stroke through the same valves used for entry. If the term were not so unwieldy, this process could be referred to more accurately as a "free flow, non-compressing" control, which is actually what does take place. Gas is permitted to move freely in and out of the cylinder, with no compression taking place.

At a preset pressure limit called the *cut-out point,* a pressure sensing element pilot device causes air to be admitted to the suction valve unloaders. Pressure sensing elements typically include trigger switches, pressure switches, and instrument pilots.

Figure 3-53 shows a typical pressure switch. Fitted with bellows and spring combinations, these elementary switches will usually give good service in their intended operating ranges.

Two typical suction valve unloaders are shown in Figure 3-54.

The unloader mechanism, attached directly to the suction valve, consists of valve fingers operating in the unloader cylinder directly on the valve strips. The spring-loaded fingers are forced to the unloading position either by the movement of the diaphragm stem through a piston and power spring in the unloader cylinder or directly by the stem.

This mechanism, when actuated, holds the suction valve strips in the open position allowing the air or gas to pass freely in and out of the suction valves without compression in the cylinder, thus reducing the delivered capacity of the cylinder and unit.

Air pressure applied to the diaphragm through action of the control unit, causes the diaphragm stem to move downward forcing the valve fingers, through the unloader mechanism, to open the valve strips and hold them solidly against the valve guard.

When the control air is released from the diaphragm, the diaphragm spring retracts the diaphragm stem from the unloader mechanism returning it to its original position. This removes the force from the unloader fingers

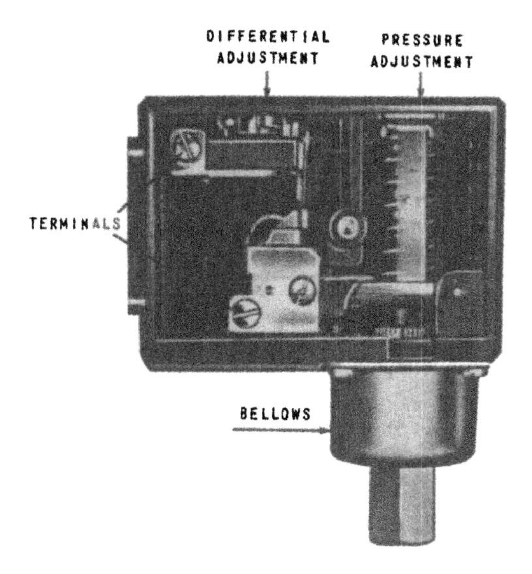

FIGURE 3-53. Pressure switch.

allowing them, by action of the return springs, to be withdrawn from the valve strips, thus allowing the valve to operate in its normal manner.

Typical arrangements of a diaphragm-operated suction valve unloader are shown in Figure 3-54.

When new unloader fingers or a complete unloader assembly is applied to a valve or a new valve or valve guard is applied to the complete assembly, the assembly should be checked to ensure that all strips will be held firmly against the valve guard, when the valve fingers are in the unloading position (strips fully depressed).

This required condition can be checked, with the valve and unloader assembly on the bench, by placing strips of tissue paper between the guard and each strip. Then depress the strips, by forcing the unloader fingers down solidly against the valve guard. With the strips held in this position, all tissue papers should be tight, unable to be withdrawn, because they are clamped between the guard and strips.

If any of the tissues are able to be withdrawn, the valve should be disassembled from the unloader and the ends of the fingers filed and the assembly rechecked as above until the clamping of all strips is accomplished.

The diaphragm head is typically mounted on supporting studs extending from the valve cover. The projection of these support studs is often made adjustable to permit the proper setting of the diaphragm in relation

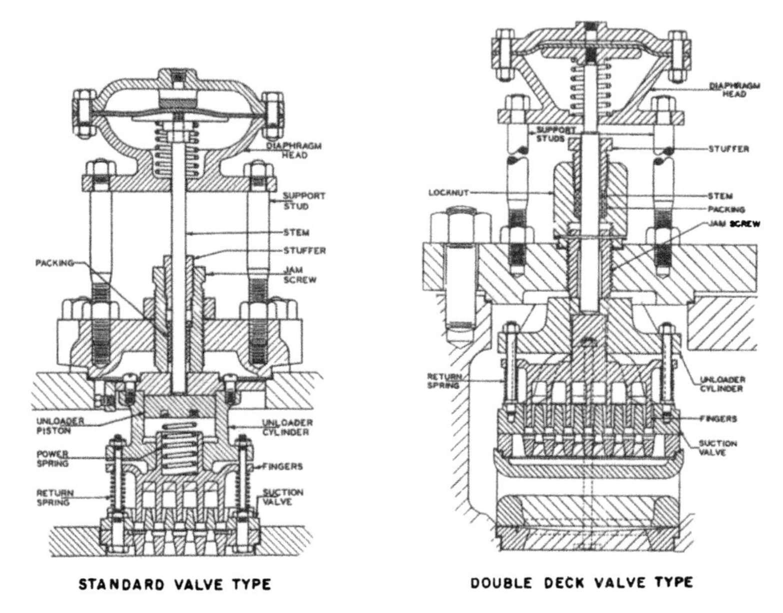

FIGURE 3-54. Suction valve unloaders.

to the valve unloader mechanism. The proper setting is such that the end of the diaphragm stem has a maximum of 1⁄16″ clearance with the unloader piston or fingers (in the case of the double-deck type) when these are in their extreme upper positions assembled in the cylinder.

The following procedure is suggested in order to check and establish the proper setting of the diaphragm stem with the unloader. Remove the valve cover with the diaphragm head mounted thereon without loosening the jam screw. Measure the distance the diaphragm stem projects beyond the end of the jam screw. Also measure the distance, through the hole in the unloader cylinder cover from the top of the cover to the top of the piston or finger. These distances should be equal or the diaphragm stem projection distance slightly less (1⁄16″ maximum).

If such a condition does not exist, the support studs should be adjusted, lengthened or shortened, to obtain this condition. With this setting made, the valve cover can be reinstalled. Upon reinstallation, the jam screw should be loosened and retightened after the valve cover is securely tightened on its seat gasket.

Whenever an unloading valve assembly is replaced in the cylinder suction port, the gasket should be in good condition and the valve should seat solidly. The unloader contains set screws or latches to facilitate holding of the assembly on position while the cover is being applied. Proceed to apply and tighten the valve cover and then tighten the unloading valve assembly on its seat and seat gasket by the aid of the jam screw and jam screw lock nut.

CAUTION: Always remember that whenever valve covers (whether unloading valves or otherwise) are removed, the jam screws and lock nuts must be readjusted.

A stuffing box arrangement is usually provided in the jam screw to prevent gas leakage along the diaphragm stem. The stuffing box may contain several rings of packing with tension thereon applied by adjustment of the stuffer.

Only sufficient tension should be applied to this packing to prevent leakage along the stem. Excessive tightening of the packing will cause binding of the stem and restrict proper action of the unloader. Suitable packing is often placed between the jam screw locknut and the valve cover to prevent gas leakage at that point.

START AND STOP CONTROL

In those installations where air requirements are intermittent, as in the case of an air starting compressor for a diesel engine, start and stop control is desirable. The compressor in this instance shuts itself down when the demand is low and automatically restarts on an increase of air requirements.

Start-stop controls are composed of the same elements that make up two-step bypass capacity control. These controls have, in addition, a time-delay relay and an automatic water valve.

In this instance, the contacts of the pressure switch will be wired in the pilot circuit of a magnetic starter as well as in the solenoid valve electrical circuit. Thus, an increase in pressure will break the starter circuit, stopping the compressor motor and, at the same time, causing the compressor to unload as it slows to a halt.

Because the compressor is starting and stopping when the load is extremely close to its peak, it is desirable to have the compressor completely unloaded at these instances. The incorporation of the pressure switch starter and pressure switch solenoid circuit answers the unloading problem for stopping, but the problem of starting still remains. For this phase of the operation, the time-delay relay is often incorporated into the starter circuit. This allows the unit to unload for only a few seconds of starting, or long enough for the motor to attain full speed.

Dual Control

Sometimes, the conditions of service within a plant favor both types of control, bypass and start-stop. For instance, the air requirements may be reduced to only a small fraction of the compressor capacity on Saturday and Sunday. Plant-specific conditions obviously call for a combination control employing the best elements of the two previously mentioned control types, bypass and start-stop.

Three-Step Variable Control

This control, which uses suction valve unloading, regulates capacity in steps of 100%, 50% and 0% of full capacity and is activated by some sort of governor. The actual unloading of the compressor can take place electrically *and* pneumatically, or pneumatically only.

Figure 3-55 shows a pneumatic control panel including controls for electric auxiliaries. Included are switches and pressure gauges for indicating the intercooler and system and oil pressures.

Normally, for a 100 psig air system, the pressure switches would be set to completely unload at 100 psig. When the pressure falls to 95 psig, the compressor would load to 50% capacity, and if the pressure continues to fall to 93 psig, the compressor would be completely loaded.

Rising pressure would cause the compressor to unload to 50% capacity at 98 pisg and completely unload for automatic initial unloading. Thus, whenever the compressor is started, it will always start unloaded. It is also unloaded when stopped.

Figure 3-56 shows diagrammatically the sequence of operation of unloading with the three-step control.

Clearance Control

Although bypass control constitutes the major portion of the capacity control devices, other methods are available and are often used. There are many instances when bypass control does not offer the necessary flexibility for a given control application.

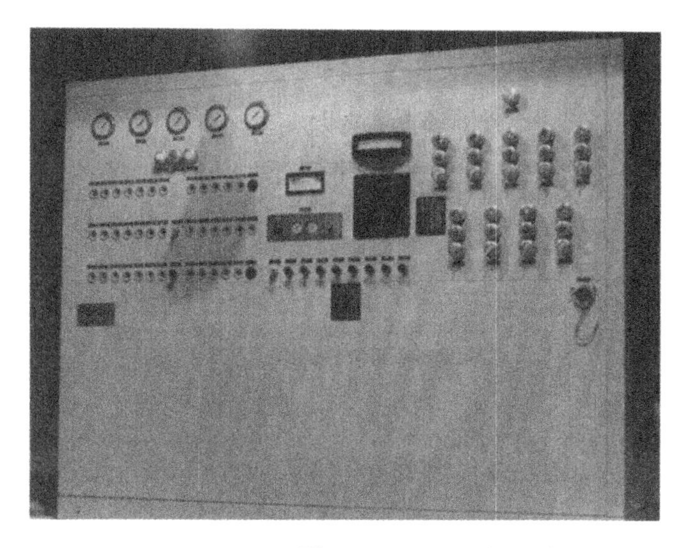

FIGURE 3-55. Three-step control panel.

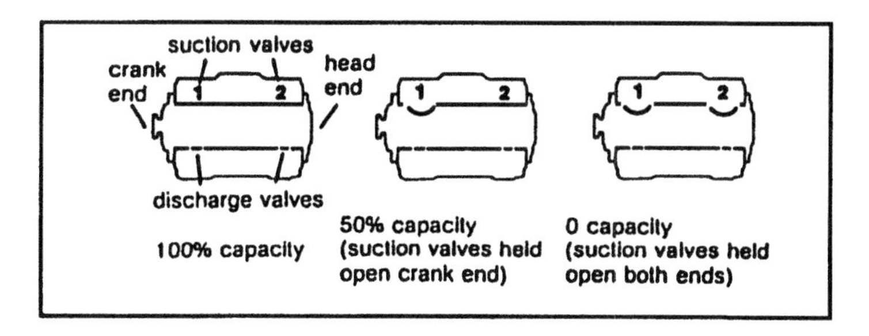

FIGURE 3-56. Sequence of operation with three-step control.

As an example, consider a simple single-acting compressor cylinder in which compression takes place on one end of the stroke only. Using suction valve unloaders permits operation at two points only, 0% or 100%.

Suppose that a particular application requires an additional step so that the compressor can operate at a point offering 50% capacity. This control would then be a three-step type, permitting operation at any three points, 0%, 50% or 100% of full capacity. By supplementing the bypass control with a clearance pocket, intermediate unloading can be accomplished.

The clearance pocket, when open, increases the cylinder clearance volume, which causes the compressor capacity to decrease. An indicator card taken from a cylinder end with a 50% capacity reduction (see Figure 3-57) shows this effect more clearly.

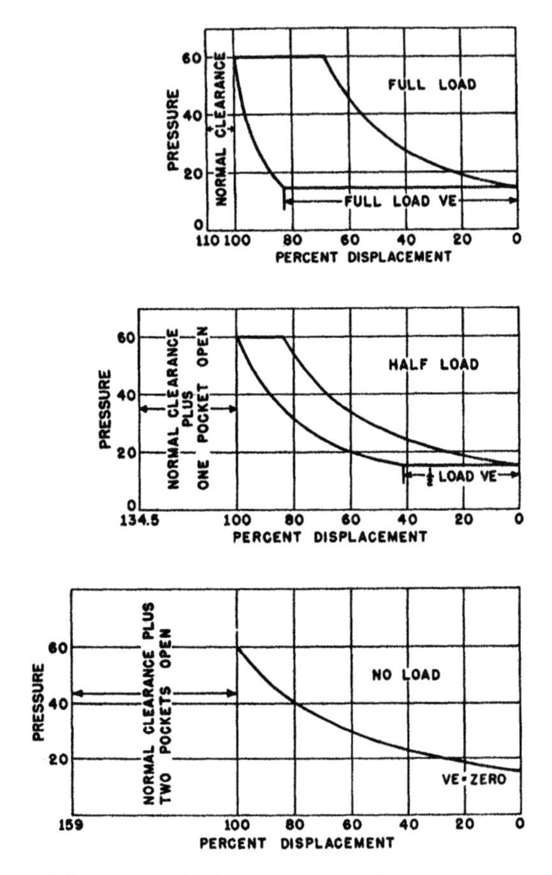

FIGURE 3-57. The effect of adding clearance for capacity control purposes.

This effect of clearance pockets on the capacity of the compressor offers many possibilities.

Clearance pockets, operated manually or automatically (see Figure 3-58) are mounted in pockets cast in the cylinder or in the cylinder head. The valve plug is located over the port opening into the cylinder bore.

Manually operated clearance pocket valves may offer a fixed clearance increase or a variable amount of clearance increase.

In automatically operated clearance pockets, the valve is held closed by constant pressure from the intercooler and compression of the spring. Control pressure applied to the underside of the piston forces the piston upward, opening the valve. When the valve is opened, the compressed air or gas passes into the clearance pocket space on the compression stroke. The capacity of the cylinder is thus reduced.

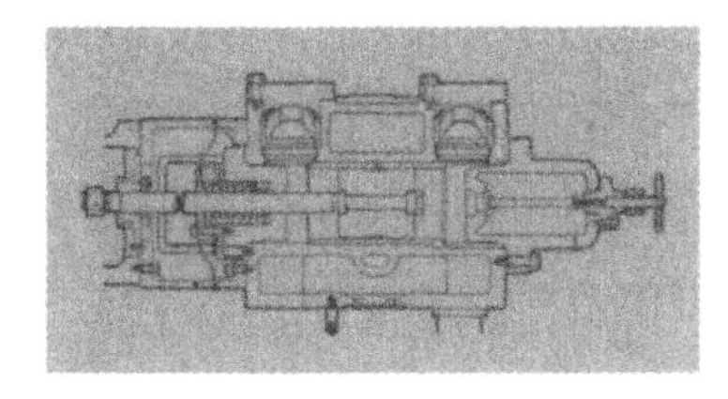

Manually operated, fixed-volume clearance pocket.

Upon manual opening, the fixed-volume clearance pocket reduces compressor capacity a specific amount in a single step by adding a fixed amount of clearance to the cylinder. In reverse, reducing the clearance will increase the compressor capacity.

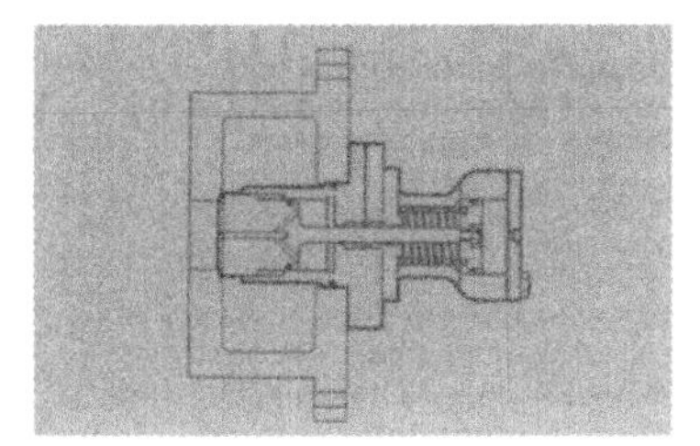

Pneumatically actuated, fixed-volume clearance pocket.

Pneumatic control is often used for remote, semi-automatic operation. The pneumatic signal can be manually controlled by three-way valves or automatically controlled by electrically actuated solenoid valves.

FIGURE 3-58. Fixed-volume clearance pockets.

FIVE-STEP CAPACITY CONTROL

When more control is required, it is possible to use suction valve unloaders and a clearance pocket in each cylinder to regulate the capacity in steps of 100%, 75%, 50%, 25%, and 0% of full compressor capacity.

With a five-step capacity control, the compressor does not normally operate over the entire range of control. Only a sudden change in demand would cause the compressor to unload completely. Because the compressor usually floats between two steps of unloading, the pressure fluctuation of the system is less with five-step control than with a three-step control.

Figure 3-59 shows the control sequence of a compressor with five-step capacity control.

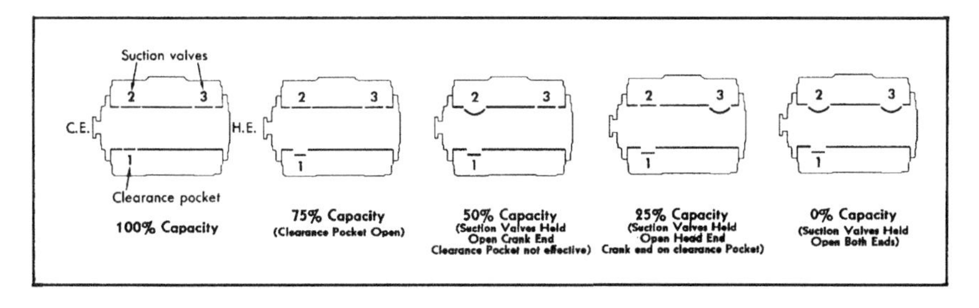

FIGURE 3-59. Sequence of operation with five-step control.

Figure 3-60 shows a typical pneumatic-electrical setup of suction valve unloaders and clearance pocket on a compressor, allowing five-step capacity control.

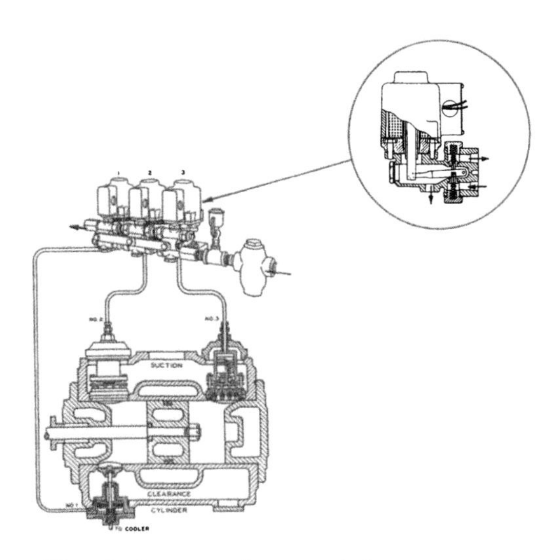

FIGURE 3-60. Electrical-pneumatic control elements associated with five-step control.

A second reason for using more steps of control is the savings in horsepower. Assume a typical plant air demand is 584 cfm with a 100 psig discharge pressure. The plant has a compressor with a capacity of 814 cfm driven by a 150 hp, 585 rpm motor. Table 3-11 shows the power consumption of the compressor equipped with two-, three-, or five-step capacity control.

As can readily be seen from Table 3-11, approximately 1% in power is saved when using a five-step control instead of a three-step control. This amount represents a considerable savings when evaluated over a year of operation.

REVERSE FLOW CONTROL

The reverse flow regulation used by Hoerbiger is a stepless capacity control system that can be used on reciprocating compressors. It allows close adjustment of the output of a piston compressor to the demands of a process. It can follow variations of gas demand, gas composition, and

TABLE 3-11
POWER CONSUMPTION OF COMPRESSOR AT DIFFERENT LOAD CONDITIONS

Control Step	A Percent Of Time At Each Control Step	B Capacity, cfm	C Brake Horsepower	D Motor Efficiency	E Kilowatts Required*
			2-Step Control		$\left(\frac{A \times C \times .746}{D}\right)$
100%	70	570	151	90.4%	87.2
0%	30	0	18.9	65.0%	6.5
	100	570			93.7
			3-Step Control		
100%	55	448	151	90.4%	68.5
50%	30	122	83.1	89.0%	20.9
0%	15	0	18.9	65.0%	3.3
	100	570			92.7
			5-Step Control		
100%	13	106	151	90.4%	16.2
75%	60	366	118	90.4%	58.4
50%	21	86	83.1	89.0%	14.6
25%	6	12	49.8	82.0%	2.7
0%	0	0	18.9	65.0%	0
	100	570			91.9

**Kilowatts required = bhp × .746/motor efficiency*

even pressures and still automatically match the output of the compressor to process needs.

Principle of Operation

An unloading device (see Figure 3-61), is fitted to each suction valve.

At partial loads, this device does not allow the suction valve to close when the piston is in its bottom dead-center position, but delays this closing in a controlled way. Thus, a certain amount of gas, which is steplessly adjustable, is allowed to return towards the suction manifold before the compression starts.

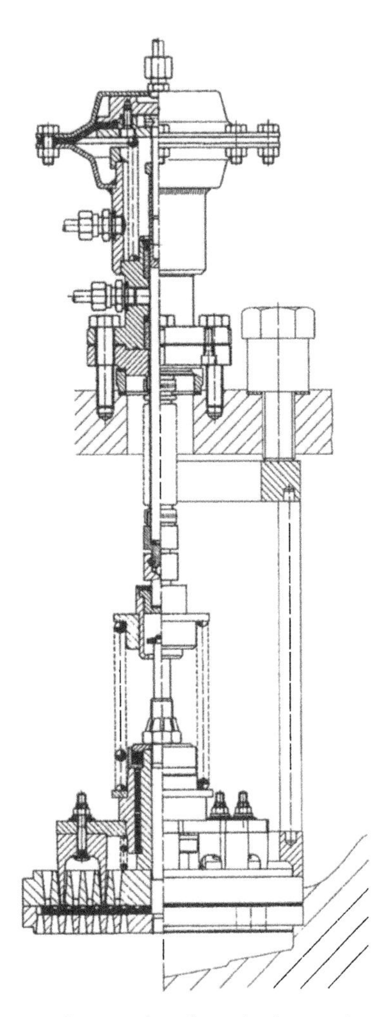

FIGURE 3-61. Reverse flow unloading device, schematic view (*Source: Hoerbiger Corporation of America, Pompano Beach, Florida*).

As the piston accelerates, it pushes gas in a "reverse flow" back out the suction valve faster and faster, which increases the drag forces on the valve plate. These forces will eventually overcome the unloader forces and the compressor valve will close. At that point, compression will start with the reduced capacity output associated with that closing point.

By predicting the drag forces based on application data supplied and controlling the unloading pressure, reverse flow regulation becomes possible.

FIGURE 3-62. Reverse flow unloading device in suction passage of a compressor cylinder (*Source: Hoerbiger Corporation of America, Pompano Beach, Florida*).

Figure 3-62 shows this control in the suction passage of a compressor cylinder.

MAINTENANCE OF COMPRESSOR CONTROLS

Compressor control systems that have been properly set initially are generally trouble-free and require little maintenance. Probably the greatest problems are caused by dirt and moisture entering the pneumatic system and plugging orifices. It is, therefore, important to maintain and change any strainers and filters that are included in the system.

Table 3-12 represents an operation chart for a two-stage compressor equipped with five-step control. It includes checks to assist in troubleshooting.

Table 3-12
Operation Chart for Two-Stage Compressor With Five-Step Control

Load Position			Trouble Checks	
Lights	Capacity	Operation	Low Intercooler Pressure	High Intercooler Pressure
No Lights	Zero	All suction valves held open Clearance pockets closed		High pressure discharge valves faulty (Pressure continues to build up)
One Light	25%	Head end suction valves held open Crank end suction valves closed Clearance pockets open	Crank End Low pressure discharge valves faulty Crank End Low pressure suction valves faulty or unloader stuck open	Crank End High pressure suction valves faulty or unloader stuck open Crank End High pressure discharge valves faulty
Two Lights	50%	Head end suction valves closed Crank end suction valves held open Clearance pockets closed	Head End Low pressure discharge valves faulty Head End Low pressure suction valves faulty or unloader stuck open	Head End High pressure suction valve faulty or unloader stuck open Head End High pressure discharge valve faulty
Three Lights	75%	Head end suction valves closed Crank end suction valves closed Clearance pockets open	If low also at 25% capacity but good at 50%, same check as for 25% capacity If good at 25% but low at 50% also same check as for 50% capacity	If high also at 25% but good at 50%, same check as for 25% capacity If good at 25%, but high at 50% also, same check as for 50% capacity
Four Lights	100%	Head end suction valves closed Crank end suction valves closed Clearance pockets closed	If also low at 25% and 75% capacity but good at 50%, same check as 25% If also low at 50% and 75%, same check as for 25% If low also at 50% only, same check as for 50%	If also high at 25% and 75% but good at 50%, same check as for 25% If also high at 50% and 75% same check as for 50% If high also at 50% only, same check as for 50%

Note: Faulty valves refer to incorrect installation, bad gasket, or broken strips.

Compressor Cylinder Cooling

Benefits of Cylinder Jacket Cooling

The major benefits of water cooling come from its role in reducing gas temperature rise during compression. This is important. Gas discharge temperature is a major factor in compressor maintenance.

High temperature results in:

1. Less effective lubrication
2. Deposits on valves
3. Shorter valve life
4. Higher cylinder maintenance costs
5. Increased risk of fires in discharge piping

Another benefit of lower gas temperatures is higher efficiency and lower power requirements. Water-cooled units also can be designed for higher single-stage pressures, that is, 125 vs. only 100 psig in air compressors exceeding 100 horsepower. Additionally, water cooling maintains a more even temperature in the cylinder, thus reducing distortion.

Multistage compression with intercooling between stages permits control of the compression ratio and limitation of gas discharge temperature to acceptable levels; for example, approximately one percent (1%) specific power saving for 10°F intercooling, based on air compressed to 100 psig in two stages.

More specifically, efficiency gains can be attributed to:

1. Dissipation of frictional heat generated by piston rings.
2. A flow of relatively cooler gas through the cylinder and jacket cooling that prevents the temperatures of rubbing surfaces from rising too high.
3. Water jacketing that allows the existence of a high viscosity lubricating oil film on the cylinder wall to reduce friction and wear.

It should be noted that whenever the cylinder is completely unloaded, flow of fresh gas to carry the heat away is eliminated. Jacket cooling may become essential in these circumstances.

AVOIDANCE OF OVERCOOLING—AN IMPORTANT ISSUE

While low gas temperature is usually beneficial, cooling should be controlled to avoid harmful condensation of vapor that is always present in air and in many gases. This is important and must be emphasized, as it is often overlooked. Condensation can impair lubrication or cause formation of rust in the cylinder bore. This condition can accelerate wear of the cylinder and piston and, in non-lubricated units, Teflon rings and wearing bands.

Condensation in cylinders can usually be prevented by regulating the cooling water rate to control the water outlet temperature. The correct water outlet temperature is a function of inlet gas temperature, moisture content, and discharge pressure. No one value is best for all conditions. An outlet water temperature of approximately 100°F to 130°F is a good average figure for many installations.

Unduly low jacket cylinder cooling water temperatures will result in the following difficulties:

1. Piston ring wear
2. Cylinder bore wear
3. Wear of valve seats
4. Valve sealing element wear and breakage
5. Packing wear and leakage
6. Piston rod wear

CYLINDER COOLING SYSTEMS

Series and Parallel Systems

Coolers and cylinders may be arranged for series or parallel flow.

Series Flow

In series flow, the water passes through the intercooler or aftercooler first and then through the cylinders. One of the two coolers, but *not* the jacket, should receive the coldest water. In the intercooler, this results in the greatest power savings. In the aftercooler, it permits the greatest moisture removal.

Usually, there is no harm in cooling air to as low a temperature as possible in the coolers. However, with some gases, it is necessary to limit the cooling in order to avoid condensing the gas or one of its constituents in the intercooler.

In multistage compressors, the possibility of moisture condensation in the cylinder is greater at high pressures than at low pressures. Therefore, in a two-stage unit, the water should flow through the low-pressure cylinder jacket first, then through the high-pressure cylinder jacket.

One benefit of series flow is that it requires less water than parallel flow. Also, if the water source is cold, the temperature rises as the water first passes through the cooler. This is beneficial in avoiding over cooling of the cylinders.

Often the water requirement of the cylinders is less than that of the cooler. It may then be advantageous to divide the water flow leaving the cooler, directing one part through the cylinders and the remaining portion through a bypass around the cylinders. The temperature of the water leaving the cooler can be controlled by a thermostat in the bypass line. Similarly, a thermostat in the water line leaving the cylinders could be used to control the temperature of water leaving the cylinder.

Parallel Flow

Parallel flow is preferred in cases of warm water, or low water pressure, or both. With warm water (approximately 95°F and higher), there is little danger of causing condensation, and the problem is to achieve effective cooling, assuming sufficient water is available. Cylinder cooling is improved slightly because the water is not preheated in the cooler. Also, due to lower pressure drop in parallel circuits, higher flow rates are possible. Higher flow rates may be needed with warm water. Lack of water pressure differential available to the compressor frequently necessitates parallel flow. This may be due to low inlet pressure or high back pressure.

All components should be considered when estimating water pressure drop in a compressor system. This includes not only coolers, cylinders, and piping, but also automatic valves, thermostatic control valves, hand control valves, and sight flow indicators. Although sometimes overlooked, accessories often constitute a large part of the total pressure drop.

Open Water Systems

While its use in new installations is declining, open raw water systems are still in wide use today. Usually, a city water supply, river, pond, lake, or stream is used and it is normally quite cold. Although this cold water provides the greatest possible amount of cooling, it is often necessary to restrict the water flow so as not to drop the cylinder operating temperature too much. Excessively low water temperatures can prove bothersome when handling partially or completely saturated gases.

As discussed earlier, inlet passages colder than the incoming gas can cause condensation and liquid knockout, which may result in serious damage. Occasionally, attempts to prevent condensation by restricting water flow may cause an appreciable water temperature rise across the cylinder and, if the water is hard, scale deposits begin to form. These deposits always build up most at the hot spots in the cylinder where the greatest heat transfer is usually desired. Thus, frequent descaling operations are required in an effort to curtail performance and maintenance drawbacks.

Thermostatic and Thermosyphon Systems

These systems may be used where a small amount of cooling is advisable but the expense of a system involving a pump does not seem justified. They make use of the natural thermal circulation that results when the cooling water is heated in the cylinder. In these systems, the jackets are filled with water, oil, or ethylene glycol, and the heat of the cylinder is distributed by convection currents. The thermosyphon system is used with gases having a "k" value of 1.26 and below and with cylinder discharge temperatures over 210°F and up to 250°F.

Closed Water System

The most common form of compressor cooling water system involves a closed, soft or treated water arrangement. The water coming from the compressor cylinder or cylinders is cooled by either an air radiator system, a cooling tower, or by a water-to-water heat exchanger. Figure 3-63 shows a schematic arrangement of a closed cooling system using a water-to-water heat exchanger.

Figure 3-64 shows piping schematics for thermostatic, thermosyphon and forced cooling systems.

Cooling Water Recommendations

Cooling water recommendations have been developed for intercoolers, cylinder jackets, and aftercoolers of typical air compressors:

In general, the following guidelines should be considered:

GPM Per 100 CFM of Actual Free Air	
Intercooler Separate	2.5 to 2.8
Intercooler and Jackets in Series	2.5 to 2.8
Aftercoolers:	
80–100 psi, Single-Stage	1.8
80–100 psi, Two-Stage	1.25
Two-Stage Jackets Alone (Both)	0.8
Single-Stage Jackets:	
40 psi	0.6
60 psi	0.8
80 psi	1.1
100 psi	1.3

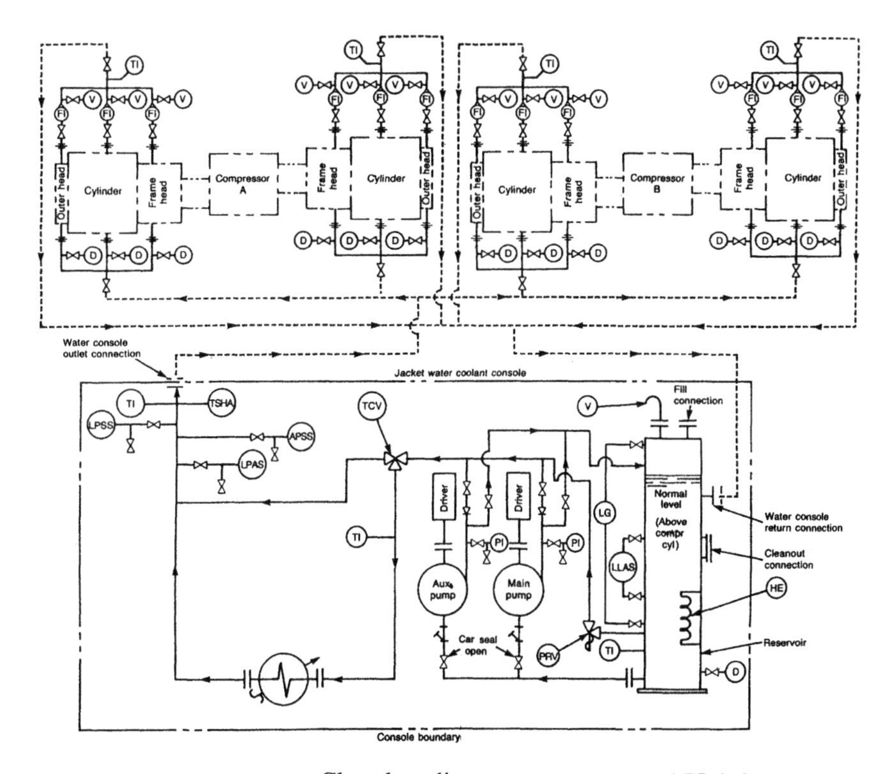

FIGURE 3-63. Closed cooling water system per API 618.

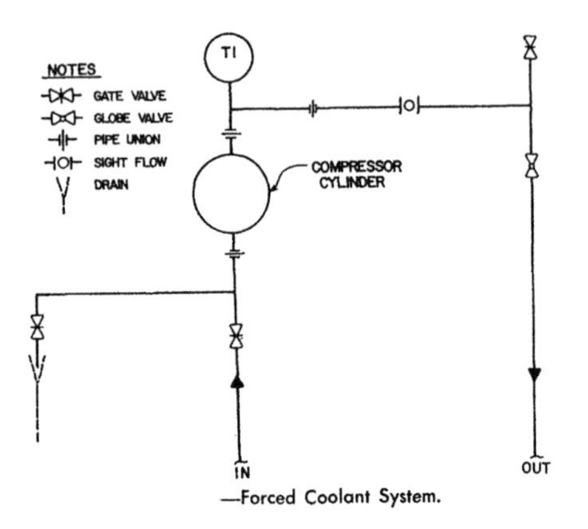

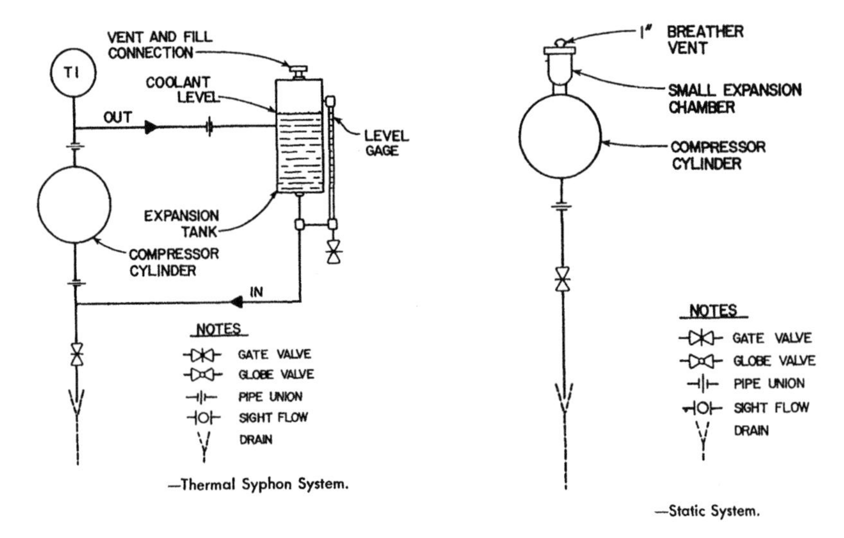

FIGURE 3-64. Piping schematics for different types of cooling systems.

1. The cooling water should be passed through the intercooler first and then this warmer water to the cylinder in a series arrangement.
2. Quantity of water should be regulated so that outlet temperature is not more than 15°F above the inlet water temperature.

$$\Delta t = t_o - t_i = 15°F$$

Cooling system flows (Rule of thumb)

Intercoolers – GPM* = BHP/4 (base 20°F rise)
Aftercoolers – GPM* = BHP/6 (base 30°F rise)
* Multistage Machine—use BHP of last cylinder

3. Water quantity should maintain discharge temperature well below 400°F, preferably 300°F.
4. Inlet water temperature to the cylinder jackets should never be less than that of incoming gas temperature.

   Jacket water temperatures should be 15°F to 20°F above the incoming gas temperatures to prevent condensation.
5. Cooling water should be shut off when unit is shut down. Remember that condensation may occur on cylinder walls whenever wall temperatures drop below gas temperatures.

   CAUTION: Do not start compressor without turning on cooling water.

MAINTENANCE

The amount of maintenance required by a closed system is very small relative to other types of systems. If the cooling system is stretched to its limits, the maintenance requirements may increase. Frequent cleaning of the heat exchanger and replacement of the coolant with one having better heat removing capacity may be required.

Obviously, system components such as pump impellers, pump seals, and thermostatic elements need periodic maintenance or replacement, but typical time intervals should be years, not weeks or months.

NON-LUBRICATED COMPRESSOR MAINTENANCE

Chapter 2 showed how the design of the non-lubricated or oil-free compressors differs from the more conventional cylinder-lubricated compressor. The design should make oil-free compressors reliable and give trouble-free operation. However, these machines are likely to require more frequent maintenance than lubricated compressors if they are to give dependable long-term service.

It is important to understand the basic difference between lubricated and conventional non-lubricated cylinders. In reciprocating compressors, pistons work against pressure and must have a sliding seal to allow the piston to compress the gas without leakage past the piston. The piston rings provide this sealing. Rings are made with "spring," which tends to push out against the cylinder wall, making a tight sliding fit. The piston rings float in the ring grooves of the piston and seal only. They do not support the piston off the cylinder wall. The piston is supported off the cylinder wall by only the liquid lubrication film.

The conventional non-lubricated piston and piston ring has no liquid oil film to support the piston off the cylinder wall; therefore, the metallic piston must be kept away from the metallic cylinder wall by other means, or serious damage could result.

The piston is kept off the cylinder wall by using guide rings, which are referred to as "bull" rings or rider rings. The outside diameter of the piston is smaller relative to the lubricated piston to allow clearance between the piston OD and the cylinder wall. This allows the rider ring to wear somewhat before approaching metal-to-metal contact.

Conventional non-lubricated compressors require more attention to installation and maintenance than oil-lubricated ones. When servicing a non-lubricated compressor, be aware of its weaknesses. The greatest drawback of the non-lubricated compressor is its sensitivity to dirt, abrasives and moisture.

Contamination and failure risks are reduced by the following:

1. Air (or gas) filtration. An efficient filter capable of filtration down to 10 microns is required, and it must be carefully maintained. It is typically sized larger than one for the lubricated compressor.
2. Suction and interstage piping. These must be maintained clean of rust and scale. If a problem of rusting persists on typical air compressors, the piping should be replaced with aluminum or stainless steel.
3. Moisture. Moisture may act as a lubricant, however, more care is required to assure that the system remains corrosion-free. Cylinder ports of air compressors are treated with phenolic or epoxy paints to prevent rusting.
4. Cylinder cooling. Teflon is a poor conductor of heat. It softens, thus extruding more easily, when hot. Friction heat is generated during operation and is not transferred through the Teflon. It is important that good jacket water cooling is maintained and scale deposits are minimized. It is also strongly recommended that an automatically controlled solenoid-operated valve be placed in the piping to shut off water flow when the compressor is out of service for an extended period of time.

5. Alignment. For both piston and packing rings, it is very important that true alignment be maintained. This ensures that, in the reciprocating motion of the piston and rod assembly, the piston and/or rod does not rise and fall. Such action would result in high loadings at the sealing surfaces.
6. Surface finishes. These finishes must be maintained to prevent abnormal and rapid wear of the piston rings, rider bands, and packing rings. It has been shown that Teflon performs optimally on cylinder bore surface finishes of 8 to 16 RMS and 8 to 10 RMS surface finish of piston rods.

Teflon wears and fills the uneven metallic structure resulting in Teflon-against-Teflon contact. It is not unusual that an initial set of rings wears rapidly because of this phenomenon. Surfaces that are too smooth do not allow Teflon to deposit and may result in higher ring wear.

Any cylinder that has been rebored and honed should be given an additional final honing using Teflon blocks substituted for the abrasive stones. This impregnates the cylinder with Teflon for initial break-in.

A daily log of compressor pressures and temperatures should be maintained. Inspection of such critical areas as cylinder bores, piston and rider rings, and piston rod surfaces is recommended on a regularly scheduled basis.

When replacing piston rings and rider bands, it is important that the side clearances and end gaps are checked. Remember that the coefficient of expansion of Teflon, which determines these clearances, is seven times that of cast iron.

LABYRINTH-PISTON COMPRESSORS*

Low maintenance costs and reduced downtime make labyrinth-piston compressors attractive for virtually any gas in chemical and petroleum-refining processes—even severe fouling services. They are especially well-suited for oil-free gas compression, where the common alternative—a compressor with dry-running piston rings—often has maintenance costs that are higher than those for an equal-size lubricated-piston machine.

*Bloch, H. P., "Consider a Low-Maintenance Compressor," *Chemical Engineering,* July 18, 1988.

The following section discusses some actual operating data on maintenance expenses, downtime events, availabilities, and spare-parts consumption for labyrinth-piston compressors. It also outlines the routine-maintenance program followed by one plant that has had particularly good experience with these units.

Labyrinth-piston compressors, as illustrated in Figures 3-65 and 3-66, have labyrinth grooves machined in the periphery of the piston and in the cylinder wall, and a similar labyrinth tooth design between the piston rod and packing gland. A so-called "distance piece," which is a space that may be either open to the atmosphere or closed (see Figure 3-66), separates the oil-free compression space from the lubricated crankcase. Four designs are typically available, as shown.

The grooves in the piston and cylinder rod provide a contactless seal between those parts. The seal is made up of a large number of throttling points and volume chambers arranged in series. Each throttling point acts as a small orifice, where pressure energy is transformed into kinetic energy. The gas velocity then decreases in the subsequent volume chamber, and the kinetic energy is transformed into heat and vortex energy. This

FIGURE 3-65. Cutaway illustration of labyrinth-piston compressor with gas- and pressure-tight crankcase. (*Source: Sulzer Roteq, Winterthur, Switzerland and New York, New York*)

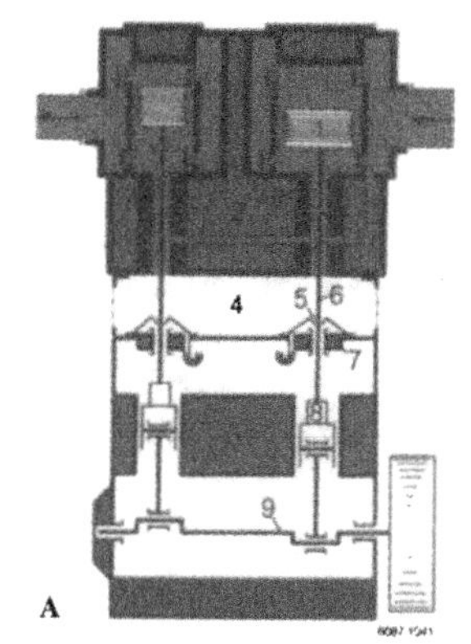

1 Labyrinth piston	4 Distance piece	7 Guide bearing
2 Cylinder	5 Oil scrapers	8 Crosshead
3 Labyrinth-piston rod gland	6 Piston rod	9 Crankshaft

A. Labyrinth-piston compressor with open distance piece and non-pressurized crankcase. Typically used for compression of gases, where a strict separation between cylinder and crankcase is essential and where process gas is permitted in the open distance piece (for example, for O_2, N_2, CO_2, process air; generally in the industrial gas industry).

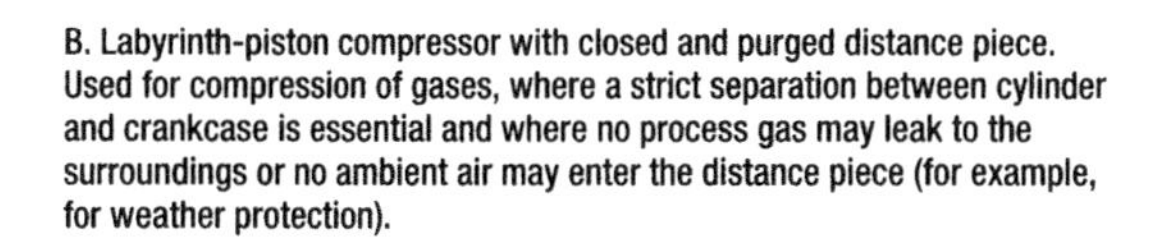
B. Labyrinth-piston compressor with closed and purged distance piece. Used for compression of gases, where a strict separation between cylinder and crankcase is essential and where no process gas may leak to the surroundings or no ambient air may enter the distance piece (for example, for weather protection).

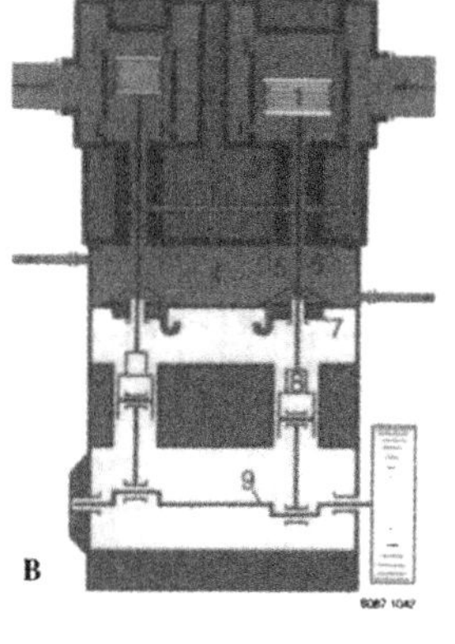

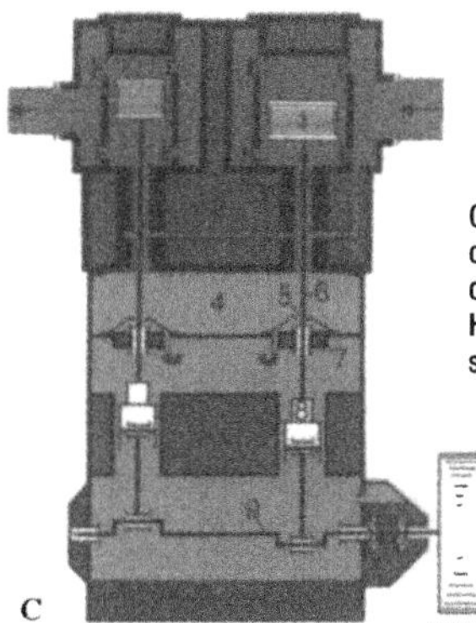

C. Labyrinth-piston compressor with gas-tight crankcase and mechanical crankshaft seal. This design is used for compression of gases that are compatible with the lubricating oil (for example, for hydrocarbon gases, CO, He, H_2, Ar) and where no process gas may leak to the surroundings. The suction pressure is limited by the design pressure of the crankcase.

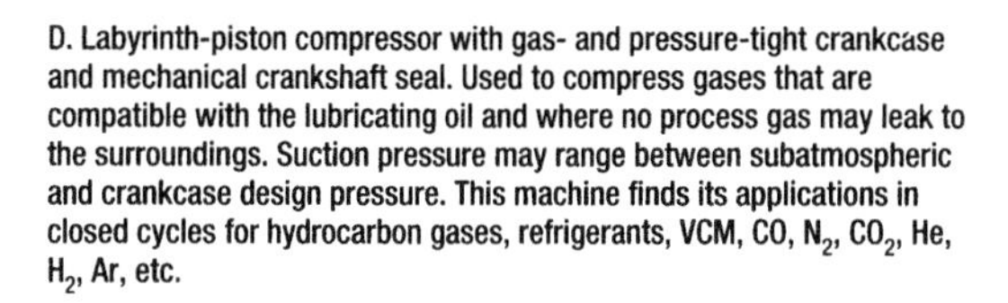
D. Labyrinth-piston compressor with gas- and pressure-tight crankcase and mechanical crankshaft seal. Used to compress gases that are compatible with the lubricating oil and where no process gas may leak to the surroundings. Suction pressure may range between subatmospheric and crankcase design pressure. This machine finds its applications in closed cycles for hydrocarbon gases, refrigerants, VCM, CO, N_2, CO_2, He, H_2, Ar, etc.

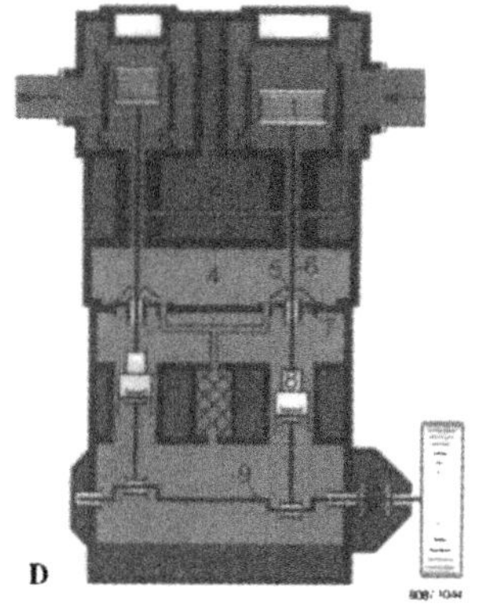

FIGURE 3-66. Labyrinth-piston compressors are available in four typical configurations (Source: Sulzer Roteq, Winterthur, Switzerland and New York, New York).

design reduces the pressure along the piston and piston-rod gland and, thus, achieves the sealing effect.

A "running-in process" and self-centering system provide for optimum piston clearance, high efficiency and economical operation. Compressed gas is delivered with unchanged purity, and the risk of carryover of abrasives from the piston, rider, or packing rings is virtually eliminated, either because there is no rubbing or because these components simply are not part of labyrinth compressors.

MAINTENANCE COSTS REVEALING

Six plants owned by a large corporation and six owned by other firms in North America and Europe were investigated. Not surprisingly, no two locations used identical data-collection and data-reporting methods. Some of the plants monitored repair cost and downtime data; whereas, others kept track primarily of spare-parts usage, or perhaps, only the number of outages. However, all of the available statistics did prove relevant.

Exceptionally detailed records have been kept for six machines in fouling polypropylene services at a U.S. Gulf Coast plant (designated Plant A) that was commissioned in 1983 (Table 3-13). As illustrated in

TABLE 3-13
EXPERIENCE SUMMARY FOR LABYRINTH-PISTON RECIPROCATING COMPRESSORS AT PLANT A

Compressor	Run time before preventive maintenance, h	Maintenance cost, $			Maintenance cost per run-hour, $
		Labor	Material	Total	
A-1	7,284	1,800	7,410	9,210	1.26
A-2	8,997	8,500	6,280	14,780	1.64
A-3	8,545	7,200	9,700	16,900	1.98
A-4	7,000	1,800	5,600	7,400	1.06
A-5	8,300	3,700	6,100	9,800	1.18
A-6	7,545	6,000	17,200	23,200	3.07
Total for plant	**47,671**	**29,000**	**52,290**	**81,290**	

Average annual maintenance expense: \$15/hp, or \$11/kW

Equipment data:

4 throws, 300-mm (12-in.) stroke, 2 stages, 920 kW, 1,234 hp
4,600 m^3/h (2,700 cfm)
P_s = 1.31 bar abs. (19 psia), P_D = 20.7 bar abs. (300 psia)
(conversions are approximate)

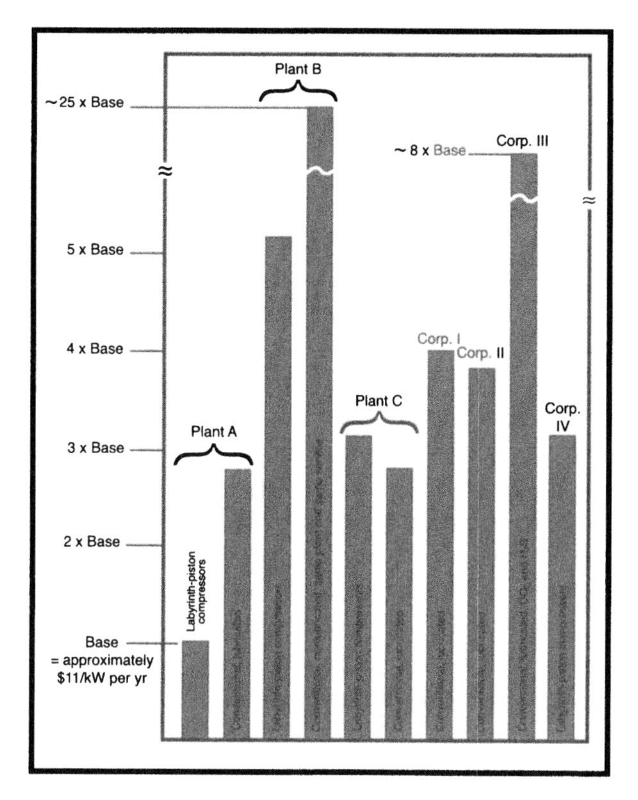

FIGURE 3-67. Average maintenance costs for reciprocating compressors at various North American Plants (1986) (*Source: Bloch, H. P. "Consider a Low Maintenance Compressor,"* Chemical Engineering, *July 18, 1988*).

Figure 3-67, the annual maintenance expenses (approximately \$15/hp, or \$11/kW) at this well-managed location were the lowest of the plants studied. The other installations tabulated in Figure 3-67 include both lubricated and non-lubricated compressors in severe service.

Figure 3-67 deserves a closer look. The labyrinth machines at Plant A had maintenance expenditures roughly one-third those reported for conventional reciprocating compressors at the same site. The comparison between labyrinth and conventional units at Plant B involves the exact same service in the same size range. There, the maintenance cost for lube-free conventional machines exceeded that of the labyrinth machines by almost five to one.

Plant C, however, was an anomaly. There, the labyrinth compressors exceeded the reported average cost of conventionally lubricated units in plants owned by the same corporation by about 18%. Plants in this study that were owned by other corporations spent at least 1.33 times the base amount of roughly $11/kW per year (for example, the amount spent for maintenance of labyrinth machines at Plant A), and as much as three times the base amount for maintenance of conventional reciprocating compressors.

The number of downtime events per year (see Table 3-14), compressor availability (see Figure 3-68) and consumption of spare parts (see Table 3-15) were also studied. The data in Table 3-14 were gathered from six plants in North America and Europe belonging to the same corporation. For the two plants reporting the most frequent downtime events (Plants B and C), the actual availability for each machine was calculated (Figure 3-68); however, downtime data on the very troublesome ring-equipped non-lubricated reciprocating compressors originally installed at Plant B were unavailable. Only one of the five compressors (which was at Plant B) exhibited an availability below that reported for conventional lubricated reciprocating compressors in the U.S. Gulf Coast plants (for example, 96.36% vs. 98.5%). All other machines exceeded the average.

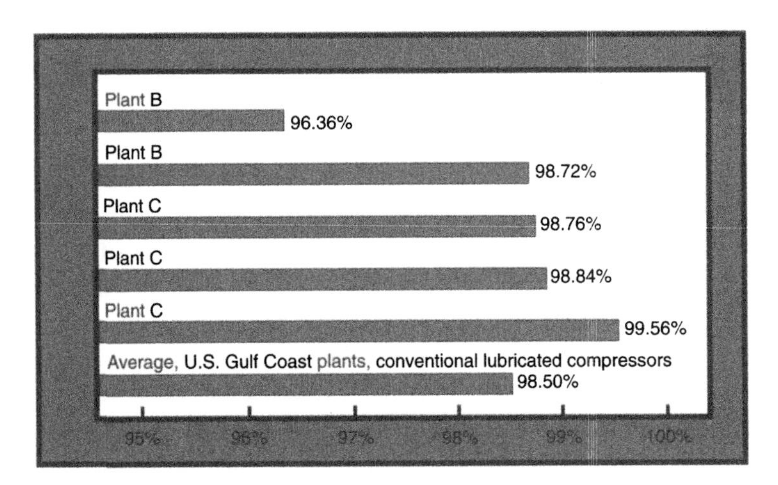

FIGURE 3-68. Availabilities of individual labyrinth-piston compressors reported by two North American plants (*Source: Bloch, H. P. "Consider a Low Maintenance Compressor,"* Chemical Engineering, *July 18, 1988*).

Finally, the consumption of major spare parts (excluding valve components, for which no records had been kept) by 14 labyrinth-piston compressors in ethylene service in Europe was examined (Table 3-15). From 1977 through 1980, these machines had been in operation for a combined total of

TABLE 3-14
NUMBER OF DOWNTIME EVENTS PER LABYRINTH-PISTON COMPRESSOR

Plant	Number of machines	Year of installation	Downtime events per machine per year
A	6	1982	1.00
B	2	1984	6.00
C	3	1982	2.00
D	5	1961–74	0.67
E	3	1976	0.35
F	12	1966–77	0.43

TABLE 3-15
CONSUMPTION OF SPARE PARTS BY LABYRINTH-PISTON COMPRESSORS

Location	Number of units	Total operating hours	Shaft seal	Crosshead body	Main bearing	Pin bearing	Connecting-rod bearing	Guide bearing	Piston rod	Piston I	Piston II	Crosshead pin
I	3	98,500		1		4		2	3		1	4
II	3	83,530	2						1			
III	2	43,000										
IV	2	33,000	1									
V	4	55,500	3			2		6	2		4	3
Total	14	313,530	6	1	0	6	0	8	6	0	5	7

313,530 hours. Although comparable statistics for conventional lubricated compressors were not available, it is unlikely that their repair history would have been more favorable than that of the labyrinth-piston compressors.

ROUTINE MAINTENANCE IS SUFFICIENT IN FOULING SERVICE

The excellent performance of the labyrinth-piston compressors in fouling service at Plant A can be attributed to the routine maintenance the units receive. Six four-throw, two-stage machines are given routine overhauls after every 8,000–10,000 operating hours. Three of the six machines are always in service, and there are times when all six machines are online.

Overhauls are generally completed in three to four working days, with three or four machinists working 12-hour shifts. Typical tasks include:

FIGURE 3-69. Amorphous polymer collected on piston rod.

- Replacement of valves with a spare set.
- Replacement or refitting of rod packing by lapping the ends of each 120° segment.
- Removal of amorphous polymer near the guide bearing (Figure 3-69).

FIGURE 3-70. Cleaning of valve unloader covers.

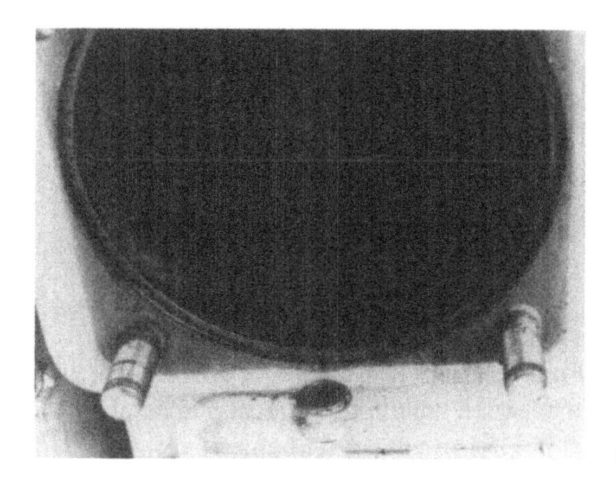

FIGURE 3-71. Cleaning of valve ports.

FIGURE 3-72. Routine cleanup of labyrinth piston.

FIGURE 3-73. Component appearance after routine cleanup.

- Cleaning of valve unloader covers (Figure 3-70).
- Cleaning of valve ports (Figure 3-71).
- Cleaning of the deadspace between the cylinder bores and heads.
- Occasional replacement of O-rings in the mechanical shaft-seal assembly.
- Routine cleanup of labyrinth piston (Figure 3-72).

CHAPTER 4

Overhaul and Repair of Reciprocating Compressors

RULE OF THUMB FOR GENERAL RUNNING CLEARANCES

Before an overhaul is attempted, it is important to read and understand the instruction manual supplied by the manufacturer of the compressor. It is particularly important to establish the recommended clearances given for the various components of the compressor.

If the manufacturer's data for running clearances are not available, the following may be used as guides and good rules of thumb. It must be remembered that repair and overhaul procedures may not lead to restoration of the original dimensions. It is important, however, to always maintain the proper tolerances and clearances between mating parts.

Piston (cast iron) to cylinder bore or liner:
.00125″ per inch of bore diameter
Example: 20″ diameter cylinder
20 × .00125″ = .025″ clearance

Piston (aluminum) to cylinder bore or liner:
.003″ per inch of bore diameter
Example: 20″ diameter cylinder
20 × .003″ = .060″ clearance

Main bearing and crankpin bearing to journal clearance:
a. Cast iron or steel-backed shells
.00075″ per inch of journal diameter

b. Aluminum bearing shells
.001″ –.0015″ per inch of journal diameter

Crosshead pin to crosshead bushing clearance:
.0005″–.0015″ per inch of pin diameter

Crosshead pin to crosshead clearance:
.0005″–.002″ per inch of pin diameter

Crosshead to crosshead guide clearance:
.00075″–.001″ per inch of crosshead diameter

Piston ring end gap:

cast iron	–.003″	per inch of cylinder diameter
carbon	–.003″	″
bronze	–.004″	″
Teflon®	–.024″	″
phenolic	–.005″	″

Piston ring side clearances (average):

cast iron	–.003″ to .004″	per inch of ring width
carbon	–.003″	″
bronze	–.004″	″
Teflon®	–.010″	″
phenolic	–.015″	″

Minimum clearance between rider ring and cylinder bore in middle of piston adjacent to piston rings:

cast iron piston –.00125″–.0015″ per inch of cylinder diameter
aluminum piston –.002″ per inch of cylinder diameter

Estimating Clearances by Calculating Thermal Expansion

Formula
$\Delta L = \mu (\Delta t) L + 20\%$
$\mu = .000006''$ per inch per degree F
(coefficient of expansion of cast iron or steel)
Example: Crosshead to crosshead guide clearance:
Crosshead diameter – 8½″
$\Delta t = 75°F$ (ambient) to $170°F = 95°F$
$\Delta L = (.000006)(95)(8.5) \times 1.2 = .006''$

Example: Piston to cylinder bore clearance:
piston diameter = 20″
discharge temperature = 280°F
ambient temperature = 80°F
material of piston = cast iron

$\Delta L = (.000006)(280 - 80)(20) \times 1.2 = .029''$

CHECKING BEARING CLEARANCES

The use of feeler gauges to check bearing clearances, while common, is subject to error due to the problem of inserting the gauge in the limited space and also because of the round surfaces being checked. It is therefore recommended that other methods be used.

Defining Clearances by the Use of "Lead" or Plastigage

The clearance in a babbitt bearing, or the clearance between piston and cylinder head, can be determined by opening up and inserting a soft lead wire, then measuring the thickness of the lead with a micrometer after the lead has been compressed in the bearing or between piston and cylinder head. Fuse wire is normally used, if it is soft. Plastigage is used in the same manner and is preferred because it is softer and will not embed in the soft babbitt of the bearing shell.

Measuring Clearances With a Dial Indicator

The most accurate method of measuring clearances is with a dial indicator. This is also a quick method that does not require any disassembly. Here is how it works.

Clamp the dial indicator on the connecting rod, with the pin resting on the crankshaft and parallel with the connecting rod. Place a bar under the lower bearing cap and bounce the bearing up and down to read the clearance directly on the dial indicator. To avoid errors in the readings, the crankshaft should be prevented from turning.

Depending on compressor type and design, some ingenuity may be required to place the dial indicator in the optimum position. The indicator should read bearing clearance by getting only the relative movement between the bearing and the journal, without lost motion being added to the reading.

Crosshead pin clearance can be checked with a dial indicator by employing a procedure similar to that for the bearings. Figure 4-1 shows a method of checking main bearing clearances with a dial indicator.

COMPRESSOR ALIGNMENT

Chapter 1 discussed the importance of proper foundations and grouting to maintain alignment and proper elevation of the compressor/driver as well as to minimize vibration and prevent its transmittal to external structures. It is not uncommon for compressors to "come loose" on the foundations because of deterioration of the grout that supports the compressor on the foundation.

A "loose" compressor often causes and transmits vibration and movement of the machine and associated piping. Even more important, looseness allows for misalignment of the compressor crankshaft running gear system and the compressor cylinders relative to compressor frame or crankcase.

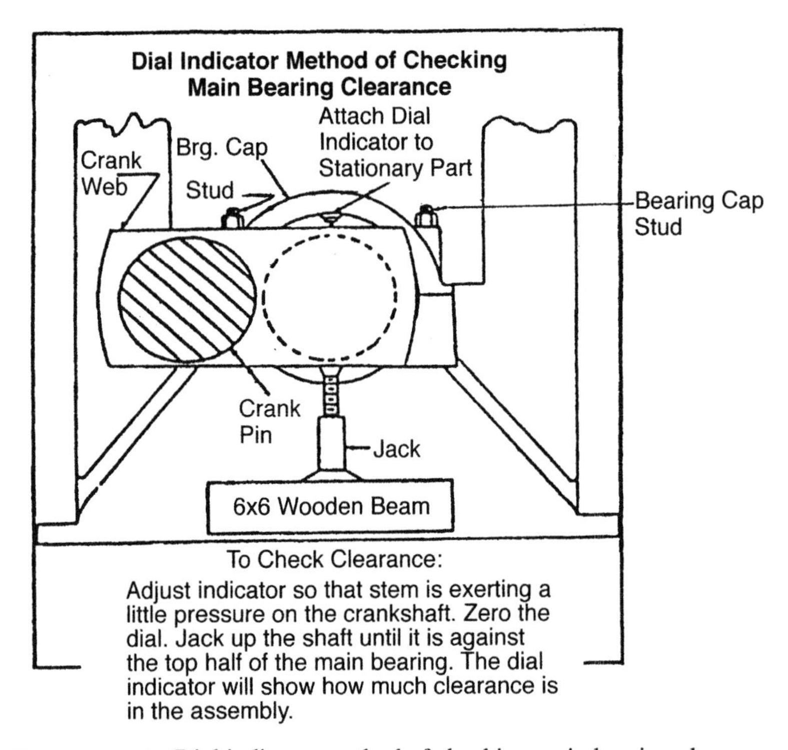

FIGURE 4-1. Dial indicator method of checking main bearing clearance.

Fortunately, we can employ one of several methods to check the alignment of the compressor crankshaft system and the relationship of the compressor cylinders to the frame. When foundation or grout failure occurs, more serious distress may follow. The compressor crankcase or frame takes on a new alignment pattern because the foundation no longer furnishes precise support.

WEB DEFLECTION MEASUREMENTS*

Web deflection measurements are thus required for determining deviations in the crankshafts of large industrial/marine engines and reciprocating compressors. These measurements and their interpretation are vital in disclosing conditions that can lead to disastrous and expensive crankshaft failures. Accurate periodic measurements, combined with careful analysis, will permit such conditions to be corrected before they cause problems.

Web deflection is defined as any movement of the crank webs from their ideal position during the entire rotation of the crankshaft (360°). Consider the crankshaft in Figure 4-2, whose centerline, perhaps because

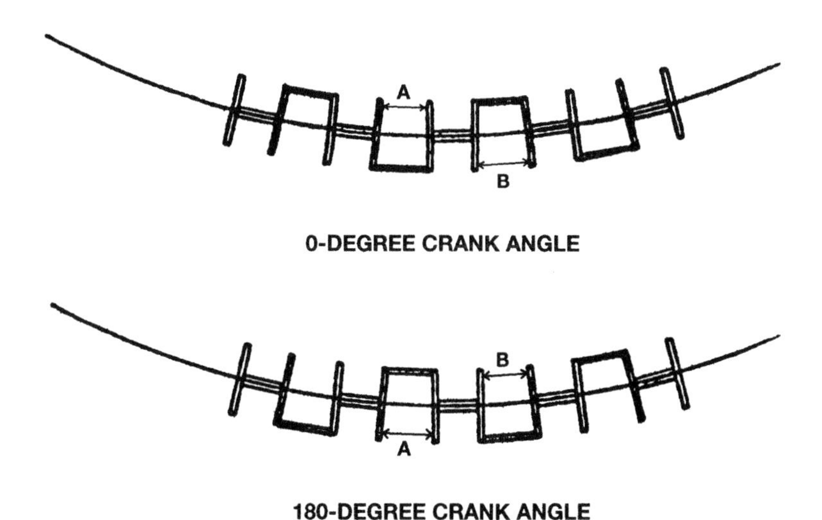

FIGURE 4-2. Crankshaft deflection mode.

*Contributed by Jim McNabb (Southern California Gas) and Leon Wilde (Indikon Company, Inc., Somerville, Massachusetts).

of faulty equipment support, has been forced into a curve. At zero degrees of rotation, the web opening A is less than ideal, while that at B is greater than ideal. After rotating the crankshaft 180°, the condition is reversed. Because the centerline is the same, the crankshaft must now bend in the opposite direction. The A opening has now increased, while B has decreased. This flexing of the crankshaft as it rotates represents cyclic stresses that, if of sufficient magnitude, can lead to fatigue failure. It can be appreciated that because deflections are small (less than 5 mils or .005″), accurate measurement is essential. Permissible deflections vary with crankshaft stroke. Figure 4-3 can be used for general guidance.

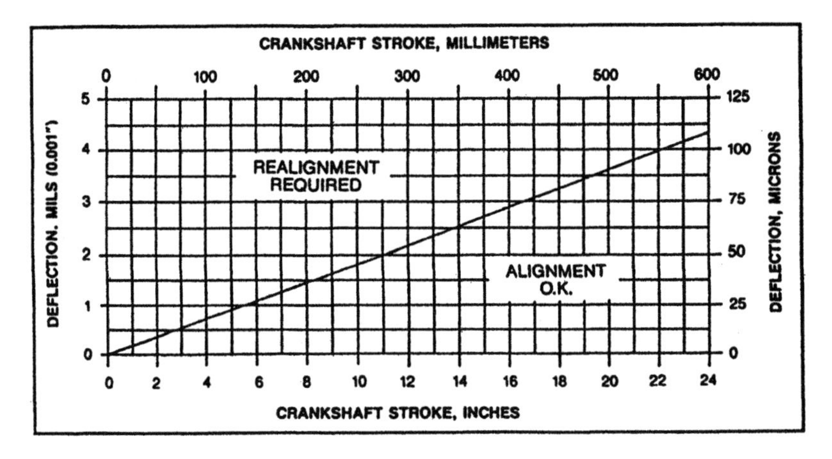

FIGURE 4-3. Permissible crankshaft deflections.

Both dial indicator and electronic instrument-based measuring methods are available to the compressor and engine maintenance professional. Using traditional dial indicator methods, Figure 4-4, crankshaft deflection is determined with a *strain gauge* or crankshaft deflection gauge. It is placed between the throws of the crankshaft and, when the shaft is rotated, *deflections* or bending of the throw can be measured.

An electronic digital web deflection indicator system (see Figure 4-5) was introduced in 1983. This system makes deflection measurements safer, easier, and more accurate than the dial indicator that had been used in the past.

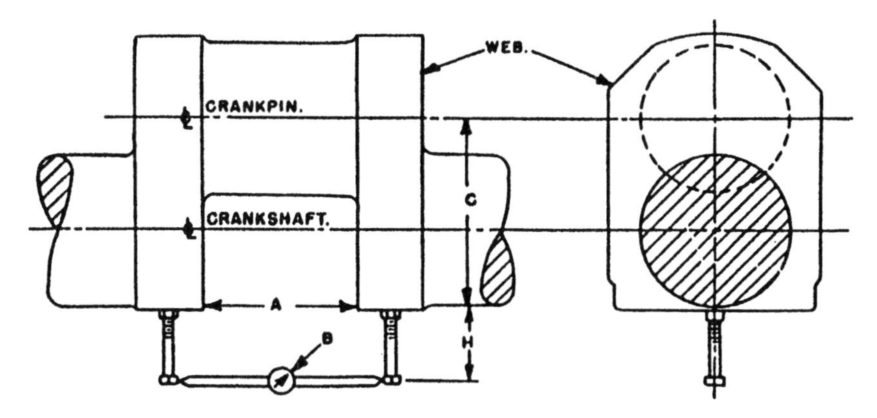

FIGURE 4-4. Web deflection being measured with dial indicator (strain gauge).

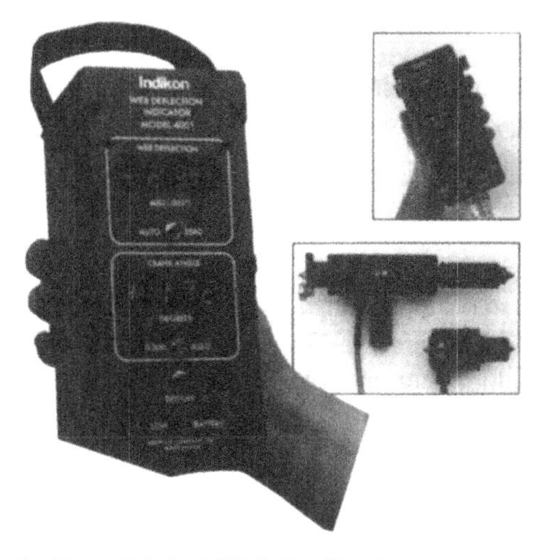

FIGURE 4-5. Indikon Digital Web Deflection Indicator (*Source: Indikon Company, Inc., Somerville, Mass.*).

The two components making up digital web deflection measuring systems are shown in Figure 4-6, which shows a single crank. One of the most important advantages of the Indikon system is that it permits taking readings from outside the engine. In addition, an angle transducer also allows the crankshaft angle to be read simultaneously.

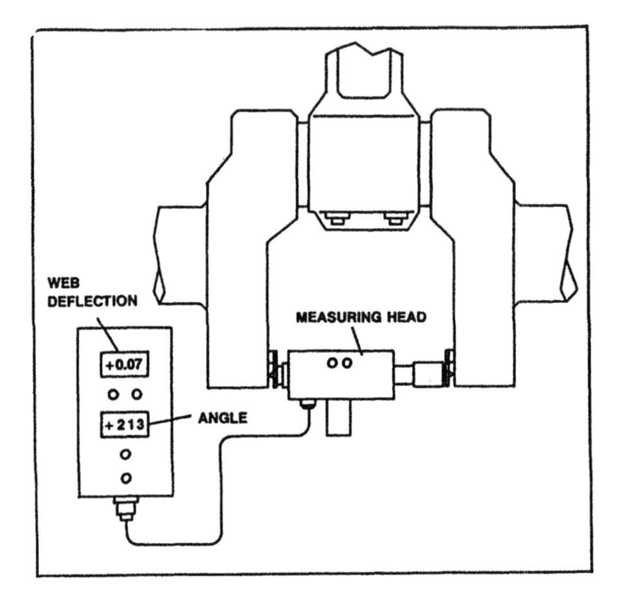

FIGURE 4-6. Schematic illustrating the two components of an electronic web deflection measuring system (*Source: Indikon Company, Inc., Somerville, Mass.*).

CAUSES OF WEB DEFLECTION

Several factors may contribute to web deflection and must be examined before, during, and after taking web deflection measurement. Some of these factors include, but are not limited to:

- compressor support—grouting conditions
- anchor bolts—torque value
- sole plate and chock adjustment and torque
- shim packing, alignment of any compressors, distance pieces, and cold supports
- rod run-out, proper frame alignment
- main bearings, clearances
- good lubrication, oil analysis
- compressor load, overload
- soil conditions, fill

All of these factors, either solely, together, or in any combination, will contribute to web deflection and must be addressed and corrected before accurate conclusions may be reached concerning the actual crankshaft condition.

MEASUREMENT CONSIDERATIONS

The following factors should be considered for accurate web deflection measurement.

1. Punch marks

   Crisp, properly aligned punch marks on the webs are critical. A hollowed-out punch mark, instead of a precise, conical punch mark, may cause the measuring tool to move as much as one mil (.001″). Good punch marks must be readily identifiable and used for each future web deflection measurement. Lining out "extra or old" punch marks is achieved simply by using a small cold chisel and brass hammer. (Never grind out punch marks because this removes weight from the balanced webs.)
2. Equipment condition

   Web deflection measurements should be taken after a compressor has operated for three to four days (72 to 96 hours) nonstop. This allows adequate time for the entire compressor, foundation, frame, bearings, and crankshaft to come up to normal operational temperature. This condition allows complete thermal expansion to occur and will reflect the normal operating mode of the crankshaft. "Hot" web deflection is recognized as the best method for measuring web deflection. "Hot" is usually represented by machine temperatures of 120°F and higher ("cold" is usually 119°F and less).

   The instrument to be used for measuring deflection should also be acclimated to compressor temperature. A cold instrument in a hot crank will expand and reflect false readings.
3. Instrument

   With each type of measuring instrument, appropriate precautions are necessary.

   a. Dial indicator

   The dial indicator (strain gauge) has been the standard precision instrument used in taking web deflection in the past. This instrument is very accurate provided the following conditions are met:

   - Calibrate yearly or as often as necessary, such as after a severe blow to the instrument.
   - All extensions for measuring varied throw widths are straight.

- Needle on dial face moves with ease. There should be no needle movement "hang-up" or drag after preloading the gauge, approximately one to one and one-half revolutions.
- Needle points on instrument and all extensions are sharp and crisp so as to correctly fit the web punch marks.
- The dial face must remain visible to the observer throughout the entire crankshaft rotation.

b. Indikon Electronic Web Deflection Indicator

As shown in Figure 4-6, this system consists of two units: a measuring head and a digital indicating instrument. After selecting the proper extensions, place the measuring head between punch marks.

Then take the following steps:

- Move the threaded adjustment to obtain the proper preload. Red and yellow lamps on the measuring head indicate the required direction of adjustment. When both lamps are out, the adjustment is correct and the transducer is within its measuring range.
- Attach the magnets at the center points to the adjacent web. This drives the centers in synchronism with the crankshaft as it is rotated, preventing the center points from rotating or "squirming" in the punch marks. Inaccuracies are thus minimized if the punch marks are not perfect.
- Because one magnet also drives the angle transducer, position it so that the correct starting angle is obtained (usually 0° and in the direction the crank is rotated when operating).
- Verify that the counterweight hangs freely to assure correct angle readings.

Once the measuring head has been placed in position, readings of both web deflection and crank angle may be taken from the digital indicator.

Checking for proper operation is easily accomplished before use. For deflections, move the measuring plunger inward by hand while observing the change in distance readings. For angle, rotate the transducer magnet and observe the readings.

A rigid calibration stand is essential because the relatively large seating force (approximately 5 pounds) can cause spurious deflections in conventional outside micrometers. A complete calibration stand, or fabrication drawings for one, may be obtained from the manufacturer.

4. User skills

It is important to ascertain that the engineer, technician, or mechanic is skilled in web deflection measurement procedures and knowledgeable enough to correctly interpret the information. Choosing one instrument over another becomes a matter of skill level and personal preference.

Regardless of the instrument used in taking web deflection readings, a well-trained technician or mechanic holds the key to accurate web deflection measurements. The more accurate the measurements taken and the more precise the reading, the greater the amount of information that will be available to determine the condition of a compressor crankshaft. Combining this information with all other factors and comparing it with previous records can be the difference between a budget-depleting crankshaft failure or an online profit-producing machine.

Electronic Web Deflection Indicator

A well-designed electronic web deflection system gives at least the same degree of accuracy as the dial indicator, and sometimes more.

It can provide measurement readings of deflection to within one hundredth of a thousandth of an inch (10 microinches). Being a digital instrument, the system eliminates all guesswork as to measurement readings, while also providing movement indication of either positive (+) or negative (–) changes. The system provides up to a 15-foot cable between the measuring instrument and the receiving digital unit. This means that all readings can be taken outside the crankcase. For safety reasons and just plain ease and comfort, this is a marked advantage. Remember, "hot" web deflection occurs when unit temperature is 120°F and above. The monitoring of web deflection readings during rotation of the crankshaft is greatly facilitated with the digital readouts.

A major attraction of any well-designed electronic unit is ease of placement in the webs. The preloading monitor lamps, digital angle and deflection readouts, and being able to stay outside of the crankcase all make the unit easy to train on and operate with confidence. An experienced technician or mechanic using the electronic unit can take a series of web deflection readings on several units in approximately half the time than if using a conventional dial indicator approach.

For the system operator, accurate, unambiguous readings and independence of operator technique are additional advantages. System-wide, historical records allow the operator to monitor the performance of a single machine over a period of time as well as to compare similar machines within a system. In addition, maintenance procedures and repair methods, such as grouting, can be evaluated.

INTERPRETING WEB DEFLECTION READINGS

In order to interpret the readings, some experience and sound thinking are required, but the effort will be worth it. Because problems with the lower end of the machine are never the same, the best way to deal with instructions on analyzing the data is to examine several hypothetical cases. The reader can better follow the examples and subsequent problems by using a model crankshaft made from wire or a paper clip.

The readings listed in the table of Figure 4-7 will be used as Case 1 and were obtained from a machine with a 22-in. stroke. The compressor builder assigned a maximum deflection figure of .004 in. The –.005 at the 180° position for No. 3 throw is above the specified limit, indicating that something is wrong. The 90° and 270° positions are normally used to determine whether the main bearings are out of alignment in a hori-

Crank Position	Throw					
	1	2	3	4	5	6
"Down Position" 0°	000	000	000	000	000	000
90°	000	.00025	-.001	000	000	-.0005
180°	000	-.001	-.005	-.002	-.00025	-.001
270°	000	000	-.001	-.00025	000	-.00025

FIGURE 4-7. Example web deflection readings. (*Source: Bloch, H. P. and Geitner, F. K.* Major Process Equipment Maintenance and Repair, *Houston, TX, Gulf Publishing Company, 1985*)

zontal plane. However, when the 180° position has excessive deflection (caused by one journal being low), it carries up to the 90° and 270° positions, which in this case results in the –.001 reading. Furthermore, if the bearing saddles were out of alignment in a horizontal plane, the signs at the 90° and 270° positions would be reversed. Therefore, nothing is wrong with the horizontal alignment. (Actually, the 180° position readings are the most significant, because rarely will main bearing saddles be found in sidewise misalignment.)

Returning to the example, because the No. 3 throw 180° reading is the only one that is excessive, it is apparent that the bearing to the right of No. 3 throw is wiped. Note that this low bearing has caused distortion in the shaft past No. 3 as indicated by the –.002 reading of No. 4 throw. In this case the correction is simple, because it is only a matter of replacing the bearing.

Figure 4-8 shows a set of crankshaft deflections that will be used to explain Case 2. Here, the 180° deflections get worse from No. 1 to No. 3

CRANKSHAFT DEFLECTIONS—CASE NO. 2

CRANK POSITION	THROW					
	1	2	3	4	5	6
0°	000	000	000	000	000	000
90°	–.00025	–.0005	–.0015	–.001	–.0005	000
180°	–.001	–.003	–.006	–.004	–.0015	000
270°	000	–.00025	–.0015	–.001	–.00025	000

Figure 2—A set of crankshaft deflections illustrating six different possible types of deflection problems that might be encountered.

CASE NO. 3—SAME AS ABOVE EXCEPT ALL SIGNS (+)

CASE NO. 4

CRANK POSITION	THROW					
	1	2	3	4	5	6
0°	000	000	000	000	000	000
90°	–.00025	–.0015	000	+.00025	+.0015	+.00025
180°	–.001	–.004	–.0005	+.001	+.005	+.001
270°	–.00025	–.001	000	000	+.001	000

FIGURE 4-8. Crankshaft deflections illustrating different types of deflection problems that might be encountered (*Source: Bloch, H. P. and Geitner, F. K.* Major Process Equipment Maintenance and Repair, *Houston, TX, Gulf Publishing Company, 1985*).

throws and better from No. 3 to No. 6, and all signs are minus (–). A condition such as this means that the shaft is in a continuous bow. This can be verified by bending your wire model crankshaft into a bow and by rotating it as is done in taking the readings. It will be seen that all signs would be (–), and the highest separation of the webs would be in the middle throw. This situation is not characteristic of one or more bearings being wiped, because it is improbable that both end bearings would be wiped, leaving the center high. A typical cause for this condition is for the bond between the frame and grout at each end of the engine to have broken loose. The horizontal couple forces cause the frame to move relative to the grout, which, over a period of a year, can actually wear it down.

If this is the problem in Case 2, it can easily be checked by inserting long feelers (about 8 in.) between the frame and grout. If the feeler thickness is too great (up to .025 in.), the situation is actually worse than the deflections indicate because the frame is not supported. There are many installations in which feelers can be inserted all the way at the end of the frame, but the gravity of the circumstance is determined by how far the feelers can be moved from the end toward the middle once they are inserted. Regardless, the deflections are excessive in Case 2, and if there is a loosening of the grout, with frame movement, the unit may have to be regrouted. A common error is to tighten the foundation bolts to restrict movement. Such tightening is useless because, once the bond is broken, the foundation bolts cannot hold the engine down. The amount by which the maximum deflection can be exceeded will be discussed in subsequent paragraphs.

If the inspection just described indicates that the bond between the frame and grout is satisfactory and the grout has not broken up, then the bowed condition of the shaft could be caused by a change in the shape of the foundation. There is a possibility that it may be cracked. This can be verified by a thorough examination of the foundation. Almost all concrete structures have hairline cracks that should be ignored. However, open cracks, regardless of the width, are a good indication of trouble. A sketch showing the exact location of the open cracks is sometimes useful in correlating their location to the crankshaft deflections.

In Case 3, if the deflections were exactly the same as Case 2 but the signs were all plus (+), then the grout or foundation is in a bad sag. Comments for this condition are the same as Case 2.

In Case 4, the changes in signs of the deflections show the shaft to be in a reverse bend. This could be caused by bad bearings, grout, foundation or frame. In this case, as well as in the preceding three, the analysis should not be confirmed or acted on until all main bearings have been inspected.

Maximum Deflection Specifications

The number of variables involved and the complexity of the problem make it impossible for an equipment builder to predict the deflection at which shaft failure will occur. Therefore, a very tight maximum figure has to be assigned to any shaft so that all situations will be covered. It is for this reason that failures have happened to shafts with deflections slightly above specifications while other machines have run for years with deflections much higher than compressor builders' limits. Furthermore, some locations make it difficult to keep a machine level enough to stay within the limits. The problem is to decide how far one can go beyond recommendations. The following discussion might help in making that decision.

In Case 1, the change in deflection from throws No. 2 to No. 3 is very abrupt. In that case, the web stress is very high, and it is recommended that the specified maximum deflection not be exceeded. This can be demonstrated by holding adjacent main bearings of the wire-model shaft and creating a bending motion. This would break the shaft quicker than by holding it at the end main bearings.

Case 4 is also an undesirable situation in that there is a reverse bend, or "S," indicated by a change from plus to minus signs. The stress concentration in the throw between the change of signs can become pronounced if the deflection is much above the manufacturer's standards.

Case 2, which is a bow (all plus), should allow more deviation from standards than the other examples, because the stress concentration, as in the case of the sag, is not as dangerous. Also, a bow is better than a sag because in the former the deflection is minus. Where a minus reading is involved, the webs are inward from the neutral position when the throw is up. On vertical compressors, the up position occurs when the peak pressure exerts maximum force on the journal and tends to spread the webs apart. Because the webs are already inward, the peak pressure does not contribute as much to web stress as it does in the situation of a plus reading, where the webs are spread apart before the maximum force is exerted.

It can be seen that it is difficult to assign a maximum deflection to any reciprocating machine, but if the value specified by the builder is not exceeded under any conditions, experience has shown that the shaft should not break. It is always wise to consult the manufacturer when deflection limits are reached.

COMPRESSOR CYLINDER ALIGNMENT

The relation of the compressor cylinders to the frame or crankcase should be examined with laser-optical instruments or by the tight wire method. This checks the relationship of the compressor cylinder bore, packing case bore, housing bore, and crosshead guide bore to the crankshaft. Refer to Figure 4-9 for reference points for wire alignment, as given by Dresser-Rand for the Model HHE Compressor.

If it is found that the alignment of the compressor cylinder is not correct or that the crankshaft strain gauge readings are beyond acceptable limits, the compressor frame may have to be regrouted and perhaps some of the foundation cap removed and repaired. This is not an uncommon repair and is done with epoxy materials that are far superior to the sand and cement used on older installations.

Cracks in the foundation or grout under the compressor base and crosshead guides are repaired by pressure injection. These sorts of repairs

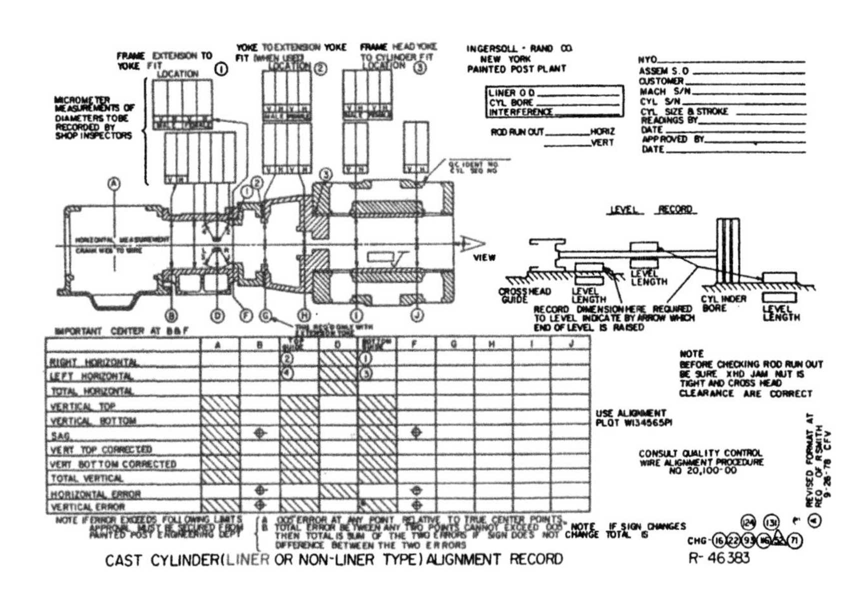

FIGURE 4-9. Reference points for wire alignment.

must be made by companies that have experienced personnel and who can determine whether a compressor foundation is bad and whether it is out of alignment.

Operating a compressor with bad foundation/grout and misalignment is extremely risky. Failure, damage, and degradation typically involve:

- Main bearings
- Crankshaft breakage
- Piston scoring
- Packing failures
- Piston rod scoring
- Crosshead shoe wear

PISTON ROD RUNOUT

Piston rod runout is a measure of how well the cylinder is lined up with the frame and, also, how well the internals of the compressor (piston, rod, crosshead) are lined up with each other and the frame and cylinder combination. This runout is checked and adjusted with all parts "cold"; the ambient temperature should be in the 60°–68°F range.

Prior to performing runout measurements, cylinder and support, cylinder frame head, and doghouse nuts should be properly set or torqued. Oil scraper and packing should be loosened to prevent binding. Runout should be set during erection; followup measurements are needed whenever regrouting, major parts replacement, overhaul, etc., are performed.

The following procedure is a guide for obtaining and interpreting runout readings using dial indicators.

- *Initial setup.* Use a dial indicator with facilities to be securely fastened to a non-moving surface in or on the housing. The compressor piston is to be set at its maximum outboard stroke by barring the unit. The dial indicator traveling pin is to be positioned anywhere along the rod where it is convenient.

  To measure horizontal runout, set the pin on the horizontal centerline of the piston rod. For vertical runout, the vertical centerline is used. Both vertical and horizontal runout are extremely important and both are to be measured.

When setting the traveling pin in the desired position, depress the dial indicator so the needle makes at least one revolution, lock the indicator neck, and set the dial face so that the needle reads zero.

- *Reading the runout.* Bar the unit over 360° (one complete revolution). The dial will go to a limit when the unit has been barred 180° (the piston should be at its maximum inboard stroke). This reading, whether plus or minus, is the total runout. The unit is then barred over the remaining 180° (piston is now at its maximum outboard stroke) to ensure that the needle returns to the zero setting. If the needle does not return to zero, redo the measurement after checking the indicator arrangement for tightness and making sure it is well secured at its base.
- *Vertical runout.* This reading generally indicates how well the compressor piston and rod are in line with the cylinder and frame. We like to see a cold runout of a minus on the piston end. The expansion of the piston while in service is thus anticipated.
- *A positive reading means that the piston is high* in relation to the crosshead, and a minus means that the piston is low, as desired, in relation to the crosshead. If the piston is high, or if it is lower than the desired limit, adjustment must be made in accordance with the compressor manufacturer's instruction manual.

Figure 4-10 shows allowable piston rod runout (vertical and horizontal) as a function of cylinder running clearance (Cooper-Bessemer). Cylinders may be operated if they fall within these limits. If, after a period of time, runout exceeds allowable, this is an indication that wear has occurred and maintenance is required.

Proximity Probe Method of Monitoring

A more efficient method of measuring rider band wear is to use a proximity probe, mounted vertically to the packing case, to measure the position of the piston rod. Probe gap voltage can then be read with an electronic rod drop readout device. Monitors such as Bently-Nevada's Six-Channel Rod Drop Monitor use a Keyphasor®, probe to provide a once-per-revolution crankshaft pulse. This pulse is used as a reference, so an instantaneous rod position can be displayed. There are three main advantages to this method of rod drop measurement.

First, by taking readings at only one point in the stroke, the effects of scratches, wear, or rod coatings are minimized. Second, the most effec-

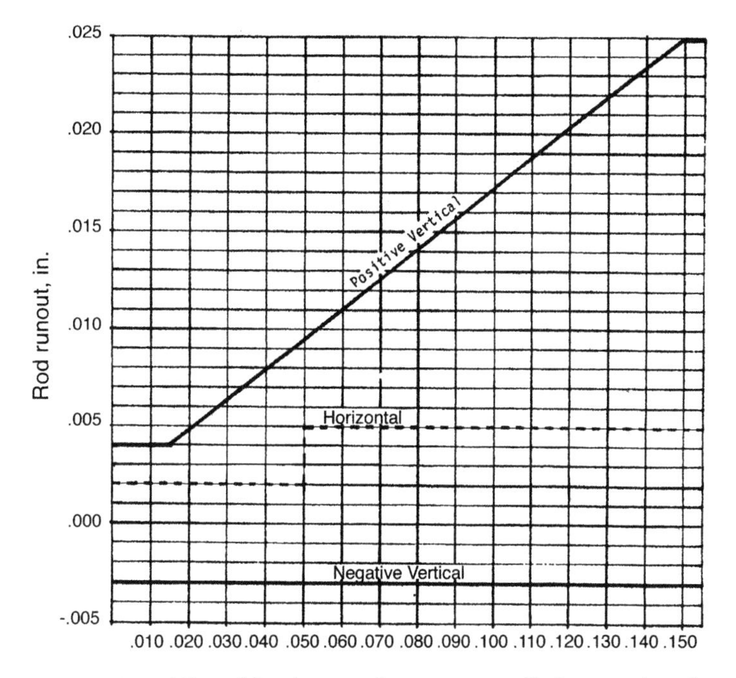

FIGURE 4-10. Allowable piston rod runout vs. cylinder running clearance.

tive point in the stroke can be selected for taking the reading. Typically, this will be just before, or just after, bottom dead center (BDC), when the dynamic forces on the piston and crosshead have a minimal effect on the reading.

Third, since we know where the center of the piston is at this instant, the monitor can correct for the geometry of the machine and display the amount of actual rider band wear.

Bently-Nevada's Model 3300/81 Monitor provides two levels of alarm, Alert and Danger, for each channel. This monitor eliminates the need to periodically stop the machine and inspect for rider band wear. In addition, the monitor provides the information to develop a wear trend as part of a predictive maintenance program, unlike other rod drop indicators that provide only an alarm on excessive rider band wear.

Preliminary Checks if Runout Exceeds Limits

If the guidelines above have been closely followed and runout cannot be set at the proper limits, the following checks are recommended.

- First, ensure that all the housing nuts are torqued properly.
- Second, ensure that the packing and scraper assembly are not binding the rod. If the proper runout still cannot be attained, rotate the piston and rod assembly 90°. Make sure the frame and outer end piston clearances are still adequate to prevent contact with their respective heads. Relock the crosshead jam nut; take both vertical and horizontal runouts again and record them.

Always refer to the instruction book for acceptable piston rod runout readings and the methods of correcting or determining unacceptable readings.

Table 4-1 gives shimming guidelines for crosshead shoes.

TABLE 4-1

PISTON MATERIAL	TOTAL THICKNESS OF SHIMS BETWEEN CROSSHEADS & SHOES IN INCHES Individual Shim Thickness = .005"									CROSSHEAD SHOE	CROSSHEAD DIAM. IN.
C.I.	.020"	.020"	.025"	.025"	.030"	.030"	.035"	.035"	.040"	TOP	25
	.045"	.045"	.040"	.040"	.035"	.035"	.030"	.030"	.025"	BOTTOM	
AL.	.020"	.025"	.025"	.030"	.035"	.040"	.045"	.050"		TOP	
	.045"	.040"	.040"	.035"	.030"	.025"	.020"	.015"		BOTTOM	
C.I.	.015"	.020"	.025"	.025"	.030"	.030"	.035"	.035"	.040"	TOP	19-22½
	.035"	.030"	.025"	.025"	.020"	.020"	.015"	.015"	.010"	BOTTOM	
AL.	.020"	.025"	.025"	.030"	.030"	.035"	.040"	.045"		TOP	
	.030"	.025"	.025"	.020"	.020"	.015"	.010"	.005"		BOTTOM	
C.I.	.020"	.020"	.025"	.030"	.030"	.035"	.035"	.040"	.040"	TOP	12¼-17½
	.020"	.020"	.015"	.010"	.010"	.005"	.005"	.000"	.000"	BOTTOM	
AL.	.020"	.025"	.025"	.030"	.035"	.040"	.040"	.040"		TOP	
	.020"	.015"	.015"	.010"	.005"	.000"	.000"	.000"		BOTTOM	
C.I.	.020"	.020"	.025"	.030"	.030"	.035"	.040"	.040"		TOP	10½
	.020"	.020"	.015"	.010"	.010"	.005"	.000"	.000"		BOTTOM	
AL.	.020"	.025"	.030"	.035"	.035"	.040"	.040"			TOP	
	.020"	.015"	.010"	.005"	.005"	.000"	.000"			[illegible]TOM	
PISTON DIAMETERS - INCHES	0 - 5 5	5 - 10 10	10-15 15	15-20 20	20-25 25	25-30 30	30-35 35	3[illegible] 40 [illegible]	40-45 45		

SHIMMING OF CROSSHEAD SHOES FOR MORE ACCURATE PISTON ROD RUN-OUT

FOUNDATION PROBLEMS AND REPAIRS

As has been pointed out in Chapter 2, reciprocating compressors experience alternating movement of the reciprocating parts. The resulting shaking or inertia forces must be contained by mounting these machines on properly designed foundations.

Regrouting of oil soaked concrete is very often required on old compressor installations. The new grout material should be an epoxy.

When regrouting on oil saturated concrete, the expected results should be comparable to the properties of good concrete because these properties were the criteria for the original design.

Experience has definitely shown that the best method of preparing a concrete surface for bonding is through mechanical scarification to remove surface laitance. This can be accomplished by chipping away the surface at least ½ inch. Sandblasting the surface on reciprocating equipment foundations is not acceptable. At one time, acid washing was widely accepted as a means of surface preparation, but this practice has not proved reliable.

Concrete can absorb oil and once oil has been absorbed, a gradual reduction in both tensile and compressive strengths will follow. Given enough time, the compressive strength of the concrete may be reduced to the point where it can be crumbled between the fingers. Fortunately, the deterioration process is slow and it may take many years for complete degradation. When total deterioration has occurred, the damaged concrete must be replaced with either new concrete or epoxy grout.

There are measures that can prevent this problem, such as sealing the concrete with an epoxy sealer to provide an oil barrier. Sealing of the foundation is usually done at the time of original construction. Concrete foundations that are oil-soaked but have not undergone total loss of strength can be salvaged with proper regrouting techniques. When contaminants such as oil or grease are present, special consideration should be given to surface preparation and epoxy thickness.

Proper epoxy grout thickness is important. It should be recognized that in solid materials, forces resulting from compressive loading are dispersed throughout the solid in a cone-shaped pattern with the apex at the point of loading. Consequently, the weaker the concrete, the thicker the epoxy covering should be in order that loads can be sufficiently dispersed before they are transferred to the concrete.

Other techniques may further enhance the remedial measures. For example, tensile loads can be transferred by means of reinforcing steel to locations deep in the foundation where good concrete still remains.

A severely oil-degraded foundation may be capped with a thick layer of epoxy grout laced with reinforcing steel properly placed and bonded deep in the foundation. This technique is similar to the technique used when a dentist caps a weak tooth. If a weak material can be contained, its strength may be maintained.

Investigating Whether Regrouting is Necessary

Measurements and analysis will help determine if regrouting is needed. Here is one proposed sequence.

1. Make precision laser profile readings of compressor base or main bearing machined areas from permanent benchmarks and record for comparison with future readings.
2. Take hot crankshaft web deflection readings of the compressor and record for study and future comparison.
3. Check for proper compressor rod runout and record readings.
4. Check for proper foundation bolt material and proper torque values. Make record of any broken bolts as well as torque value. Foundation bolt material should be AISI 4140 (B-7) or comparable high tensile strength steel.
5. Inspect main bearings and record clearances.
6. Make visual inspection of compressor movement at corners during operation and note condition on outline diagram of compressor foundation.
7. Make visual inspection of grout and concrete foundation and note condition on outline diagram of compressor foundation.
8. Compare readings and conditions with previous records. Compare crankshaft web deflections with manufacturer's suggested maximum permissible. Discuss essential corrective measures and develop a schedule for implementing corrections.

Effects of Postponing Regrouting

Postponing routine maintenance of operating equipment, particularly when regrouting is needed, will usually result in the foundation failing.

The most serious type of failure is foundation cracking at a location in a plane parallel to the crankshaft. These cracks may be caused by inadequate design, or by operating conditions that exert excessive forces on the foundation. Unless these foundation cracks are repaired at the time of regrouting, grout life will be greatly reduced (usually to about 10% of its normal life).

Lateral dynamic forces are generated by compressor pistons. Theoretically, if a machine were perfectly balanced, only dead weight forces would be exerted on the foundation. In such a condition, anchor bolts wouldn't be needed. In reality, a perfectly balanced reciprocating machine has never been built.

After establishing the fact that unbalanced forces do exist on well-designed and maintained equipment, consider what happens when maintenance is postponed. Suppose there are lubricating oil leaks that puddle on the foundation shoulder. If any movement exists between the machine and grout, oil will penetrate voids caused by the movement, and hydraulically fracture any remaining bond between the machine base and grout. As movement between the machine and grout increases, forces exerted on the foundation increase at an exponential rate because of change in direction and impact.

Repairing Foundation Cracks

The notch provided in the top of a foundation for the oil pan creates a perfect location for stress risers. A moment is created by lateral dynamic forces multiplied by the distance between the machine base and transverse reinforcing steel in the foundation below. The possibility of foundation cracking at this location increases as the depth of the notch increases. The farther the distance between the horizontal forces and transverse reinforcing steel, the greater the moment.

Figure 4-11 illustrates a method of repairing such cracks by drilling horizontal holes spaced from one end of the foundation to the other end. A series of holes is placed at an elevation just below the oil pan trough. A high tensile strength alloy steel bolt is inserted into each hole and anchored at the bottom of the hole.

Alternatively, the long bolt can extend all the way through.

Next, a small diameter copper injection tube is placed in the annular space around the bolt. The end of the hole is then sealed and the nut tightened to draw the two segments of the block back together.

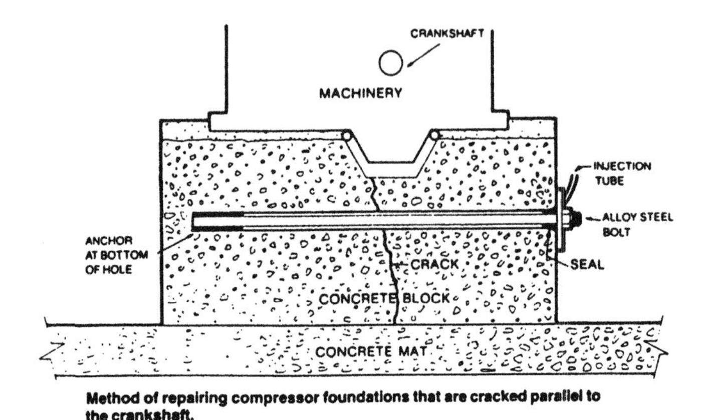

FIGURE 4-11. Method of repairing compressor foundations that are cracked parallel to the crankshaft.

An unfilled or liquid epoxy is now injected into the annular space around the bolt. Air in the annular space around the bolt is pressed into the porous concrete as pressure builds. After the annular space has been filled, injection continues and the crack is injected from the inside out.

This repair method places the concrete in compression. Note that it would otherwise be in tension. This compressive condition must be overcome before a crack could possibly develop again and, as a result, the repaired foundation is usually much stronger than the original foundation. This technique is often used when the concrete in the foundation is of poor quality.

Anchor Bolt Replacement

When anchor bolt failure is such that complete replacement is necessary, it can be accomplished using techniques consistent with the illustration in Figure 4-12.

Complete replacement of an anchor bolt is possible without lifting or regrouting the machine. This is accomplished by drilling large diameter vertical holes, adjacent to the anchor bolt to be replaced and tangent to the base of the machine. Once the cores have been removed, access is gained to concrete surrounding the anchor bolt.

After the surrounding concrete is chipped away, a two-piece and sleeved anchor bolt is installed. After the replacement anchor bolt has

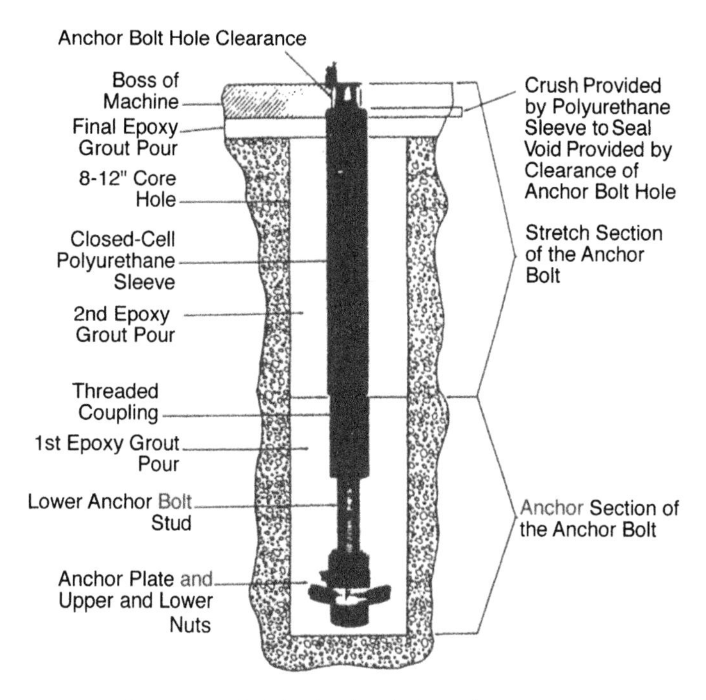

FIGURE 4-12. Anchor bolt replacement.

been installed, epoxy grout is poured to replace the concrete chipped from around the original bolt and the concrete removed by the coring.

COMPRESSOR BEARING MAINTENANCE AND REPLACEMENT

Premature bearing wear or failure is usually caused by a combination of factors. Mechanical overload, dirt, and incorrect installation are sometimes involved. But misalignment, improper lubrication, and overheating are by far the major causes of serious journal, bearing and mainframe damage. Such damage can be difficult to spot and often goes undetected for a time. Even though oil flow to one bearing may drastically deviate from the norm, oil pressure and temperature indicated on gauges may still appear normal if the problem is limited to a single oil line.

That's just one example of the many reasons manufacturers' recommendations on periodic inspection of bearings, crankshaft, and other components are so important. Whenever there is a need to replace a main or crankpin bearing, the analyst should ask why the bearing required replacing. Always investigate possible causes and determine if other components have been affected as well. Careful inspection of the area around a replaced bearing is an essential step in preventive maintenance.

Crankshaft straightness, bedplate saddle alignment, and the bore diameter of the connecting rod bearing bores are potential trouble spots. The localized overheating that often accompanies bearing failure creates dimensional changes that can lead to distortion or cracking in related parts. Be sure to consult the manufacturers' manuals and check all related components for damage whenever bearing problems occur.

Regardless of other factors involved, most severe compressor alignment deficiencies will ultimately lead to destructive overheating of major components.

Bearing Maintenance

Many serious problems can remain hidden long enough to cause damage. Good bearings may actually conceal a bearing saddle problem. A bent crankshaft can, in some cases, give "zero" deflection readings. Sometimes the lack of normal wear can indicate a problem. Thus inspections should be made a regular part of the maintenance program. Be suspicious when anything out of the ordinary is seen—even if the overall condition of the machine seems acceptable.

Expect wearing parts to wear out. In one catastrophic failure event, a compressor broke a crankshaft. There was no abnormal crankshaft deflection at the last inspection, and it had run for over 15 years on the same bearings.

During disassembly, it was discovered that, while the bearing surfaces of the mains appeared to be in good condition, their backsides had become worn into the bedplate. Large compressor components such as crankshafts are subject to considerable flexing despite their mass. They should be supported along their entire length, or else severe damage can result. In this case, the worn bearings failed to provide that support. The distortion and fretting on the backs of the bearing shells also impaired heat transfer.

These bearings should have been removed, inspected, and replaced thousands of hours earlier. Spending about $12,000 on new bearings to protect a $400,000 crank would have been a wise trade-off. This unfortunate story is instructive. Compressor bearings should be replaced every 5 to 8 years, or at least thoroughly inspected.

It is vital to install bearings correctly. Improper installation is the most common cause of bearing failure. Tightening bearings for correct crush is of prime importance; otherwise, bearings may come loose and fail from constant pounding action or poor heat transfer. Check with the bearing supplier for more specific information on bearing crush.

We've already commented on the importance of determining the cause and extent of damage when a bearing fails. Damaged or failed main bearings or rod bearings should prompt checks of the crankshaft, main bearing caps, bedplate saddle bores, and connecting rod bearing bores.

Check Top Clearance for Main Bearing Alignment

The clearance between the upper main bearing shells and main bearing journals gives preliminary indications of bedplate alignment. All top clearances should be within approximately .001 inch of one another (see Figure 4-13). A tight top clearance may indicate a low saddle. If so, jack down on the crankshaft and check the top clearance for settling.

Check Crankshaft for Deflection and Runout

Deflection readings give some indication of whether the crankshaft is straight, properly supported and running in correct alignment. Deflection is normally tested with a deflection gauge. Deflection gauges measure changes in the spread between adjacent crank webs while the crank is rotated (see Figure 4-4 and 4-6).

Whenever a main bearing fails or any time bearings are replaced, it is vitally important to check both runout and deflection. Reading deflection alone does not give the full picture because zero deflection is no guarantee that the crankshaft is straight. Actually, some nominal deflection on a

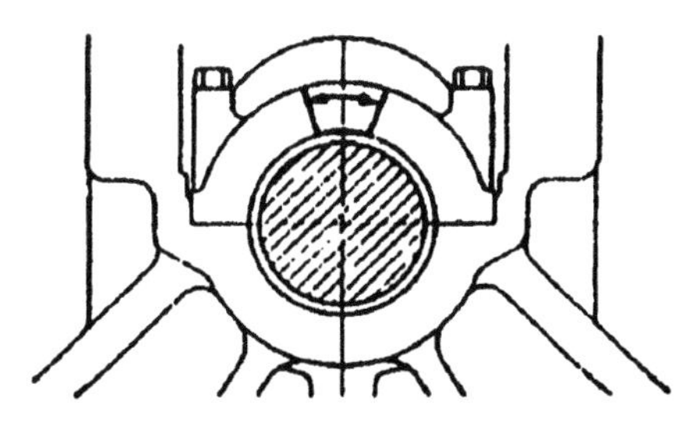

Figure 4-13. Compressor main bearing.

straight crankshaft is expected. If there is no deflection, there may be a problem.

As mentioned earlier, the deflection gauge should be read with the crankshaft jacked downward after 180° rotation. This allows deflection to be seen that otherwise might be missed if the crankshaft is bent just enough to be held "straight" by the main bearings.

Journal runout is normally checked with a dial indicator on the side of the crankshaft at the level of the main bearing parting line to cancel out the effects of crank weight. With the bearing removed and the dial indicator against the main bearing journal at the bearing oil groove location, rotate the crankshaft and observe the readings at 90° intervals to get the complete picture. If the crankshaft and bedplate are both straight and properly aligned, the readings should be within about .001″ of instruction manual specifications.

Crank Bearing Failures

Failure of these bearings is actually more common than loss of mains. When it occurs, always check the connecting rod bearing bore for signs of distortion and destructive overheating.

Remove the connecting rod and check the bore for roundness with the bearing removed. Before reinstalling the rod, temporarily install the new bearings and torque the bolts, then check the rod bore for roundness again. An out-of-round bore may show up during either, but not necessarily in both, checks.

When installing replacement bearings, always be sure to properly torque the caps to ensure proper bearing crush for good fit and effective heat transfer. The crosshead pin bore of the rod should also be inspected for roundness and the clearances checked whenever the connecting rod bearing is replaced. It is good practice to confirm that the centerline is parallel through both bores of the connecting rod.

Bearing Assembly Checklist

This checklist itemizes the most important items that should be verified prior to and during assembly of crankshaft bearings.

- Ensure bearings are the correct design.
- Bearings should be free from burrs and thoroughly cleaned.

- Ensure housing, crankshaft, and oilways are thoroughly cleaned.
- Ensure any locating dowels are satisfactory and free from burrs. The locating tab should be correctly positioned.
- Ensure the bearing has a positive free spread.
- Clearly mark on the bearing its location within the compressor.
- Clearly mark on one end of each bearing housing its position within the compressor.
- Do not apply oil between bearing back and housing bore.
- Apply a liberal coating of oil between bearing bore and crankshaft surfaces.
- Tighten bolts in the correct sequence to the correct torque or stretch, as defined in the equipment instruction manual.
- Check that the shaft can rotate freely.

Remanufacturing of Shell-Type Bearings

Remanufacturing or rebabbitting of shell-type compressor bearings should be done only to specific styles.

The old bearing shells with thick babbitt, usually .032″ to .064″, can be rebabbitted by a shop experienced in this type of work. Centrifugal casting or spraying of the babbitt has been found superior to pouring babbitt in these shells.

The newer shells with reduced babbitt thickness (.007″–.015″), can be rebabbitted by centrifugal casting if the backing is made of cast iron or steel. Bearings made with bronze backing should never be rebabbitted because the bearing shell crush has been destroyed.

Shells made of the tri-metal construction should not be remanufactured or rebabbitted.

Any bearing shell that is rebabbitted should be checked by ultrasonics to be sure the area between the babbitt and the backing is 100% bonded.

Compressor Component Repair and Rebuilding

In the overhaul and maintenance of the reciprocating compressor, many repairs can and are made to the major components. When making repairs, several things must be considered to prevent jeopardizing the integrity of the part and to ensure that it performs as good or better than the original installed in the compressor by the manufacturer. Here are questions to ask:

- Is the repair made with sound engineering judgment?
- Is the repair done with care and proper manufacturing techniques, such as, squareness, parallelism, and finishes?
- Are the materials used in making the repair equal or better to the ones available from the manufacturer?

CYLINDER REPAIR AND MAINTENANCE

A regular schedule for inspection of cylinder bore and piston should be determined as soon as the projected usage and service conditions can be established. At the outset, make frequent inspections and keep a detailed record of observations. Tabulate the information, and once a pattern is recognized, adjust the time span between inspections and service accordingly.

Thoroughly inspect all components for wear and damage. Inside and outside micrometers should be used when checking cylinder bores, piston diameters, and rod diameters. Cylinder bores should be measured at 6–12 and 3–9 o'clock, at both ends and the middle. Visually inspect the bore, counterbore, and valve ports for cracks. Use dye penetrant, if cracks are suspected.

Problems generally found with compressor cylinders include:

1. Cylinder bores worn out of round, especially on horizontal machines. (Wear is usually in an hourglass shape from front head to back head.)
2. Cylinders worn to an excessive (oversize) bore dimension.
3. Water passages fouled with deposits or water treatment chemicals.
4. Air passages fouled with dirt and carbon deposits.
5. Lubricating oil passages clogged.
6. Valve seat and cover gasket surface pitted or eroded.
7. Head-to-cylinder water ports eroded.
8. Cracks in cylinder bore.

INSPECTION OF CYLINDERS

Cylinders are inspected either visually or, with greater precision, by measuring bore size.

Visual inspection should be done at every shutdown, for example on the occasion of a valve change-out. This can be accomplished by looking through the valve ports and inspecting the bore for any sign of scuffing or scoring. Also, the amount of cylinder oiling should be determined, that is, determine whether the cylinder is being over-oiled or not enough oil is being fed to the cylinder. Adjustment to lubricator feed can be done at this time.

A more detailed inspection should be performed at every overhaul. At that time, the piston and piston rods are removed, exact measurements of the bore are made, and the general condition of the bore is determined.

TYPES OF INSPECTION

There are two types of inspections.

1. Visual. This inspection determines whether any roughness, scuffing, or scratching has taken place.
2. Measurements. Measurements are taken to determine what wear has taken place and whether that wear is within limits allowing operation to continue. If measurements show that the wear is beyond acceptable limits, the measurements determine what repair procedures should be implemented.

CYLINDER BORE MEASUREMENTS

Compressor cylinder bore measurements are taken in the vertical and horizontal directions and in at least three locations along the bore, each end of ring travel, and in the center (see Figure 4-14).

All measurements are taken with a micrometer and should be recorded on appropriate inspection report forms. One form should be used for each compressor cylinder and all pertinent data entered.

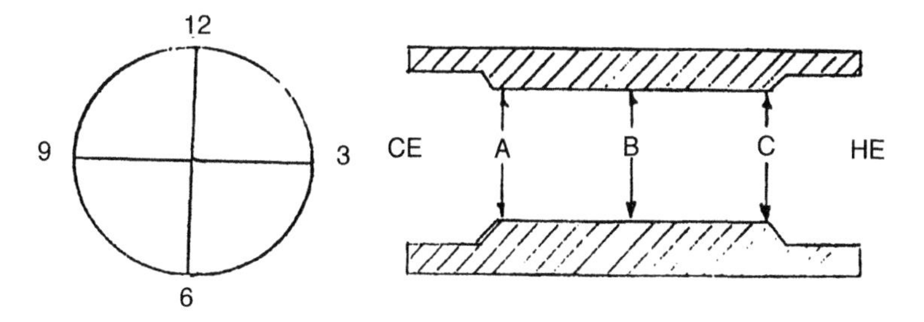

FIGURE 4-14. Measurement locations along cylinder bore.

Types of Cylinder Bore Wear

Cylinder bores usually wear in the following ways.

1. Compressor cylinder bores generally wear in a barrel-shape manner, that is, they wear more in the center than at either end of the compressor stroke (see Figure 4-15).
2. Bores may also wear "out of round," that is, larger in the vertical than horizontal direction, or the reverse.
3. Taper wear may also be evident, that is, larger at the head end than crank end, or the reverse.

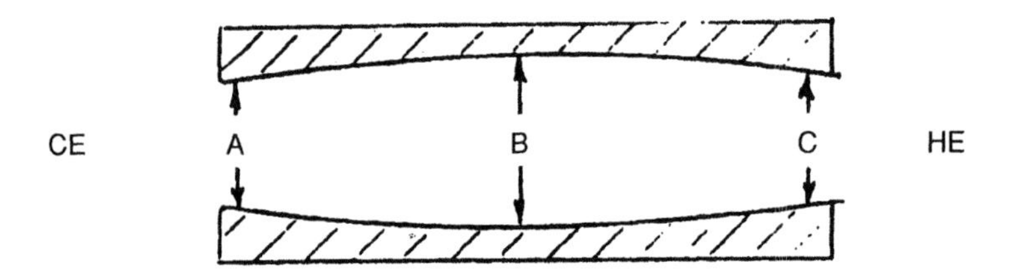

FIGURE 4-15. Barrel-shaped bore wear pattern.

Repair Procedures

When a compressor cylinder has been scored, scuffed, or worn so that operation should not be continued without correction, there are three repair procedures that can be undertaken. The proper action to take depends on the condition.

Honing

When cylinder bore inspection shows that scuffing or scratches have occurred, the problem can be corrected by proper honing. If the bore has deep scuffing, it is recommended that it be bored first, then honed. Correction by honing may be feasible, but could take a great deal of time. A proper honed finish on cast iron cylinder bores is achieved by honing with a coarse J 13 stone, finishing with a fine J 45 stone, and then crosshatching the bore with a coarse stone, if it is a lubricated cylinder.

Cylinder bore finish should be a minimum of 16 RMS for lubricated compressors and a minimum of 8 RMS for non-lubricated cylinders or lubricated cylinders used at operating pressures above 2500 psi.

Note: A compressor cylinder bore that is "out of round" or tapered cannot be corrected by honing. It must be rebored. Honing will only result in a smooth "out of round" or "tapered" bore.

Cylinder Reboring

Correcting an out-of-round, tapered, or barrel-shaped bore is done by reboring. After the cylinder has been corrected by boring, all tool marks must be removed by honing as previously described.

- Rebore cylinders when the maximum measured bore diameter is .00075″ times the nominal bore diameter. This difference applies to measurements within any circumference, plane, or to any difference in measurements along the bore axis.

  Example: Nominal bore—10.000″ diameter

  Maximum allowable bore difference = 10 (.00075)
  = .0075″

  Therefore, if the minimum bore diameter is 10.0025″, the maximum measured at any point should not exceed 10.010″.

  Note: When compressing a gas with a molecular weight less than 17 or over 700 psi differential across the piston rings, reduce the multiplier to .0005″ times nominal bore diameter.
- Maximum out-of-round diameter
  1. Metallic rings—.001″/inch of bore diameter up to maximum of .015″ regardless of bore diameter.
  2. Teflon rings—.002″/inch of bore diameter up to a maximum of .018″ regardless of bore diameter.
- Taper

  Maximum taper for length of bore is .0005″ per inch of stroke.
- Ridges

  Rebore all sizes that have a ridge of .007″ to .010″ at the end of piston ring travel.

Quantity of Material to be Removed When Reboring

- Solid Bore Cylinders

  The maximum diameter to which a compressor cylinder can be rebored is usually limited by the counter bore. Generally, the counter bores are ⅛″ to ¼″ larger than the cylinder bores.

  While a cylinder may have sufficient strength to be bored larger than the counter bore diameter, it may be impractical to do so because there would be no way to center the heads without major redesign (see Figure 4-16).

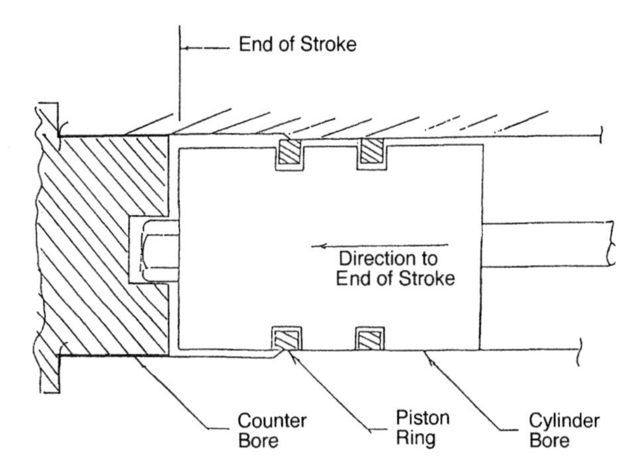

FIGURE 4-16. Cylinder bore and counter bore configurations.

- Cylinders with liners

  On cylinders with liners, the liners can be honed, but not much boring can be done. The amount will depend on the liner thickness, which typically varies from .375″ to .750″. When replacing a cylinder liner, it is important that the cylinder bore is not out of round or tapered. The OD of the liner requires full support. Unsupported liners will distort, and the liner bore (ID) will take an out-of-round or tapered shape.

  The cylinder may be rebored to .060″ over the original liner diameter. Obviously, the amount removed from the cylinder bore must be added to the liner OD.

Oversize Piston Rings

How much honing or reboring can be done before oversize rings are needed? This is determined by considering piston ring gap. When honing or reboring a cylinder, oversize rings become advisable when the ring gap becomes three times normal due to the bore increase.

The ring gap will be increased 3.1416 times the amount of material removed from the cylinder bore.

Example:

Cast iron ring gap .003 inches per inch of diameter. Hence, a 10-inch cylinder would have a normal piston ring gap of 0.030 inches.

Since $3 \times 30 = .090''$ or $.090 - .030 = .060''$, an increase of .060″ would make the gap three times normal.

Therefore, .060 divided by $3.1416 = .019''$ would represent the maximum allowable increase in cylinder bore size, without resorting to oversized rings.

Table 4-2 can be used as a guide for the use of oversize rings, although the example above is more conservative.

TABLE 4-2
GUIDE FOR OVERSIZE RINGS

Nominal Bore Diameter, In.	Max. Bore Diameter Increase Allowed Before Oversize Piston Rings are Required, In.
3.99	.015
4.000 – 5.999	.015
6.000 – 7.999	.020
8.000 – 9.999	.030
10.000 – 14.999	.040
15.000 – 17.999	.050
18.000 – 22.499	.065
22.500 – 29.999	.085
30.000 – 34.999	.100
35.000 – 42.000	.120

Installation of Sleeves or Liners

If the practicality of overboring a cylinder is in question, or if it is desired that the bore size remain the same as original, installing a special liner may be considered. The amount of material that can be removed from a cylinder for overboring remains the same, but a suitable liner may be made for the cylinder. It must be remembered that installing a liner will not strengthen the cylinder (see Figure 4-17).

A liner or sleeve can be manufactured from centrifugally cast, high-strength iron with the outside diameter slightly larger than the cylinder bore. The inside diameter should not be finished to size until it is installed into the cylinder. After installation into the cylinder, it can be bored to finished size and honed to achieve the proper finish.

The interference fit between the liner OD and the cylinder bore should not be excessive, or distortion of the liner will take place and the stress on the cylinder may exceed the material strength. Generally for liners with inside diameters up to 16″, an interference fit of line-to-line to a maximum of .002″ is used.

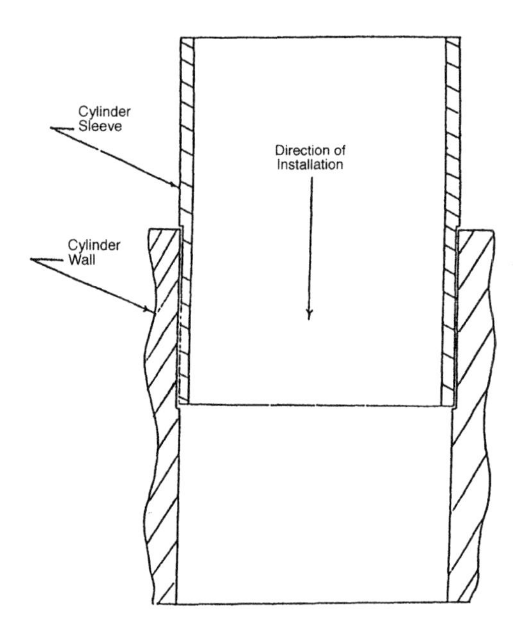

FIGURE 4-17. Installing a cylinder liner.

Even though the cylinder may have enough material that can be removed to accept a liner, this option is not always available. In some cylinders, liners are confined by specially configured head geometries. Thus, each case must be looked at and evaluated separately.

Metal Stitching

When cracks are discovered, welding is not usually recommended. Cast iron is difficult to weld. Sometimes, the welded cylinder can be damaged beyond repair. Wherever possible, a procedure called metal stitching is recommended. This is a cold repair. When done correctly, metal stitching can return the equipment to its original strength characteristics and integrity (see Figure 4-18).

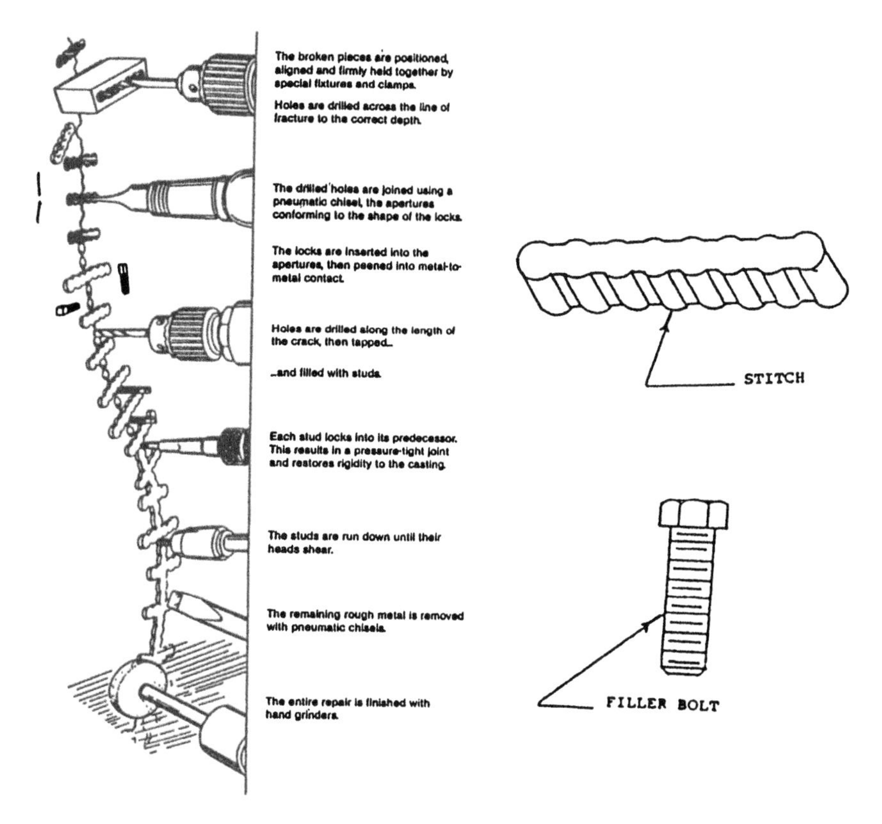

FIGURE 4-18. Metal stitching sequence. (*Source: In-Place Machining Company, Milwaukee, Wisconsin.*)

Cylinders crack for various reasons, and overheating or freezing are among the most common. If a machine is inadvertently started with no cooling water flow, do not apply the cooling water with the machine running. Shut the machine down and let it cool gradually, without water flow. After it has sufficiently cooled, it may be restarted with the water circulating. If shut down in time, the worst that usually happens is scoring of the piston and cylinders. If cold water is applied too soon, the casting will crack. A high temperature discharge switch may prevent this damage.

In the case of freezing, the protective measures are obvious. Cooling water should never be left in cylinders that are not in service in subfreezing temperatures. Simply opening drain valves does not guarantee that all water has been removed. Discharge lines can be plugged. Note, also, that some cylinders have several drain lines for each head and the cylinder.

Metal Spray

Worn or scuffed cylinder bores on compressors on low pressure air service may be restored to original size by building up the bore with metal spray. The cylinder must first be prepared by boring and undercutting, degreasing, and spraying. As much as 3/16″ on the diameter has been replaced in this manner.

Inspection of Valve Seats

At the time of inspection and taking measurements of the cylinder bore, an in-depth visual inspection should be done of all the valve seats, paying particular attention to the condition of the gasket seating area.

Valve seats become damaged by not properly torquing the valve jam screws or other devices that hold the valves in position. If the jam screw is not torqued to a high enough value, the valve may become loose during operation and "pound" the seat, or if over-torqued, distortion or cracking of the seat may occur.

Valve seat damage may occur also when the valve seat gasket is not in the proper position to seal, and wire drawing might then take place. A folded or wedged gasket can cause indentations in the seating surface.

Many so-called valve problems are not due to bad valves, but rather to bad valve seat gaskets or valve seats in such poor condition that seating cannot take place and leakage occurs. In this situation, it is common to try to correct the leakage problem by tightening the jam screw, even to the point of breaking the valve seat in the cylinder.

Another common cause of cylinder valve seat breaking is not backing off the valve jam screws when the valve cover is unbolted. The cover is commonly reinstalled using impact wrenches and with the jam screw not backed off, too much force is put on the valve, causing the seat to crack.

If visual inspection indicates possible damage to a valve seat, a dye-penetrant inspection should be done to determine if the seat is cracked. When a valve seat is found damaged, steps should be taken immediately to remachine the seat to restore the seating surface. This can be done by the use of portable boring equipment. It may not be necessary to remove the cylinder from the compressor.

Gaskets

When reassembling the compressor cylinder assembly, new gaskets should be used throughout. It is particularly important to replace the gaskets for the front and rear heads, water jacket covers, clearance pockets, and valve covers.

Proper materials must always be used. Room temperature vulcanizing cement (RTV) or similar material should not be used in place of cut gaskets for any joint where fits are employed since the thickness of the RTV will vary and cause misalignment. The only place where RTV may be considered for use is on water jacket covers.

Water Jackets

After years of operation, water jackets become dirty and fouled, particularly if jacket water quality is not good and the jacket cooling effect may have deteriorated considerably. The inside of the jacket walls may be covered with sand, lime, or magnesium, which interferes with normal heat transfer and cooling.

The solution to this problem is to clean not only the cylinder jackets but also the water cooling passages in the front and rear heads. The only way to effectively do this is by chemically cleaning. Figure 4-19 shows the correct hook-up for chemically cleaning cylinders and jackets.

Chemical cleaning is recommended at every fourth or fifth overhaul for normal water conditions and more frequently with inadequate water conditions.

CLEANING COMPRESSOR CYLINDERS & WATER JACKETS

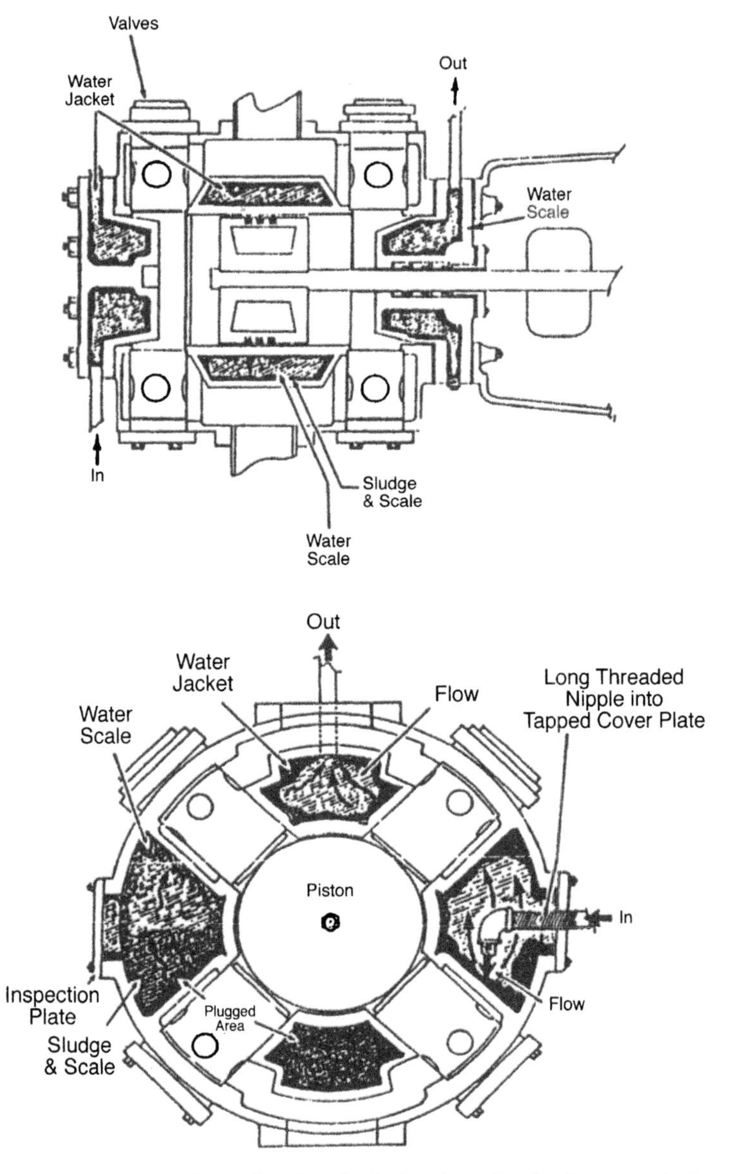

FIGURE 4-19. Hookups for chemical cleaning of cylinder water jackets.

NUT AND BOLT TORQUING

It is important to apply the proper torque to all bolts/nuts used on the cylinders to prevent distortion, damage to the components, and breakage of the castings. This is particularly important on the valve jam screw and valve cover bolting.

INSPECTION FORMS

It is important that forms are available at the time of inspection so that the measurements taken may be immediately recorded. Typical inspection forms for compressor cylinders are shown in Figure 4-20.

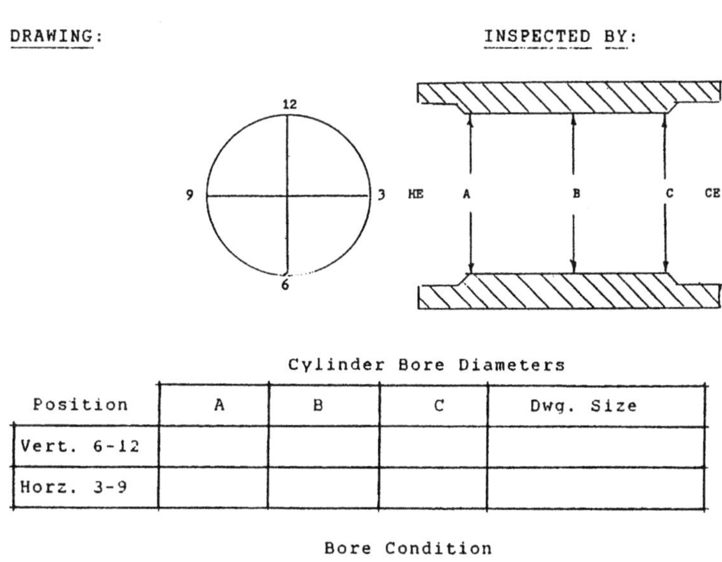

DRAWING:

INSPECTED BY:

Cylinder Bore Diameters

Position	A	B	C	Dwg. Size
Vert. 6-12				
Horz. 3-9				

Bore Condition

	Measured	Limit
Oversize		
Out of Round		
Taper		

REMARKS:

FIGURE 4-20. Inspection form.

COMPRESSOR PISTON MAINTENANCE

Figure 4-21 illustrates the typical piston and groove dimensions and areas that will be discussed in this section.

CLEARANCES

As has been previously pointed out, it is important to maintain the proper tolerances and clearances between mating parts after each overhaul.

Piston-to-Cylinder/Liner Bore

The piston-to-cylinder or liner bore clearances vary with piston design and piston ring configuration.

- Oil-Lubricated Cylinders with No Rider Band
  The clearance between the piston-to-cylinder bore or the liner bore on an oil-lubricated cylinder can be determined as follows:
  Cast iron piston = .00125″/inch bore diameter
  Example: 20″ diameter cylinder
  20″ × .00125″ = .025″ clearance
  Aluminum piston = .003″/inch bore diameter
  Example: 20″ diameter cylinder
  20″ × .003″ = .060 clearance
- Oil-Free Cylinder Rider Band
  See Table 4-3, page 269.

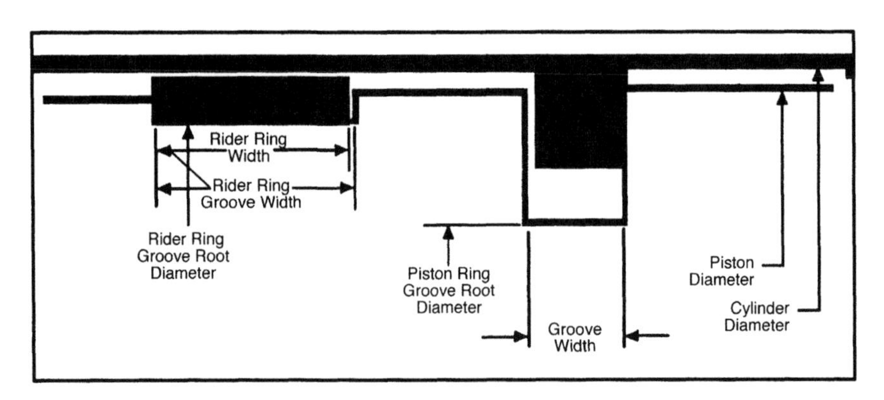

FIGURE 4-21. Typical piston and groove dimensions.

Table 4-3 shows the clearance between piston and cylinder (liner) bore on oil-free cylinders (conventional non lube) or lubricated cylinders with rider ring (band).

TABLE 4-3
CLEARANCE BETWEEN PISTON AND CYLINDER

Cylinder Diameter	Clearance
2-½″– 5″	.125″
5″–8″	.156″
8″–12″	.188″
12″–16″	.219″

Rider Ring Diametrical Clearance

The clearance between the rider ring and cylinder (liner) bore on pistons equipped with rider rings can be determined as follows:

Diametrical clearance = cylinder bore diameter × k + .005″
Where k = expansion factor based on piston material
Cast iron piston k = .0015″
Aluminum piston k = .0025″

Piston Ring Function

Two principal functions of piston sealing rings are:

- To prevent gas from blowing by the piston.
- To transfer heat from piston to the cylinder walls and hence to the water jackets.

Piston ring groove width variations can have a significant effect on these functions. To ensure proper operation, good contact is required between the ring and the side of the groove. If the grooves become tapered or edges "dragged" over as shown in Figure 4-22, poor sealing or ring fracture may be expected. If shoulders or a tapered condition have

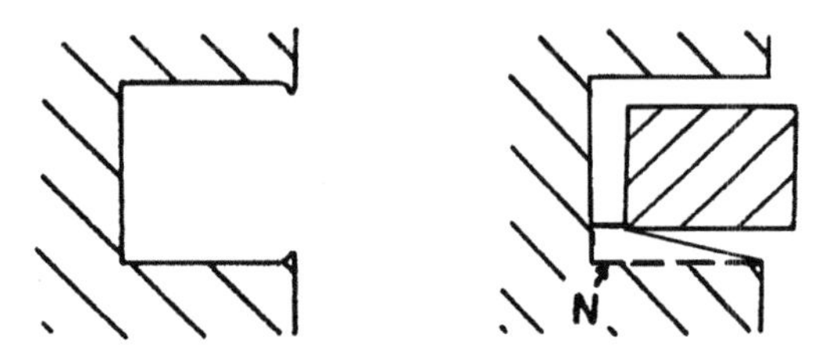

FIGURE 4-22. Unacceptable ring groove configurations.

developed, grooves should be trued up in a lathe, and rings of greater width installed. The dashed line "N" illustrates how grooves should be remachined to their full depth.

Piston Ring Groove Depth

The groove depth must be slightly greater than the radial thickness of the ring. One easy way of checking this is to put the ring in the groove and use a straight edge as shown in Figure 4-23.

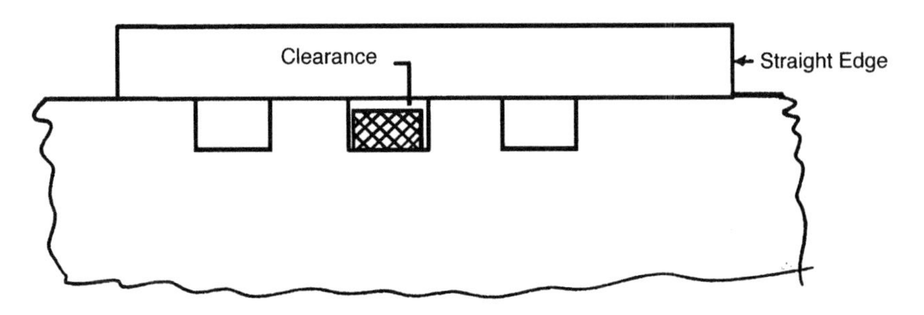

FIGURE 4-23. Groove depth must exceed radial thickness of ring.

Each ring should be tried in the groove it is to occupy and rolled completely around to make sure there is no obstruction in the bottom of the groove at any point. The ring should move freely without sticking.

Rings used in conjunction with rider rings on the piston should be checked as shown in Figure 4-24 to assure they do not protrude beyond the rider.

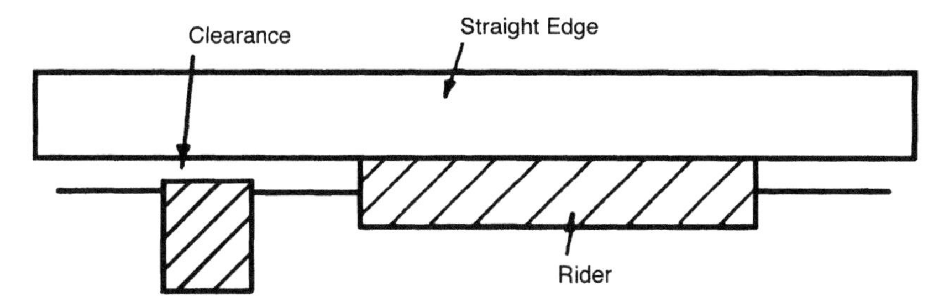

FIGURE 4-24. Piston ring periphery should be below that of rider band.

Piston Ring Sealing

For a piston ring to seal properly, it must have contact with the cylinder wall and also the side of the groove. Piston rings are made with a free diameter larger than the cylinder so that, when installed, they exert outward initial unit pressure against the cylinder wall, as shown in Figure 4-25.

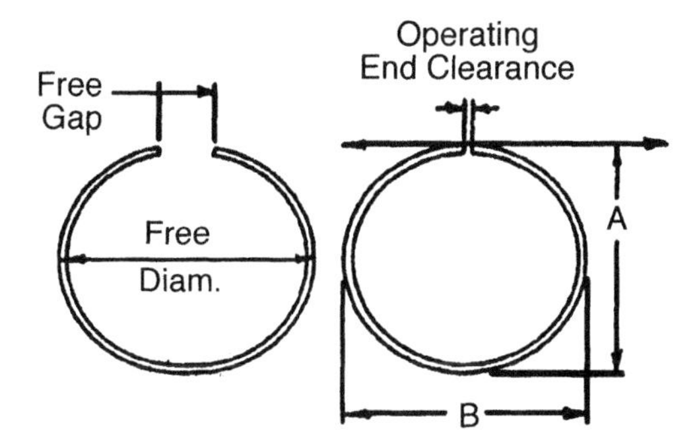

FIGURE 4-25. Free gap must exceed operating gap of piston rings.

The initial pressure varies from about zero in plastic or segmental rings to possibly several hundred psi in some situations (with metallic rings). Unit pressure will vary around the ring circumference, depending on ring circularity. Positive circularity exists when A is greater than B, and negative when A is less than B. Normally, rings have positive circularity.

During operation the primary sealing force, however, is a result of gas pressure in the ring bore and against the side. This serves to hold the ring against the cylinder and against the flat groove face as shown in Figure 4-26.

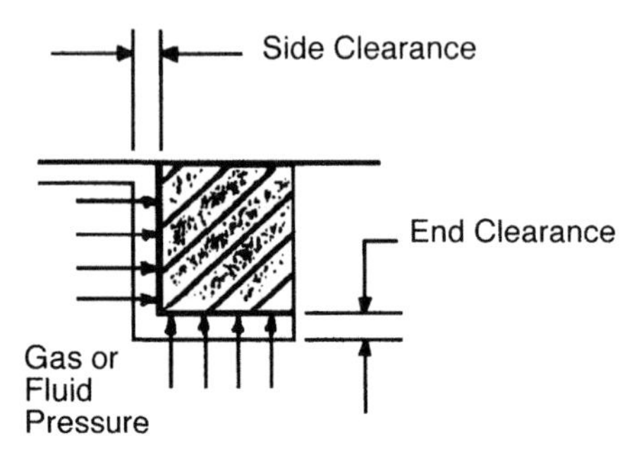

FIGURE 4-26. Gas pressure action on piston rings.

Piston Ring Clearances

The suggested minimum side clearances between ring groove (side) and end (gap) clearances for piston rings of various material are given below. Caution should be taken to make sure that each ring has the minimum end clearance. Unless there is sufficient clearance, ring ends may butt together from thermal expansion, causing ring wear, cylinder scoring, and other damage.

The end clearances shown in Table 4-4 are for butt cut or step joint rings, as shown in Figure 4-27.

For angle cut joints, multiply the value in the table for end clearance by the amounts shown in Figure 4-28.

TABLE 4-4
END CLEARANCES FOR BUTT CUT OR STEP JOINT RINGS

Material	Side Clearance (Per inch of width)	End or Gap Clearance (Per inch of cyl. dia.)
Cast Iron	.006″ (.002″ min.)	.002″
Bronze	.006″ (.002″ min.)	.003″
Teflon	.018″	.017″
Carbon	.006″ (.002″ min.)	.002″

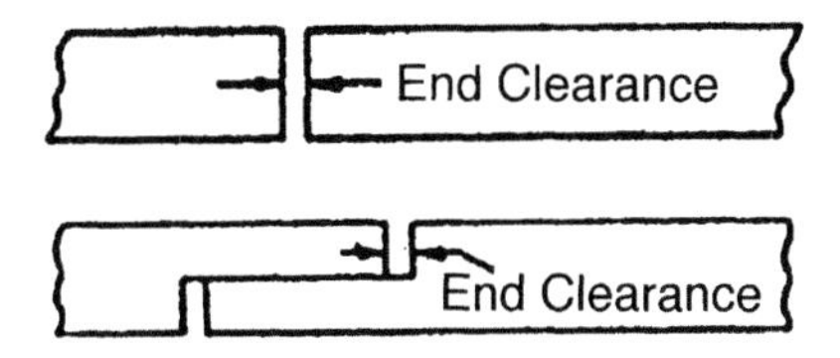

FIGURE 4-27. Butt or step joints in piston rings.

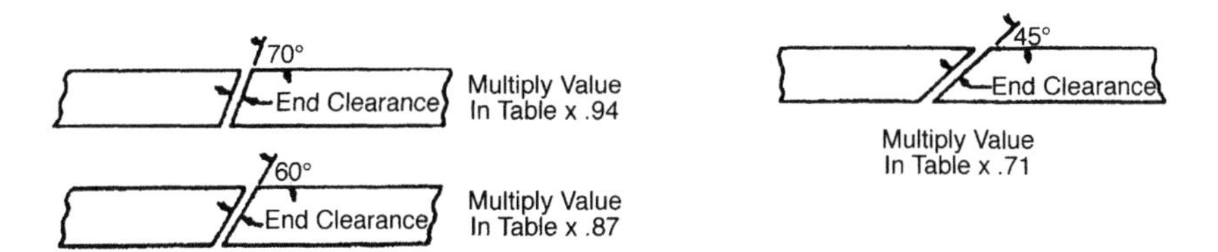

FIGURE 4-28. Angle joints and gap multipliers referred to in Table 4-4.

End Clearances

One simple method of checking end clearance is to insert each ring in the cylinder at its smallest or unworn section and check end clearance with a feeler gauge. On small bore cylinders, it may be difficult to make a measurement. In this case, a fixture or jig can be made. Turned to the exact bore measurement, it will facilitate measuring the end gap. If clearance is insufficient, the ring ends should be filed. It is better to have too much than too little end clearance. If the ring is segmental, all but one cut should be held closed and a feeler gauge inserted at the open joint. Any filing to adjust clearance needs be done at one joint only.

When checking an angle cut ring, both ends of the ring must be held against a flat surface such as the end of the piston or another ring. A jig or fixture with a square shoulder will again simplify the measuring task.

Side Clearances

The side clearances of piston rings are checked by installing the rings in their grooves on the piston. They should then be rolled around completely to make sure there is no obstruction in the bottom of the ring groove at any point, and side clearance should be checked with a feeler gauge.

The ring grooves must be in good condition, that is, not worn or tapered, if poor sealing or breakage of the rings are to be avoided. Worn ring grooves must be reconditioned.

Piston Ring Installation

Care should be taken when expanding rings over the piston. To prevent permanent deformation or breakage, they should be spread no further than necessary to clear the outside diameter of the piston.

Thin steel strips placed as shown in Figure 4-29 will help prevent damage from overstress or catching in unfilled grooves. Four strips are usually sufficient, one near each ring end and the other two evenly arranged on the opposite side of the piston.

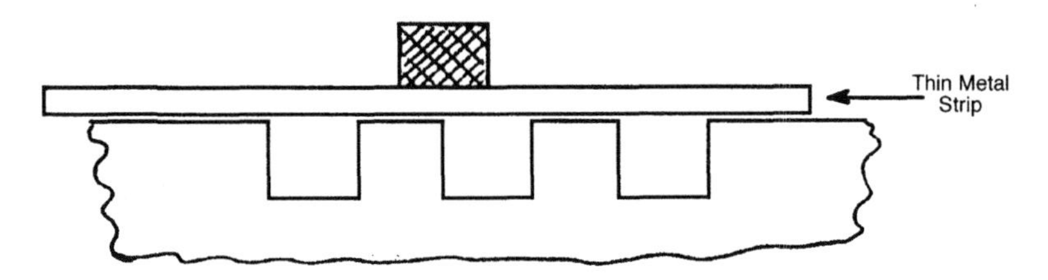

FIGURE 4-29. Thin metal strips facilitate ring installation.

Rider Rings

Rider rings or bands are designed to support the weight of the piston and piston rod assembly. These are available in a number of configurations that include the following:

- Cut-type rider rings are snapped over the piston with the same care as piston rings. Cut-type rider rings are supplied either as butt cut or as angle cut, similar to piston rings. Both types require side relief grooves to relieve the gas pressure from behind the ring so that they will not act as a piston ring. Rider rings should not seal the gas pressure.

- Solid rider rings or bands are made in one piece and are machined with an interference fit so that, once expanded over the piston end, they will contract like a rubber band to provide a snug fit on the piston. This type does not require pressure relief grooves.

The life of solid rider bands can be extended by rotating pistons 120° to 180° at each overhaul.

Rider Ring Clearances

The suggested minimum side and end clearances for rider rings of different materials are shown below.

TABLE 4-5
MINIMUM SIDE AND END CLEARANCES FOR RIDER RINGS OF DIFFERENT MATERIALS

Material	Side Clearance (Per inch of width)	End Clearance (Per inch of cylinder diameter)
Cast Iron	.001″ (.002″ min.)	.015″
Bronze	.001″ (.002″ min.)	.015″
Teflon	.012″	.035″
Carbon	.001″	.015″

The clearances shown in Table 4-5 are for butt cut or step joint rings as shown in Figure 4-27. For angle cut joints multiply value in table by the amounts shown in Figure 4-28.

The side clearance for Teflon rider rings, which must be stretched to install, applies after installation.

INSPECTION OF COMPRESSOR PISTONS

Visual inspection of the piston is recommended at every shutdown or whenever a valve is removed. While it is not possible to inspect the entire piston through the valve ports, enough can be seen to determine if any scuffing is occurring. This partial inspection gives an indication of piston problems and further corrective action can be taken, if needed.

Detailed inspection should be done when the compressor is overhauled or whenever it is necessary to remove the piston and piston rod assembly from the cylinder.

Caution: Before disassembly of a hollow piston from the piston rod, check to see that there is no pressure trapped inside the piston. All hollow pistons should be self-venting and a vent hole at the bottom of a ring groove is included for this purpose. These vent holes should be checked to be sure they are not plugged.

Detailed inspection of the piston includes:

- Cleaning piston completely including ring grooves of all carbon, dirt, etc.
- Visually inspecting all portions for signs of scuffing, breakage of ring lands, galling at hubs and piston nut washer seat, and cracking of any component part.
- Dye-penetrant inspection of ribbing on two-piece pistons and of all piston faces, particularly around the hub area.
- Complete dimensional check of piston OD, ID of piston rod boss, piston ring grooves, and dimension of OD face and hub.
- Recording all dimensions and results of non-destructive testing (NDT) inspection on appropriate forms.

Procedure for Inspection of Compressor Pistons

The following procedures should be followed when inspecting compressor pistons.

1. Check piston in lathe or place in V-blocks, then indicate inside diameter of piston boss.
2. Indicate counterbore or face of piston that bears against piston rod collar or shoulder. This must be perpendicular with inside diameter of the boss within .0005″. If indication exceeds this amount, machine counterbore or face until it indicates within .0005″.
3. Indicate counterbore or face of piston where piston nut makes contact. This must also be perpendicular to inside diameter of piston boss within .0005″. If not within this amount, machine face or counterbore piston so that it does indicate within .0005″.

4. Both counterbore or face of piston (piston rod shoulder, collar, or piston nut end) must not be galled or rough. If not in good condition, rough or marked, machine to clean up.
5. Measure the inside diameter of piston boss. If this diameter is greater than the normal diameter of the bore by .002″, the piston will not have the proper fit to the rod. Depending on the design of the piston and the boss, a repair may be made by boring oversize and installing a bushing. If the design is such that there is not enough material to install a bushing, the piston rod must be built up and ground to an amount equal to the oversize bore. Clearance between piston bore ID and piston rod OD is .001″ to .0025″.
6. On two-piece pistons, it is important to have a .002″ difference between center hub and the outside rim of the piston. Place a straight edge across the piston half and with a feeler gauge measure from the hub to the bottom of the straight edge (see Figure 4-30). This difference must be .002″. If less than that amount, machine hub to achieve the .002″ difference.

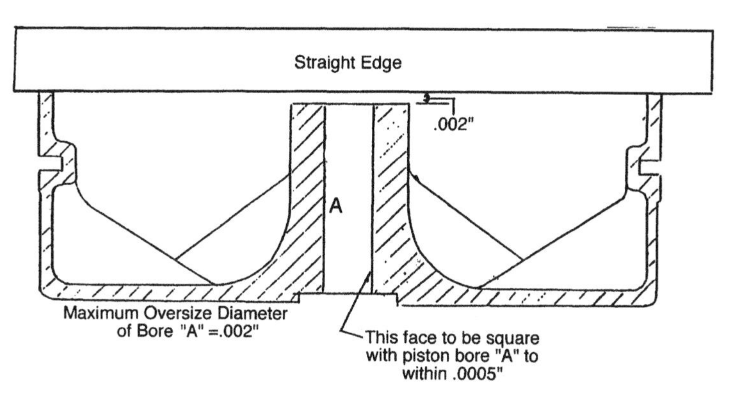

FIGURE 4-30. Verifying acceptable piston dimensions.

PISTON INSPECTION FORMS

Inspection forms should be available at the time of inspection so that all measurements and pertinent data may be immediately recorded.

Oversize Pistons

Oversize compressor pistons are required when:

1. Piston-to-cylinder bore clearance is too large to properly support rings.
2. Allowable piston rod runout cannot be obtained by adjustment of crosshead shoes.
3. Clearance between bottom of piston rod and packing flange is less than .015″.

Table 4-6 shows the maximum allowable cylinder bore diameter before oversize pistons are required.

Piston Failures

Figure 4-31 explains typical compressor piston defects and failure locations.

A. Piston mating surface fretting and wear on two-piece pistons.
B. Ring groove side wall wear.

Table 4-6
Maximum Allowable Cylinder Bore Diameter Before Oversize Pistons Are Required

	Maximum Bore Diameter Increase Allowed Before Oversize Piston is Required		
Nominal Bore Diameter, In.	Cast Iron (Riding on Bore), In.	Aluminum (Riding on Bore), In.	All Piston Materials With Non-Metallic Rings & Riders, In.
3.99	.050	.050	.015
4.000–5.999	.075	.070	.015
6.000–7.999	.105	.085	.035
8.000–9.999	.125	.085	.040
10.000–14.999	.125	.085	.050
15.000–17.999	.125	.085	.050
18.000–22.499	.125	.085	.065
22.500–29.999	.125	.085	.085
30.000–34.999	.125	.085	
35.000–42.000	.125	.085	

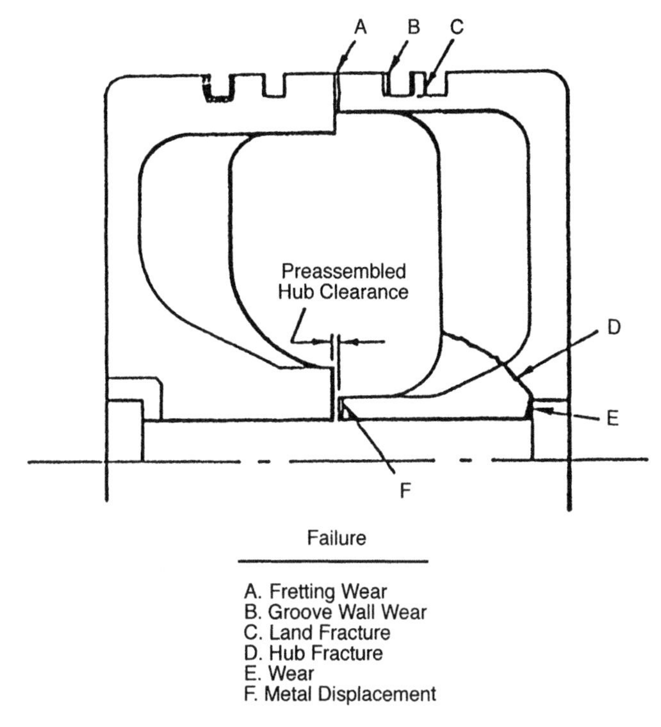

FIGURE 4-31. Types and locations of piston failures.

C. Cracked ring lands.
D. Fracture or cracking of ribs in hub area.
E. Crushing and galling of nut sealing surface.
F. Metal displacement or interference.

FAILURE OF ONE-PIECE CASTING DESIGN

Some pistons are designed as a one-piece casting and have been subject to chronic failure.

We recall an installation plagued with piston failures in the first stage of its two-stage air compressor. An examination of broken piston parts showed that there had been a core shift during casting. The problem recurred with replacement pistons for this one-piece design.

It was recommended that a two-piece piston be designed, using thin-walled, ribbed construction with the same material (aluminum). The new design incorporated clearance between hubs. Figure 4-32 shows the outer rim under compression when the piston was tightened. This keeps the joint tight. The new pistons were fabricated and installed, using the same piston rings and piston rod. The breakage problem was eliminated.

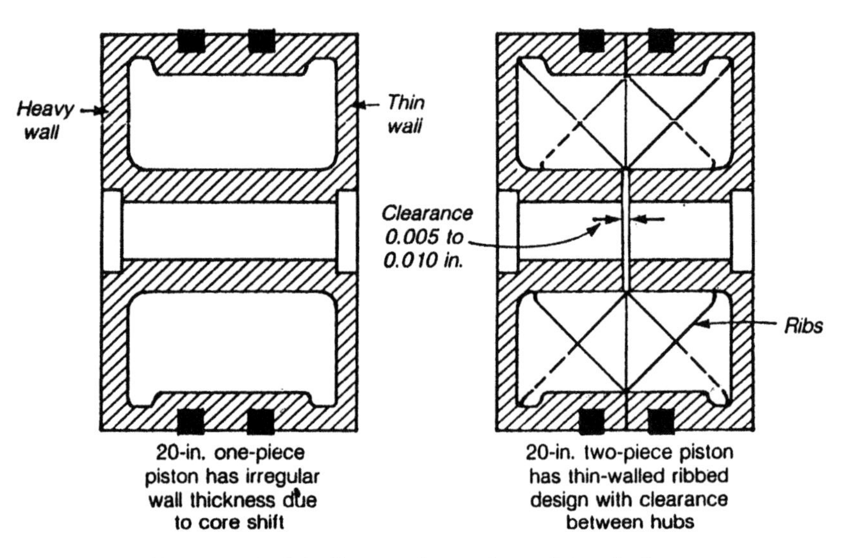

FIGURE 4-32. One- and two-piece piston designs.

Renewing Piston Rings

Piston rings must be replaced when the end gap (clearance) becomes too great and blow-by of the gas occurs. End gap increases occur when:

- Rings become worn
- Cylinder bore wear takes place
- Scuffing occurs on the surface of the ring due to lack of lubrication, or cylinder cooling is rendered ineffective, or the cylinder bore becomes rough.

Oversize Piston Rings

Oversize piston rings are required when the ring gap has tripled. Normally, the ring gap for a cast iron butt cut piston ring is .002″ per inch of cylinder diameter. Thus, a 10″ cylinder bore requires piston rings to have a gap of .020″ when new. An increase of .040″ would triple the ring gap to 0.060″. This increase in ring gap would be due to an increase in bore size of .013″ (.04″/3.1416″ = .013 in.) This three-fold increase of ring gap also governs when worn rings should be replaced even though the cylinder bore has not worn. Table 4-7 shows the maximum increase in bore size before oversize piston rings are required.

TABLE 4-7
MAXIMUM INCREASE IN BORE SIZE BEFORE OVERSIZE PISTON RINGS ARE REQUIRED

Nominal Bore Diameter, In.	Maximum Bore Diameter Increase Allowed Before Oversize Piston Rings Are Required, In.
3.99	.015
4.000–5.999	.015
6.000–7.999	.020
8.000–9.999	.030
10.000–14.999	.040
15.000–17.999	.050
18.000–22.499	.065
22.500–29.999	.085
30.000–34.999	.100
35.000–42.000	.120

Oversize rings are required to overcome an increase in ring gap due to an increase in bore dimension. Also, oversize rings are needed to overcome the inability of the piston ring lands to properly support the rings (see Figure 4-33). As the ring moves out to accommodate a larger bore size, less of the ring side is in contact and supported by the groove wall. The ring is thus susceptible to breakage.

Note: From a practical point of view, it is recommended that piston rings and rider rings be replaced whenever the piston is removed, regardless of cylinder bore size or ring wear. These are inexpensive components

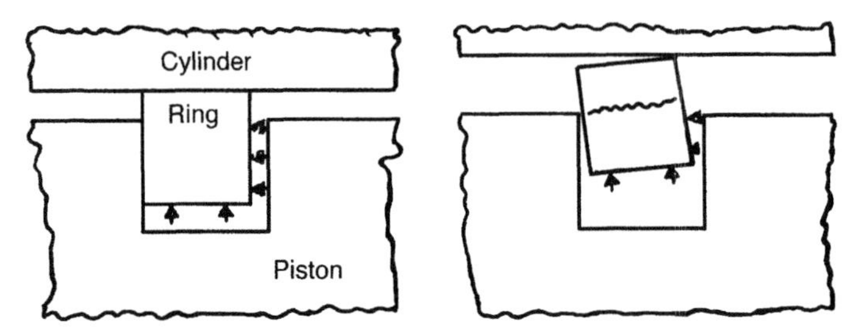

FIGURE 4-33. Effect of using standard ring in oversize bore.

and replacement is less costly than having a failure and the attendant downtime risk.

REBUILDING COMPRESSOR PISTONS

There are several types of repair and rebuilding needed with worn or scuffed compressor pistons. These include:

1. Remachine piston ring lands to correct worn grooves, then install wider piston rings.
2. Turn OD of piston to remove any scoring or scuffing. Build up OD with metal spray of proper material and remachine OD to original diameter.
3. Repair cracks in ribs or hubs in cast iron pistons by appropriate grinding, cleaning, and welding. Piston must be remachined after welding on OD, hub bore, and counterbores to restore welding-induced distortion.

   Note: It is not recommended to repair aluminum pistons by welding. Aluminum alloys used for compressor pistons are generally not weldable.
4. Reapplication of babbitt sprayed rider bands. See procedure below for proper repair.
5. Install bushing in worn hub bore and remachine to correct size to fit piston rod.
6. Install rider bands on pistons not originally so equipped to restore worn and scuffed diameter. Remachine for rider bands and install by using spray bronze or aluminum material. On some piston designs, Teflon rider bands may be installed.

REAPPLICATION OF BABBITT-SPRAYED RIDER BANDS

The following procedures should be followed when reapplying babbitt-sprayed rider bands.

Surface Preparation and Spraying

1. Clean all old babbitt from the rider ring grooves by machining to base metal. Take extreme care at this point to assure that all babbitt is removed, while keeping the depth of the groove to a minimum. If machining is not necessary, sandblast the rider groove to bare metal.
2. Thread rider ring groove in accordance with instructions given in Figure 4-34.
3. Keep the cleaned area free of all oil and spray with Metco nickel aluminide bond within four hours after sandblasting or recutting threads.
4. After cleaning, preheat piston to 175°F–200°F and keep piston between these temperatures during entire spraying operation.
5. Spray the grooves with .002″ of Metco aluminide bond, turning spray gun to 45° alternating from one side to the other. Distribute bond material evenly. Oil must not come in contact with the bond material either before or after application.
6. Apply Metco spray babbitt "A," completely building up one section to 1/16″ to 3/32″ over finished diameter. Turn spray gun at 45° angle

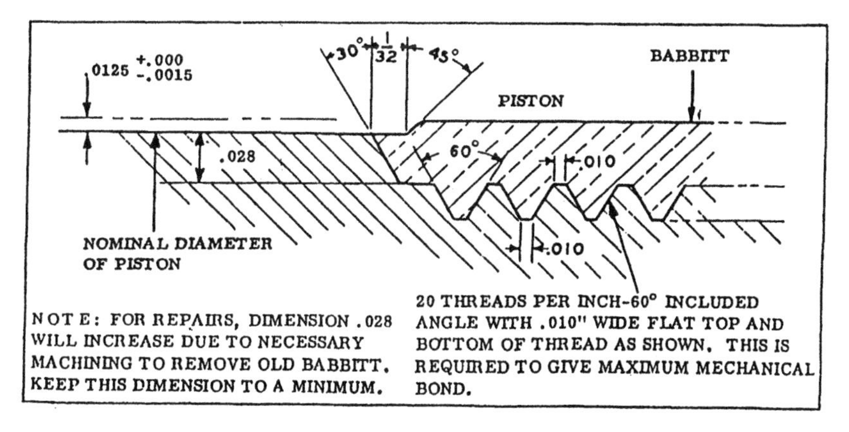

FIGURE 4-34. Repair instructions for rider bands (*Source: Dresser Rand Customer Service Training*).

alternately from one side to the other until threads are filled, and then at 90° until completely built up.

7. After one groove of the piston is completed, proceed as outlined in steps 5 and 6 above, completely building up one groove before proceeding to the next groove. Do not spray adjacent groove but alternate from one end of piston to a groove in the opposite end in order to keep the heat more evenly distributed (see Figure 4-35).

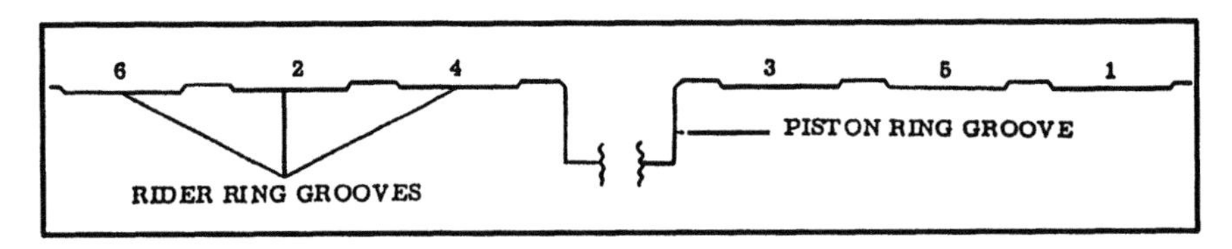

FIGURE 4-35. Rebabbitting sequence for rider bands.

Finish Machining

1. After sprayed surface has cooled, take a rough cut across bands to a diameter .002″ to .030″ larger than finished dimension of bands. Finish bevels on ends as shown in Figure 4-34.
2. Finish turn or grind sprayed surface to .025″ (+.000″/–.003″) over the nominal diameter of the piston.

   Special Note: Take all possible precautions to ensure that all oil has been removed from the rider area and that after machining or sandblasting the bond coat is applied within four hours to prevent any oxidation of the rider ring groove. The importance of this step cannot be overemphasized. The integrity of the babbitt bond is directly related to the cleanliness of the parent metal.

INSTALLING PISTONS ON PISTON RODS

Before the piston is installed on the piston rod both should be thoroughly inspected. Any deviations from standard must be corrected. Installation of a non-standard assembly will risk an unscheduled shutdown with all of its well-known consequences.

PROCEDURE FOR TORQUING PISTON NUTS

The following procedure has been developed for the torquing of piston nuts when installing cast iron, aluminum, and steel pistons, either two pieces or one piece, with straight bores. It must not be used for pistons that use piston rods with taper fit to the piston.

1. Clamp piston rod behind the collar in a suitable fixture similar to the one shown in Figure 4-36. Care must be taken not to clamp on any coated surface if the rod has been coated, because cracking or breaking the bond will result.

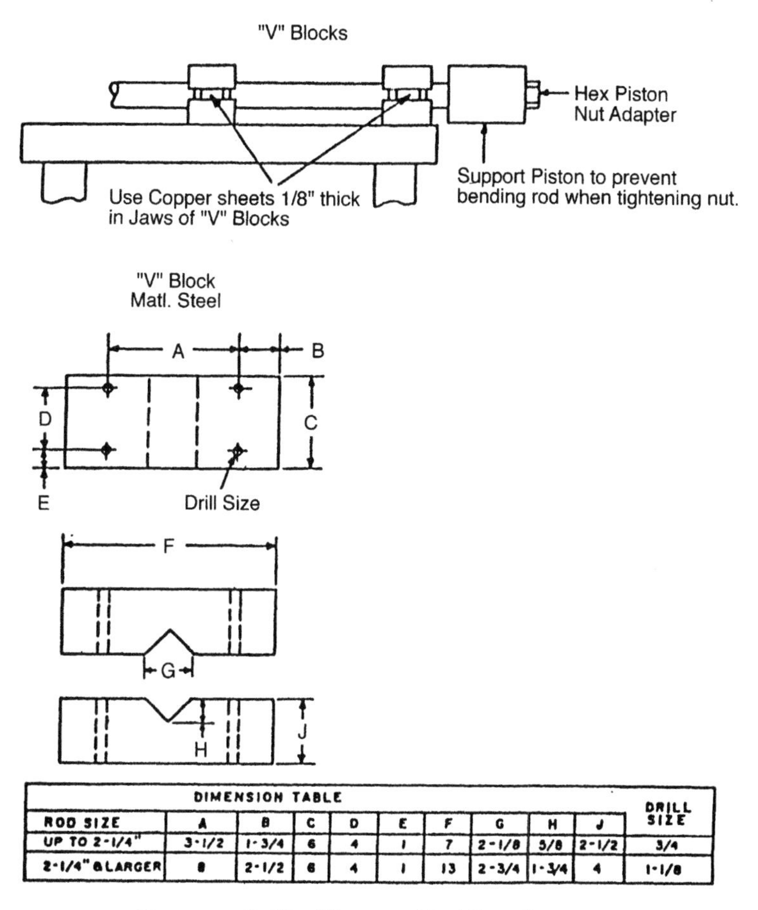

DIMENSION TABLE										
ROD SIZE	A	B	C	D	E	F	G	H	J	DRILL SIZE
UP TO 2-1/4"	3-1/2	1-3/4	6	4	1	7	2-1/8	5/8	2-1/2	3/4
2-1/4" & LARGER	8	2-1/2	6	4	1	13	2-3/4	1-3/4	4	1-1/8

FIGURE 4-36. Piston rod holding fixture.

2. The piston and piston rod should be inspected according to the instructions for Inspection of Compressor Pistons discussed previously in this chapter.
3. Blue* the piston to piston rod collar. A minimum of 75% bearing contact should be evident; if not, lap the surface of piston and collar until this minimum of 75% contact is achieved. The bearing area should be uniform around the circumference and across the bearing face.
4. Blue piston nut to piston. Again a minimum of 75% bearing contact is required. If not, lap until achieved. Remove all traces of compound by cleaning with an approved solvent.
5. Install piston on rod and coat the threads of the piston rod with an anti-galling compound. Do not use a copper based compound since seizing or galling of the threads may result. Apply a thin coating of oil to the face of the nut.
6. Torque the nut to the proper torque value as shown in the compressor maintenance manual.
7. If a suitable torque wrench is not available, another suitable method that can be used is as follows:
   - With the piston properly located on the rod, tighten the nut to 150 ft. lbs. to obtain good metal-to-metal contact.
   - Scribe a line (A) through the centerline of the piston rod, and extend it out to the piston as shown in Figure 4-37A.
   - Refer to the nomograph in Figure 4-38 to determine the number of degrees the nut must be turned in relation to the piston rod.
   - Measure from the original scribe line (A) the number of degrees the piston nut be turned in relation to the piston and scribe a line through point (B) and the centerline of the piston rod.
   - Install socket on piston nut and mark adjacent to the first scribe line (A) as shown in Figure 4-37B.
   - Tighten the piston nut until the mark on the socket coincides with the second scribe line.
8. Place assembly between centers or in V-blocks and indicate piston rod behind collar or location where piston shoulders against rod. Indicator reading should not exceed .001″. If indicator reading exceeds .001″, the piston counterbore or face is not perpendicular with piston bore or shoulder of piston rod is not perpendicular with piston diameter. Because something is not perpendicular, torquing

*"Bluing" is the application of a thin layer of dye that allows the technician to spot incorrect mating contact of mechanical parts.

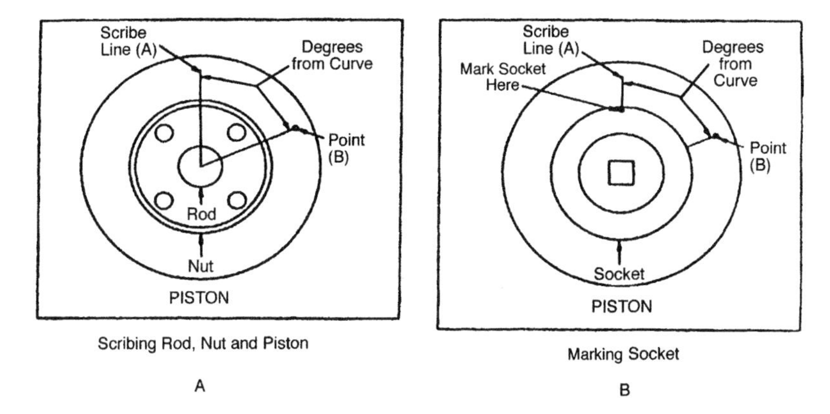

FIGURE 4-37. Alternative torque application method.

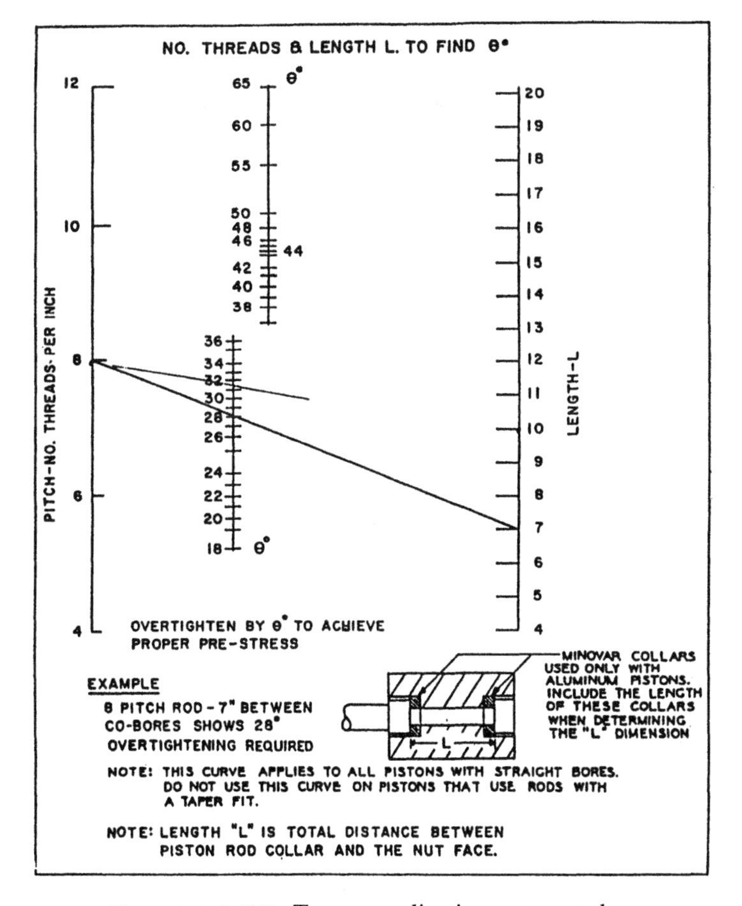

FIGURE 4-38. Torque application nomograph.

the piston is causing the piston rod to bend. Remove piston from piston rod and repeat checks described earlier under Inspection of Compressor Pistons.

This problem must be corrected before use, or a broken piston rod will almost certainly result.

SETTING PISTON END CLEARANCE

Whenever the piston rod has been removed or even turned in or out of the crosshead, it is necessary to check the piston end clearance and make any adjustments necessary to restore it to the proper amount. If the piston has not been removed from the rod, the original clearance can easily be maintained by center punching both the rod and the crosshead and making a suitable scribing tool, as shown in Figure 4-39, before removing the rod from the crosshead.

Note: The length of the scribing tool will vary depending on the size of the crosshead nut, and care must be taken to locate scribe marks so that

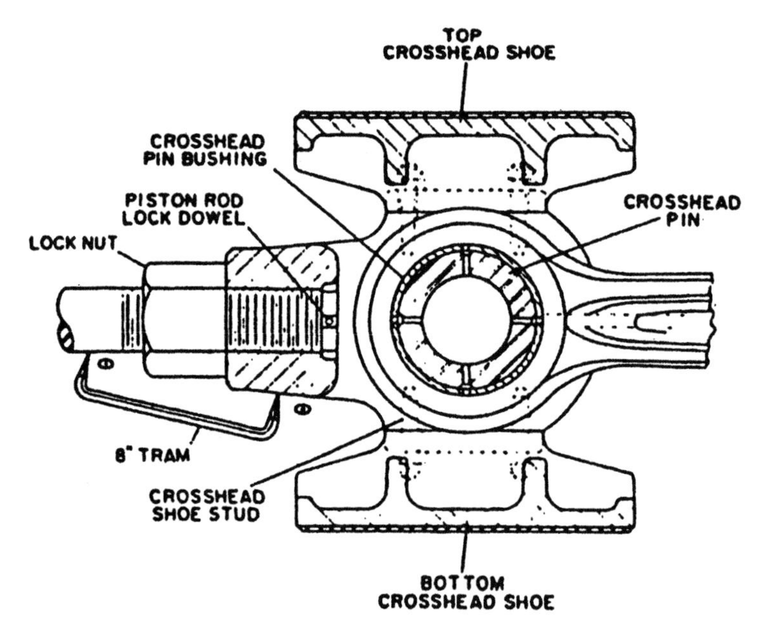

FIGURE 4-39. Tramming tool used to verify location of piston rod relative to crosshead.

during operation they will not enter the oil scraper rings and destroy the oil film.

When installing pistons or adjusting end clearances, it is desirable to have more clearance at the head end than at the crank end because expansion from running heat and normal connecting rod bearing wear tend to increase the crank end clearance. For this reason, when setting the piston end clearance, make the head end clearance approximately two thirds of the total and the crank end clearance one third.

To adjust piston end clearance, use the following procedures.

1. Lock out the electrical switch to prevent accidental start up of the compressor.
2. Remove all gas pressure from cylinders.
3. Remove one valve from each end of cylinder.
4. Bar compressor over until piston is at end of the stroke.
5. Using feeler gauge, measure the distance between the face of the piston and the cylinder head. Note this measurement.

   Caution: Be sure that feeler gauges are long enough so that your hands are never in the cylinder between piston and cylinder head. Also, never attempt this measurement with the compressor running.
6. Bar compressor over again to bring piston to the end of its stroke at the opposite end.
7. Measure this amount, using feeler gauges.

   An alternative method of checking the clearance between piston and head is to use soft lead wire. Bar over until the lead wire is flattened, and measure this thickness with a micrometer.
8. Take the difference in clearance between the two readings and screw the piston rod in or out of the crosshead to make this clearance two-thirds on the head end and one-third on the crank end.

   Note: If the piston rod cannot be turned by hand, use a strap wrench or a socket on the piston nut. Never use a pipe wrench on the rod.
9. Tighten the crosshead nut and fasten locking devices.

   Caution: Place a block of wood between the frame and the crosshead when the nut is being tightened. This will prevent the crosshead from turning and distorting the connecting rod and bearing. Add tram marks for point of reference.
10. Be sure to record the crank end and head end clearance settings for future reference.

Inspection and Reconditioning Piston Rods

Before installing the piston/piston rod assembly in the compressor cylinder or attempting to recondition a piston rod, inspect and measure it thoroughly. An important part of this inspection is checking the piston rod and the piston to be sure the assembly is able to run true to the centerline of the cylinder bore.

Inspection is done by the following method:

1. Visually inspect for signs of scuffing, longitudinal scratches, damaged or pulled threads.
2. Measure all diameters and lengths to determine actual sizes and record on suitable inspection forms.
3. Non-destructively test the rod by magnaflux or magnaglow to determine if there are any signs of longitudinal scratches or signs of cracking at threads. This is particularly important when inspecting coated or flame-hardened piston rods.
4. Determine runout by placing the piston rod between centers or in V-blocks and using dial indicators.
5. Determine whether the rod had been previously chrome plated or whether the rod had been replaced by vendors other than OEM. If so, determine the suitability of material and surface treatment.

Procedures for Inspection of Compressor Piston Rods

Use the following procedures to inspect compressor piston rods:

1. Remove piston from piston rod.
2. Place piston between centers (be sure existing centers are in good condition) or in V-blocks (see Figure 4-40).
3. With dial indicator, indicate over length of rod. Rod should not run out more than .001″.
4. Measure diameter to be sure rod is round within .001″ and not tapered more than .001″ over its length, or worn undersize.

   If rod is found to be worn not more than .005″ undersize and still round and not tapered, it may be used. If it is found to be more than .005″ undersized, or out-of-round or tapered more than .001″, it should be turned undersize to remove out-of-roundness and taper

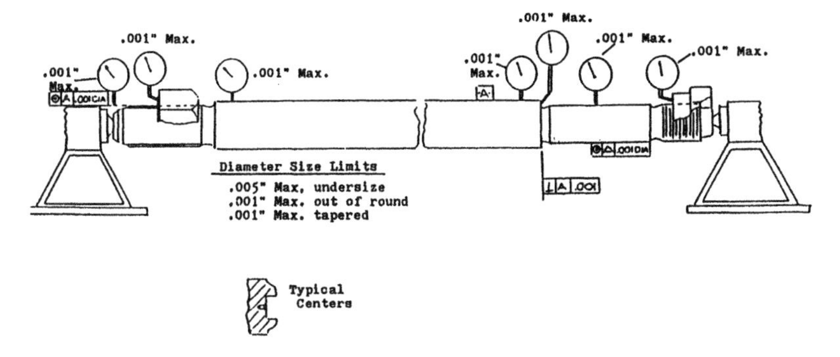

FIGURE 4-40. Verifying trueness of piston rod.

and built up to size. Material and method must be compatible with the rod base material, pressure, and gas being compressed.

5. Indicate rod in the area of piston fit. This area must be concentric with diameter of rod within .001″.
6. Measure the diameter of this fit area. If more than .002″ undersize, it must be built up and reground to size (normally no wear occurs in this area).
7. Indicate thread diameter at both the crosshead end and piston end. These must be concentric with rod diameter within .001″.
8. Indicate shoulder or collar where piston fits to rod; this must be perpendicular to diameter of rod within .001″. If indicator shows more than .001″ runout on the shoulder or collar, it must be machined so runout does to exceed .001″.
9. With the crosshead nut and piston nut screwed on rod (be sure nuts are facing in proper direction), indicate seating faces. These should not run out more than .001″. If runout is greater than .001″, the faces must be machined.

If the surface of the rod in the packing area is scuffed, scored, or scratched in any manner, the rod must be turned undersize to remove all damage and then built up to size. Both material and method must be compatible with rod material, pressure, and gas being compressed. If indicating rod shows a bend at piston fit area or at either of the thread ends, the rod should be scrapped. Do not attempt to straighten.

Reconditioning of Compressor Piston Rods

If a rod is scratched, nicked, worn, or a slight shoulder exists, it is possible to recondition. If previous inspection has revealed any indication of cracks, inclusions, or imperfections in material, the rod must be scrapped.

Note: Do not attempt to weld piston rods.

Regrinding

Piston rods may be ground undersize, and the standard packing rings can be used. Generally .002″–.003″ per inch of piston rod diameter can be ground for medium pressures, but as the operating pressure increases above 1000 psi, this should be limited to a total of .003″ under the nominal piston rod size.

On low pressure air service (125 psi), as much as .020″ has been removed and standard packing used. However, it is better practice to grind undersize and return to standard nominal size by one of several methods to be discussed next. Generally, with an undersized rod, the packing rings will take longer to break in and the leakage will be greater during the break-in period.

Metal Spray

Metal spray application is an acceptable reconditioning technique on piston rods used in pressures up to 400 psi. After proper cleaning and surface preparation, the rod may be sprayed with 420 stainless steel, which is equivalent to 413 SS. The rod is ground to normal size after spray. Surface finish should be as shown earlier.

Plasma Spray

Development of plasma spray and suitable coatings allow piston rods not only to be reconditioned but to be given a surface treatment superior to that originally on the rod. Overlays of tungsten carbide and chrome oxide can give the piston rod longer wear life. Care must be taken to check the rod for straightness after spraying. Plasma spray has superior bond strength over conventional metal sprays and is less likely to "flake" off in service.

Chrome Plating

This technique is used to restore piston rods to their original diameter, but its success depends on the quality of the plating, which varies enormously depending on the plater. *Do not use this method unless the quality of the plating produced by a vendor is truly known.*

The only chrome plating approved for piston rods is *porous-chrome,* which is a reverse etching process. The depth of the pores in the chrome produced by reverse etching is on the order of .002″–.003″ deep, and the plating has approximately 40% porosity. Porosity on piston rods enhances the ability to hold oil on the rod surface and provide adequate lubrication.

Note: Do not plate piston rods with hard chrome such as used on automobile bumpers.

For best results, the finished plating should have no more than .010″ radial thickness. Thicker plating is more susceptible to failure and should be avoided. The finish on the rod diameter prior to plating should be 32 RMS or better to assure as smooth a plating as possible. When plating a rod, it is important to avoid abrupt steps from the undersized area to the adjacent area. Two basic precautionary procedures are possible:

1. Grind the entire rod undersize and plate on top of this undersized area, allowing plate to fade to the undersized diameter outside the packing travel area. In this case, it is necessary to polish the ends of the plating to assure proper smoothness.
2. Grind a groove to the undersized diameter, then plate back to the finished size. In this case, the ends of the groove should be reduced to minimize stress.

Generally, the second method is not preferred because of the difficulty in assuring that the final grind will be absolutely concentric to the original rod diameter. It may, however, be preferred in situations where greater than recommended plate thickness must be used.

With either method, after grinding undersize and prior to plating, the rod should be carefully magnaflux inspected for grinding heat checks or other cracks. This inspection should also be repeated on final-ground, plated surfaces. No circumferential cracks or indication should be permitted.

Note: The use of bronze packing rings with chrome plated piston rods is to be avoided because it leads to scoring of rods. If the compressor uses bronze packing, chrome plating of piston rods should not be done.

Caution: All previous plating or spray must be removed before replating or spraying. DO NOT attempt to chrome plate or spray over surfaces previously coated with fuseable alloy. Also, do not attempt to recoat fuseable alloy piston rods with any coating including fuseable alloy.

Regrind piston rods after coating or plating to size. Specify 16 RMS finish for lubricated service and 8 RMS finish for nonlubricated service. Bent rods may be straightened only if they have not been coated or plated.

Manufacture of Compressor Piston Rods

Piston rods may be manufactured if proper attention is paid to selection of material, machining accuracies, heat treatment of materials, surface treatments, and surface finishes.

Note: Do not attempt to manufacture piston rod without first asking questions and consultation on the various aspects of piston rods.

Before deciding to manufacture piston rods, the following factors should be considered:

1. Determine operating conditions of compressor.
2. Material may be substituted from original design but only after consultation. Do not substitute new specifications without questions. *Ask questions!!* Many times the reasons for designs are not known and certain changes may be dangerous.
3. Coatings may be applied to piston rods that had originally been furnished without these coatings.
4. Threads on piston rods are now made either by grinding or rolling. Rolled thread is superior in that it eliminates high stresses at the root of the threads. There are few, if any, repair service facilities that have thread rolling equipment; therefore, this work must be subcontracted. Do not attempt to substitute cut threads without consultation and knowing all details of the compressor design.

In some cases, it is possible to use cut threads (on lathe, not die cut) but considerable prior investigation will be required. Modern designs make use of the higher stress levels possible with rolled threads and the unauthorized use of cut threads could cause premature failure of the rod with disastrous results.

If cut threads have been deemed adequate on a given job, careful manufacture is still essential. The tool bit must be properly ground, so that no pulls or tears occur. It must be correctly rounded at the bottom or root of the thread. Obviously, the pitch diameter must be proper.

Tightening Piston Nuts on Piston Rods

It is important to tighten the piston nut to the proper torque to prevent loosening of the piston on the rod. Loosening could result in damage to the piston, or could cause stretching of the piston rod at the thread end beyond the elastic limit of the material.

Before tightening is attempted, an adequate means of holding the piston rod/piston assembly must be available. Figure 4-36 showed one such arrangement with V-blocks to hold the rod.

If the piston rod has been coated or plated, it must be held in the area that has not been coated. The stress caused by tightening of the V-blocks could easily cause cracking of the coating or the plating.

In addition to adequate V-blocks to prevent the piston rod from turning when tightening or loosening the piston nut, the piston must be properly supported to prevent bending of the piston rod.

Caution: Never attempt to tighten a piston nut with piston rod assembly in the cylinder and fastened to crosshead. Always tighten the piston nut before installation of assembly in cylinder.

Again, under no circumstances should a pipe wrench be used to hold piston rod. Note also that copper-based compounds (such as Fel-Pro) should never be used on threads since galling may result from this practice.

Torque Values For Piston Nuts

Proper tightening of piston nut to the piston rod includes torquing the nut to the proper value. Approved torquing procedures include the use of a torque wrench or the "stretch" method.

Table 4-8 includes torque values for various size piston rod thread sizes. Instructions for tightening and determination of pre-stress value were given earlier in this chapter (see section on pistons).

TABLE 4-8
PISTON NUT TORQUE VALUES

Thread Diameter (8 TH.D.)	Foot Pounds
1″	200
1-¼″	425
1-½″	775
1-¾″	1250
2″	1900
2-¼″	2775
2-½″	3875
2-¾″	5200
3″	6850

On large diameter piston rods, it is not practical to obtain the proper torque and pre-stress of the rod by the use of torque wrenches. Here, heaters to expand (lengthen) the rod are used that allow the nuts to be easily turned the proper amount. Upon cooling, the rod "shrinks," setting up the proper pre-stress which ensures tightness.

In all cases where a nut is used, it is locked to the piston rod by means of a cotter pin. Piston rod lock nuts are usually the castellated style.

Torque Values For Crosshead Nuts

On those piston rods that use a nut tightened to the face of the crosshead, it is important that the proper torque value be achieved to prevent loosening or overstressing the rod.

Table 4-9 gives torque values for various size piston rod thread sizes at the crosshead end.

Note: It is very important to properly block the crosshead to prevent it from "cocking" when tightening the nut. Failure to take this precaution can damage the crosshead pin bushing as well as the connecting rod bearing.

OTHER COMPRESSOR COMPONENT REPAIRS

As long as the proper methods and materials are used, all components of the compressor may be successfully rebuilt or repaired.

TABLE 4-9
CROSSHEAD NUT TORQUE VALUES

Thread Diameter (8 THD.) except as shown	Foot Pounds
2-¼″	1850–1950
2-½″	2600–2700
2-⅞″	4050–4150
3-⅝″	8500–8600
4″	11400–11500
4″	11400–11500
4-¾″ - 6 THD	19200–19300
4-¾″ - 6 THD	19200–19300
5″ - 6 THD	29800–29900
5-¼″ - 6 THD	29800–29900

CONNECTING ROD REPAIRS

- Rebore the large end to re-establish the correct bearing crush and to return the bore to standard size.
- Straighten rods with heat. Cold straightening is limited by the amount of bend.
- Install new crosshead pin bushings. Fit pin to bushing.
- Straighten the rod by machining the large end of the rod true to the crosshead end. Avoid having to remove bend.
  Note: Repair by welding is not recommended.

CROSSHEAD REPAIRS

Crossheads, like other compressor components, may be successfully rebuilt if the proper materials and procedures are used.

- Non-adjustable. Rebabbitt and turn to size on babbitted crossheads. This may be done by metal spray.
- Non-adjustable, not babbitted. Repair by machining round. Consider application of babbitt to faces. This improved type of crosshead can also be welded up and turned to size.

- Rebore and bush to crosshead pin bores in crosshead. Use the same material as the crosshead for bushings or use a bronze bushing with a flange on the inside of the crosshead.
- Repair damaged crossheads by welding and remachining. This is not always possible. It is feasible only if material is steel.

Crosshead Pins

Generally crosshead pins do not wear but may become galled or scored. Reconditioning may be done in a variety of ways:

- Turn old pin down and make new pin bushing to suit.
- Grind pin true and build up by chroming back to standard size.

Crosshead Guide Repair

Crosshead guides are either built integral with the compressor frame or are separate and bolted to the frame. The guides form slides for the crossheads; they must have a good surface finish and must be in line with the centerline of the cylinder.

- Crosshead guides can be rebored in place or in the service center. This is usually a field job because it is much cheaper to machine a guide in place than it is to remove the guide from the foundation.
- Removable guides, such as tail rod crosshead guides, can be plated back to standard size.
- If a guide was overheated due to lack of lubrication, the guide will have to be checked for cracks. Try to determine if the guide is cracked before boring is started.
- Cracks or bad sections of the guide can sometimes be repaired by metal stitching. This is similar to inserting a piece of cast iron in the damaged area and remachining (see earlier discussion in this chapter).

Crankshaft Repairs

Repairs to forged crankshafts can be done in many instances, depending on type of damage, material of shaft, and size of shaft. Most crankshafts used in compressors are carbon steel forgings; however, some are cast nodular (ductile) iron.

WELD REPAIR

Crankshaft journals, mains as well as throws, can be repaired by welding and grinding back to standard. However, weld repairs are feasible only with steel shafts. This technique requires real weld repair expertise.

Because of the carbon content of these shafts, the material is difficult to weld without inducing surface embrittlement and cracking. These cracks may be microscopic and usually cannot be detected except by magnetic particle inspection or metal etch testing. If cracks are induced through welding and the shaft is returned to service, it will eventually fatigue and break.

WELD REPAIR PROCEDURE

It is important that the proper weld procedures be followed when making weld repairs on crankshafts. An approved procedure will go a long way towards maintaining the integrity of the crankshaft.

1. Cracks in the crankshaft shall be removed by grinding.
2. Crankshaft shall be inspected with magnetic particle method to ensure that the cracks are removed.
3. Crankshaft shall be preheated to a temperature between 350°F and 800°F. This preheat temperature shall also be the minimum interpass temperature.
4. If weld repair of crankshaft is necessary, weld depth shall not exceed a 0.250″.
5. Stress relief shall be performed immediately after welding at a temperature between 1000°F and 1200°F. The minimal stress-relief time shall be one hour per inch of shaft diameter.
6. Non-destructive testing (NDT) inspection.

Journals can be ground undersize and special bearings used. There is a limit on the amount of undersize a shaft may be ground. Journal grinding is often feasible in-place, with portable equipment.

Journals may be trued up, prepared for plating, and chrome plated back to size. However, chrome plating is not always accepted by insurance companies.

Chrome Plating Procedure

The following procedures should be followed when applying chrome plating to compressor crankshafts.

1. Areas to be plated shall be pre-ground, followed by magnetic particle and dye-penetrant inspection of all ground surfaces.
2. Shaft area to be chromed shall be shot peened, with care given not to damage the collar surface.
3. Maximum chrome thickness shall be 0.030″ (radial thickness).
4. Journals that are not too badly damaged may be cleaned with a belt sander and hand stoned. Extreme care must be taken because it is easy to get journal in an out-of-round condition, which will cause further bearing failures.
5. Metal spraying has been successfully done on smaller shafts, but is generally not recommended for large cranks.

Some other general repairs of crankshafts that can be done in the field or shop include:

1. Straightening
2. Machine rod size journals undersize
3. Machining damaged keyways oversize and producing new key
4. Machining outside diameter and faces of flywheel fit and bushing flywheel
5. Reaming damaged holes in shaft and flywheel and installing new oversized studs
6. Boring and tapping damaged counterweight stud holes oversize and producing new step studs

Certification

Vendors should certify that the repaired crankshaft meets or exceeds original engine or compressor manufacturer requirements with regard to structural and dimensional integrity. Alternatively, a vendor should detail all known deficiencies that could, in his judgment, affect the operational serviceability of a particular shaft. This cataloging of deficiencies must be produced during the evaluation phase with the customer, prior to the repair being authorized by the customer and accomplished by the vendor.

COMPRESSOR FRAME REPAIR

If a compressor frame needs repair, it is often the consequence of catastrophic failure of another part, such as bearing failure or breakage of a piston rod. Typical repairs include:

1. Reboring of main bearing saddles on larger machines, which can be done in place.
2. Cracked frames, which may be metal-locked (metal-stitched) and remachined.

COMPRESSOR PART REPLICATION

The previous discussions have dealt with the repair, reconditioning, and rebuilding of the major components of the compressor, which are subject to the greatest wear and/or damage through accidents. In recent years, there has been a growth of companies manufacturing and supplying replacement parts for compressors other than the original equipment manufacturer (OEM).

At one time, virtually all replacement parts were supplied by the OEM and users automatically obtained all new parts from them. This has greatly changed due, in part, to the OEM's inability to furnish parts on a timely basis, particularly in the case of emergencies, and due to the high prices that are sometimes charged for those parts.

As with every type of service, there are some very good part replicators but, unfortunately, many are not qualified and supply very poor parts. It is, therefore, advisable to select replacement parts carefully. Moreover, only qualified vendors should be selected.

Note: Lowest price may not be the best bargain.

Virtually every compressor part can be and is being duplicated. In purchasing these parts, care must be taken and basic considerations weighed:

1. The original dimension and tolerances of the part to be duplicated must be obtained.

2. Specification for materials must be the same or better than the original part.
3. Heat treating and hardness of the materials must be the same or superior in quality.
4. Weights of the reciprocating parts must be very close to original in order to minimize unbalanced forces and resulting vibrations.
5. Nothing is done that will jeopardize the integrity of the mating parts.

Any deviation from these fundamentals will result in parts that are poor in quality and that can adversely affect the operation and reliability of the compressor.

Remember, you get exactly what you pay for.

CHAPTER 5

Troubleshooting Compressor Problems

A key element in upgrading the quality of compressor maintenance is the troubleshooting capability of the maintenance personnel. Operators may also be helpful in establishing procedures and guidelines for reducing downtime, if they have a basic working knowledge of each type of compressor in the plant.

INTRODUCTION

Troubleshooting the mechanical condition and performance of machinery is a talent that falls into the natural, analytical thought process of some fortunate individuals, but, for most of us, it must be learned through experience and developed through rigid, disciplined analytical procedures.

The old approach could be paraphrased as "change a part and wait to see if it fixes the problem. If it does great, fine, if it does not, we will do something else." Unfortunately, this has caused:

- Many hours of wasted manpower
- Great losses due to replacement of unnecessary parts and components
- Unnecessary downtime and lost production
- Excessive power losses

THE ANALYST'S ROLE

The major component in a successful troubleshooting effort is the machinery analyst. The actual analysis of an operating deficiency is performed by the human mind; the analyst is therefore the key ingredient who makes it all happen.

A good analyst always:

- Deals with facts
- Collects and uses all the facts
- Never jumps to conclusions

COMPRESSOR PROBLEMS

Problems with reciprocating compressors generally fall into one of three broad categories:

- Loss of capacity
- Noise and vibration
- Failure to run

LOSS OF CAPACITY

Unless the compressor cannot deliver any pressure at all, this term usually describes a comparison between the quantity being delivered and the quantity that should be delivered.

The analyst should determine how a loss of capacity conclusion was reached. This is of prime importance on plant air machines. All too often, the addition of new users to the system or even leaks are responsible for the apparent loss of capacity. The analyst should ask: Did the loss of capacity occur quickly or is it a problem which gradually became worse? The answer might indicate whether component failure or the accumulation of wear are likely causes.

The causes and symptoms for capacity loss include the following:

1. Extreme reduction in capacity or no delivery.
   - The suction filter could be plugged or the intake line blocked. This, in effect, would unload the compressor, causing it to act as a vacuum pump, and actually pump the suction pressure below atmospheric.

- The unloader is stuck, holding a suction valve open, and not allowing compression of the gas.
- Gasket failure, at the valve seat, cylinder head, or even in the intercooler.

2. Not enough delivery. More often the complaint is not enough delivery, rather than no delivery.

   The above causes still hold true, but with the added possibility of valve element failures or valve seat gasket failure.

   It must also be remembered that as the system pressure rises, the delivered capacity of the machine is reduced. Should the compressor be set with a system pressure that exceeds its rating, not enough air will be delivered to the compressor system. Quite often, the advertised compressor capacity is erroneously interpreted as referring to the delivered pressure, rather than suction pressure.

   It should be noted that faulty valves usually are the most common cause of a loss of capacity.

NOISE AND VIBRATION

Another compressor problem is noise and vibration.

Vibration

Possible causes of compressor vibrations are:

- Inadequate foundation
- The compressor is not properly grouted
- Unloaders not functioning correctly
- Excessive pipe strain or incorrect piping design

Noise or Knocking

There is always some noise associated with a compressor in operation. Sources could include air noise from motor, gas flow in piping, clicking of valves, and slight belt slap on belted compressors.

"Noise" is often described as a mechanical, or friction-related sound. However, in compressors reports of "knocking noise" are more common. Determination of the origin of knocking sounds is generally more difficult because sound travels through the machine.

Cylinder Knocks

Most knocks will be found in the compressor cylinder. The common causes of knocking in a cylinder include:

- Loose pistons
- Insufficient head clearance
- Too great a piston-to-cylinder bore clearance
- Broken piston rings
- Loose rider bands
- Loose or broken valves
- Moisture carryover or "liquid slugging"

For cylinder knocks, simple steps can be taken to analyze the problem:

1. If the unit knocks only when idling, check piston end clearance.
2. If the unit knocks when it is loaded, check for cylinder wear.
   - The piston or piston rings may have worn a ridge in the bore.
   - Check for a loose piston. Remove a suction and discharge valve and bar the unit over. Using wood to restrain the piston, watch for movement.
   - Check end clearances with a lead wire and notice variations in readings.

Frame or Running Gear Knocks

Possible causes of knocking in compressor running gear include the following:

1. Loose flywheel or sheave
2. Loose or worn bearings
3. Crosshead pin-to-bushing clearance too large
4. Mechanical packing loose in gland
5. Excessive crosshead-to-guide clearance
6. Connecting rod hitting end of piston rod in crosshead
7. Belts misaligned, causing motor rotor to "weave" and bump
8. Unit not level, causing motor rotor to "weave" axially and bump in some manner. This problem is often related to belt alignment inaccuracies.
9. Incorrect rotation of compressor.

A knock in the compressor running gear is most frequently caused by a loose part or a bearing that has too much clearance.

Squealing Noise

Possible causes of squealing noise in compressors include the following:

1. Motor or compressor bearing too tight
2. Lack of oil
3. Belts slipping
4. Leaking gasket or joint

Failure to Run

If the machine continually blows fuses, the following should be checked:

1. Fuses too small
2. Voltage is low
3. Pressure switch differential is set too close
4. Compressor does not unload for starting
5. Motor bearings are beginning to seize

Should the motor fail to even start upon demand, the following points should be checked:

1. Pressure switch contacts not mating
2. Worn contacts
3. Motor seized
4. Thermal overloads require resetting
5. Fuses blown

Typical Compressor Problems

For possible solutions to typical problems, check Troubleshooting Chart, Table 5-1. Always be sure to study and consult the compressor instruction book. And DO NOT FORGET TO CHECK THE PRESSURE GAUGES.

(text continued on page 314)

Table 5-1
Troubleshooting Chart

The Problem / The Causes	Compressor noisy or knocks	Air discharge temperature above normal	Carbonaceous deposits abnormal	Operating cycle abnormally long	Piston ring, piston, cylinder wear excessive	Piston rod or packing wear excessive	Motor over-heating
System demand exceeds rating				•			
Discharge pressure above rating	•	•	•				•
Unloader setting incorrect		•					•
Intake pipe restricted, too small, too long		•					•
Intake filter clogged		•		•			•
Valves worn or broken	•	•	•	•			•
Valves not seated in cylinder	•	•	•	•			•
Valves incorrectly located		•	•	•			•
Gaskets leak		•	•	•			•
Unloader or control defective	•	•	•	•			•
System leakage excessive				•			
Piston rings worn, stuck, or broken	•	•	•	•	•		•
Cylinder (piston) worn or scored	•	•	•	•	•		•
Belts slipping	•						
Speed too high		•	•				•
V-belt or other misalignment	•						
Pulley or flywheel loose	•						
Motor rotor loose on shaft	•						
Foundation bolts loose	•						
Foundation uneven—unit rocks	•						
Piston to head clearance too small	•						
Crankshaft end play too great	•						
Piston or piston nut loose	•						
Bearings need adjustment or renewal	•						•
Liquid carryover	•				•	•	
Oil feed excessive	•		•			•	
Oil level too high	•	•	•				
Lubrication inadequate	•	•			•	•	•
Oil viscosity incorrect	•		•		•	•	•
Intercooler vibrating	•						

TABLE 5-1 (CONTINUED)
TROUBLESHOOTING CHART

The Problem / The Causes	Compressor noisy or knocks	Air discharge temperature above normal	Carbonaceous deposits abnormal	Operating cycle abnormally long	Piston ring, piston, cylinder wear excessive	Piston rod or packing wear excessive	Motor over-heating
Tank ringing noise (1)	•						
Ambient temperature too high		•	•				•
Ventilation poor		•	•				•
Air flow to fan blocked		•	•				
Rotation wrong		•	•				
Oil level too low						•	
Check or discharge valve defective		•					
Cylinder, head, cooler dirty		•	•				
"Off" time insufficient		•	•				
Water quantity too low		•					
Belts too tight							•
Water inlet temperature too high		•	•				
Water jacket or cooler dirty		•	•				
Valves dirty		•	•				
Air discharge temperature too high			•				
Wrong type oil			•		•	•	
Air filter defective			•		•	•	
Dirt, rust entering cylinder			•		•	•	
Packing rings worn, stuck, broken						•	
Rod scored, pitted, worn						•	
Motor too small							•
Electrical conditions wrong							•
Voltage abnormally low							•
Excitation inadequate							•
Excessive number of starts							•
Discharge line restricted		•					•
Resonant pulsation (inlet or disch)							•

(1) Consult manufacturer.
(Source: Compressed Air and Gas, *3rd Edition, Ingersoll-Rand Company, Washington, N.J., 1980, pp. 18–34 to 18–36)*

(table continued on next page)

Table 5-1 (Continued)
Troubleshooting Chart

The Problem / The Causes	Delivery less than rated capacity	Discharge pressure below normal	Receiver pressure above normal	Receiver safety valve pops	Intercooler pressure above normal	Intercooler safety valve pops	Intercooler pressure below normal	Compressor parts overheat	Outlet water temperature above normal	Valve wear and breakage abnormal
System demand exceeds rating	•	•					•			
Discharge pressure above rating	•		•	•	•	•		•	•	
System leakage excessive	•	•					•			
Intake pipe restricted, too small, too long	•	•					•	•		
Intake filter clogged	•	•					•	•		
Valves worn or broken	•	•			•H	•H	•L	•		
Valves not seated in cylinder	•	•			•H	•H	•L	•		
Valves incorrectly located	•	•			•H	•H	•L	•		
Gaskets leak	•	•			•H	•H	•L	•		
Unloader or control defective	•	•	•	•	•	•	•	•		•
Unloader setting wrong	•	•	•	•	•	•	•	•		
Piston rings worn, broken, or stuck	•	•			•H	•H	•L	•		
Cylinder (piston) worn or scored	•	•			•H	•H	•L	•		
Rod packing leaks	•	•						•		
Safety valve leaks	•	•					•			
Belts slipping	•	•								
Speed lower than rating	•	•								
Gauge defective		•	•		•		•			
Safety valve set too low				•		•				
Safety valve defective				•		•	•			
Control air pipe leaks			•	•						
Intercooler passages clogged				•	•					
Intercooler leaks							•			
Cylinder head, intercooler dirty								•	•	
Water quantity insufficient	•				•			•	•	
Water inlet temperature too high	•				•			•	•	
Water jackets or intercooler dirty					•			•	•	
Air discharge temperature too high									•	
Intercooler pressure too high									•	
Speed too high								•	•	

H (in high pressure cylinder).
L (in low pressure cylinder).

TABLE 5-1 (CONTINUED)
TROUBLESHOOTING CHART

The Problem / The Causes	Delivery less than rated capacity	Discharge pressure below normal	Receiver pressure above normal	Receiver safety valve pops	Intercooler pressure above normal	Intercooler safety valve pops	Intercooler pressure below normal	Compressor parts overheat	Outlet water temperature above normal	Valve wear and breakage abnormal
V-belt or other misalignment								•		
Bearings need adjustment or renewal								•		
Oil level too high								•		
Lubrication inadequate								•		•
Oil viscosity incorrect								•		•
Ambient temperature too high								•		
Ventilation poor								•		
Air flow to fan blocked								•		
Rotation wrong								•		
Oil level too low								•		
Check or discharge valve defective								•		
"Off" time insufficient								•		
Belts too tight								•		
Valves dirty					•		•	•		•
Liquid carryover										•
Oil feed excessive										•
Air filter defective										•
Dirt, rust entering cylinder										•
Assembly incorrect										•
Springs broken										•
New valve on worn seat										•
Worn valve on good seat										•
Rod packing too tight								•		
Control air line clogged			•							
Resonant pulsation (inlet or disch)					•	•				•

(table continued on next page)

TABLE 5-1 (CONTINUED)
TROUBLESHOOTING CHART

The Problem / The Causes	Excessive compressor vibration	Oil pumping excessive (single acting compressor)	Crankcase water accumulation	Crankcase oil pressure low	Starts too often	Compressor fails to unload	Compressor fails to start
Intake pipe restricted, too small, too long		•					
Intake filter clogged		•					
Unloaders or control defective	•				•	•	•
Unloader setting wrong					•		•
Discharge pressure above rating	•						
Piston rings worn, broken, or stuck		•	•				
Cylinder (piston) worn or scored		•	•				
Gauge defective				•			
Intercooler, drain more often			•				
Speed too high	•						•
V-belt or other misalignment	•						
Pulley or flywheel loose	•						
Motor rotor loose on shaft	•						
Foundation too small	•						
Grout improperly placed	•						
Foundation bolts loose	•						
Leveling wedges left under compressor	•						
Piping improperly supported	•						
Foundation uneven—unit rocks	•						
Liquid carryover			•				
Oil level too high		•					
Oil viscosity incorrect		•		•			
Rotation wrong							•
Oil level too low				•			
Belts too tight							•
Oil wrong type		•					
Crankcase oil pressure too high		•					
Unloaded running time too long (2)		•					

(2) Use automatic start-stop control

TABLE 5-1 (CONTINUED)
TROUBLESHOOTING CHART

The Problem / The Causes	Excessive compressor vibration	Oil pumping excessive (single acting compressor)	Crankcase water accumulation	Crankcase oil pressure low	Starts too often	Compressor fails to unload	Compressor fails to start
Piston ring gaps not staggered		•					
Piston or ring drain holes clogged		•					
Centrifugal pilot valve leaks		•					
Runs too little (3)			•				
Detergent oil being used (4)			•				
Location too humid and damp			•				
Oil relief valve defective				•			
Oil piping leaks				•			
Oil filter or strainer clogged				•			
Air leak into pump suction				•			
Gear pump worn, defective				•			
Receiver too small					•		
Receiver, drain more often					•		
Demand too steady (3)					•		
Unloader parts worn or dirty						•	
Control air filter, strainer clogged						•	
Regulation piping clogged						•	
Motor too small							•
Electrical conditions wrong							•
Voltage abnormally low							•
Excitation inadequate							•
Motor overload relay tripped							•
Fuses blown							•
Wiring incorrect							•
Low oil pressure relay open							•
System leakage excessive					•		
Bearings need adjustment or renewal				•			

(3) Use constant speed control
(4) Change to nondetergent oil

(text continued from page 307)

SOLUTIONS TO SPECIFIC PROBLEMS

The following is a more detailed listing of solutions to specific problems.

No Air Delivered

1. Intake blocked
2. Suction filter plugged
3. Unloader stuck
4. Valve damaged
5. Excessive leaks in piping
6. Gasket blown on cylinder head, valve cage assembly, or intercooler

Not Enough Air Delivered

1. Air demand exceeds rating of compressor
2. Intake partially blocked
3. Suction filter requires cleaning
4. Unloader stuck
5. Valves damaged
6. Leaks in piping
7. Gasket blown in cylinder head, valve assembly, or intercooler
8. Piston rings worn or broken
9. System pressure set too high
10. Speed too low
11. Belts slipping

Insufficient Pressure

1. Air demand exceeds rating of compressor
2. Intake partially blocked
3. Suction filter requires cleaning
4. Unloader bypassing
5. Leaks in piping and connections
6. Gaskets leaking on cylinder head, valve assembly, or intercooler
7. Piston rings worn or broken
8. Pressure gauge not properly calibrated

Machine Runs Too Hot

1. Air cooling fins loaded with dirt, or water jacket soiled (if water-cooled machine)
2. Discharge pressure set too high
3. Valve damaged, causing "wire drawing"
4. Compressor running backwards
5. Intercooler fouled and not functioning

Unit Runs Too Long To Reach Desired Pressure

1. Air demand increased
2. Intake partially blocked
3. Suction filter requires cleaning
4. Unloader partially bypassing
5. Valves defective
6. Leaks in pipe connections
7. Piston rings worn or broken
8. Compressor not running up to speed
9. Belts slipping

Receiver Safety Valve Blows

1. Unloader not functioning
2. Control air line leaking
3. Control air line strainer clogged

Unloader Operates Erratically

1. Control air line leaking
2. Control air line too small
3. Control air line strainer partially clogged
4. Unloader spring improper range

Unit Will Not Shut Off At Pressure Switch Setting

1. Contacts in pressure switch "fused" or not breaking circuit
2. Pressure switch spring broken or weak

Unit Blows Fuse

1. Fuses too small
2. Voltage is low
3. Pressure switch differential setting too close
4. Pressure switch release valve not operating (unit therefore starts against load)
5. Motor bearings beginning to seize
6. Motor leads loose
7. Motor leads too small

Motor Overheats

1. Pressure switch differential setting too close, causing compressor to cycle, resulting in excessively frequent starts
2. Voltage is low
3. Discharge pressure too high
4. Compressor running faster than rated
5. Motor needs cleaning
6. Motor bearings beginning to seize
7. Check belt or coupling alignment

Unit Will Not Start

1. Pressure switch contacts not mating
2. Worn contacts
3. Motor seized
4. Thermal overloads require resetting
5. Fuses blown

Intercooler Safety Valve Blows When Compressor Unloaded—Second Stage Not Operating

1. Broken high pressure suction or discharge valve
2. Seat leaking on high pressure unloader piston
3. Broken unloader power spring
4. Unloader power piston stuck

Intercooler Safety Valve Blows When Compressor is Loaded—Second Stage Not Operating

1. Broken high pressure suction or discharge valve
2. High pressure unloader piston stuck

TROUBLESHOOTING LUBRICATION SYSTEMS

The following lists will help identify the cause of some of the more common compressor lubricating oil systems problems.

Frame Oil System

Low Oil Pressure

1. Low oil level
2. Dirty filter strainer
3. Defective pump or relief valve
4. Relief valve on header faulty
5. Worn bearings or excessive clearance
6. Low oil viscosity
7. High oil temperature

High Oil Pressure

1. Relief valve in header faulty
2. Restriction in oil line
3. Improper grade of oil
4. Cold oil
5. Pressure regulating valve set too high

Excessive Oil Consumption

1. Oil level too high in crankcase
2. Oil is too light
3. The oil pressure may be too high
4. Piston rings and cylinder are worn

Force-Feed Lubricator

Incorrect Delivery of Lubricator

1. Feeds not vented of air
2. Low oil level
3. Plugged vent in lubricator reservoir
4. Oil check valve on cylinder faulty
5. Line leaking or kinked
6. Incorrect adjustment of pump stroke
7. Leak in line or fitting

Troubleshooting charts similar to Table 5-1 are useful aids in pinpointing problems. Two charts, Tables 5-2 and Table 5-3 have different but complementary approaches; one helps to locate the problem, and the other helps identify what caused it.

SIGNIFICANCE OF INTERCOOLER PRESSURES

Regularly observing the intercooler pressure is important in the determination of proper compressor performance and operation.

SUCTION VALVE FAILURES

As has been pointed out, capacity loss can most often be attributed to problems with valves. If a suction valve does not close and seal completely, the following occurs:

1. During the compression (discharge) stroke, some of the normally displaced gas will leak or blow by the suction valve to the suction side of the valve.
2. The volume of gas actually displaced will decrease.
3. As the compressed hot gas passes over the valve in the wrong direction, the valve heats up.
4. The inlet temperature of the gas on the suction side of the valve goes up, thus reducing the volumetric efficiency of the machine, that is, hotter than normal design suction temperatures will result.

(text continued on page 323)

TABLE 5-2
TROUBLESHOOTING CHART

Trouble	Probable Cause(s)	Remedies
Compressor Will Not Start	1. Power supply failure 2. Switchgear or starting panel 3. Low oil pressure shut-down switch 4. Control panel	1. Correct voltage or power supply 2. Check curcuitry, interlocks, relays, etc. 3a. Check switch setting b. Install momentary bypass switch on direct gear driven oil pumps 4. Check connections and settings of all devices
Motor Will Not Synchronize	1. Low voltage 2. Excessive starting torque 3. Incorrect power factor 4. Excitation voltage failure	1. Correct voltage supply 2. Unload compressor during starting 3. Adjust excitor field rheostat 4. Check field excitation system
Low Oil Pressure	1. Oil pump failure 2. Oil foaming from counter-weights striking oil surface 3. Cold oil 4. Dirty oil filter 5. Interior frame oil leaks 6. Excessive leakage at bearing shim tabs and/or bearings 7. Improper low oil pressure switch setting 8. Low gear oil pump by-pass/relief valve setting 9. Defective pressure gauge 10. Plugged oil sump strainer 11. Defective oil relief valve	1. Check oil pumps/power supply 2. Reduce oil level 3. Use frame oil heater and/or steam trace exterior piping 4. Clean/replace oil filter cartridge 5. Check oil piping 6. Set shim tabs and bearing clearances 7. Reset 8. Reset valve 9. Replace gauge 10. Clean strainer 11. Repair or replace valve
Relief Valve Popping	1. Faulty relief valve 2. Leaking suction valves or rings on next higher stage 3. Obstruction (foreign material, rags), blind or valve closed in discharge line	1. Test and reset 2. Repair/replace defective parts 3. Relieve obstruction

(table continued on next page)

Table 5-2 (Continued)
Troubleshooting Chart

Trouble	Probable Cause(s)	Remedies
High Discharge Temperature	1. Excessive ratio on cylinder due to leaking inlet valves or rings on next higher stage	1. Repair valves/rings
	2. Fouled intercooler/piping	2. Clean intercooler/piping Reduce lube rates
	3. Leaking discharge valves or piston rings	3. Repair/replace parts
	4. High inlet temperature	4. Clean intercooler
	5. Fouled water jackets on cylinder	5. Clean jackets
	6. Improper lube oil and/or lube rate	6. Use correct lube oil and correct lube rate
Frame Knocks	1. Loose crosshead pin, pin caps or crosshead shoes	1. Tighten/replace loose parts
	2. Loose/worn main, crank-pin or crosshead bearings	2. Tighten/replace bearings, check clearance
	3. Low oil pressure	3. Increase oil pressure, repair leaks
	4. Cold oil	4. Warm oil before loading unit Reduce water supply to oil cooler
	5. Incorrect oil	5. Use proper oil
	6. Knock is actually from cylinder end	6. Tighten piston nut, etc.
Crankshaft Oil Seal Leaks	1. Faulty setting of oil baffle and slinger	1. Set baffle
	2. Clogged drain hole	2. Clear obstruction
Piston Rod Oil Scraper Leaks	1. Worn scraper rings	1. Replace rings
	2. Scrapers incorrectly assembled	2. Assemble
	3. Worn/scored rod	3. Replace rods
	4. Improper fit of rings to rod/side clearance	4. Replace rings
Noise in Cylinder	1. Loose piston	1. Disassemble and tighten piston
	2. Piston hitting outer head	2. Adjust piston rod for proper end clearance
	3. Loose crosshead lock nut	3. Tighten nut
	4. Broken or leaking valve(s)	4. Repair/replace parts
	5. Worn or broken piston rings or expanders	5. Replace rings
	6. Valve improperly seated/ damaged seat gasket	6. Replace gasket and reassemble properly
	7. Free air unloader plunger chattering	7. Replace worn or broken unloader spring(s)

TABLE 5-2 (CONTINUED)
TROUBLESHOOTING CHART

Trouble	Probable Cause(s)	Remedies
Excessive Packing Leakage	1. Worn packing rings 2. Improper lube oil and/or insufficient lube rate (blue rings) 3. Dirt in packing 4. Excessive rate of pressure increase 5. Packing rings assembled incorrectly 6. Improper ring side or end gap clearance 7. Plugged packing vent system 8. Scored piston rod 9. Excessive piston rod run-out	1. Replace packing rings 2. Use correct lube oil and increase lube rate 3. Clean piping/gas supply 4. Reduce pressure and increase at more gradual rate 5. Reassemble per instructions 6. Establish correct clearances 7. Remove blockage and provide low point drains 8. Replace rod 9. Correct run-out. Reshim crosshead
Packing Overheating	1. Lubrication failure 2. Improper lube oil and/or insufficient lube rate 3. Insufficient cooling	1. Replace lubricator check valve/lubricator pumping unit 2. Use correct lube oil and increase lube rate 3. Clean coolant passages/install water filter/increase supply pressure Reduce coolant inlet temperature
Excessive Carbon on Valves	1. Excessive lube oil 2. Improper lube oil (too light, high carbon residue) 3. Oil carryover from inlet system or previous stage 4. Broken or leaking valves causing high temperature 5. Excessive temperature due to high pressure ratio across cylinders	1. Adjust lube supply 2. Use lube oil per manufacturer's recommendations 3. Install oil separators/drain system 4. Repair/replace parts 5. Clean exchangers, valves and correct cause of high pressure
Leaking Valves	1. Faulty seat gasket 2. Inadequate setscrew tightening 3. Worn valve seat, channels or plates. 4. Improper valve assembly—channels sticking in guides	1. Replace gaskets 2. Tighten setscrews 3. Replace valves/lap seats 4. Reassemble properly per "VALVES" section

(Source: Compressed Air and Gas Handbook, 5th Edition, Published by The Compressed Air and Gas Institute)

TABLE 5-3
TROUBLESHOOTING WATER-COOLED RECIPROCATING COMPRESSORS

Observed Abnormal Condition	Probable Cause or Contributing Factor
Failure to Deliver Air	Restricted suction line; dirty air filter Broken valves Defective capacity control; unloader stuck Piping system leaks
Insufficient Capacity	System demand exceeds capacity Dirty air filter Unloader stuck or defective Worn or broken piston rings Head-gasket leak Speed too low; defective capacity control System pressure set too high
Insufficient Pressure	Unloader bypassing Loose valve Pressure gauge improperly calibrated Air demand exceeds capacity Broken valves; piping system leaks Worn or broken piston rings
Compressor Overheats	Discharge pressure set too high Inadequate cooling-water supply Improper cylinder or crankcase lubrication Broken or loose valves
Intercooler Pressure Below Normal	Defective unloader in low-pressure cylinder Loose discharge valve in low-pressure cylinder Defective capacity control; unloader stuck
Intercooler Pressure Above Normal	Worn or missing valve strips in high-pressure cylinder Loose valve, leaky gasket in high-pressure cylinder Worn piston rings in high-pressure cylinder Defective capacity control; unloader stuck
Receiver Pressure Too High	Inadequate cylinder control pressure; leaks Unloader not working Improperly adjusted capacity control
Discharge Temperature Too High (Air)	Broken valves; unloader stuck or defective Improper cylinder or crank-case lubrication Inadequate cooling-water supply

Table 5-3 (Continued)
Troubleshooting Water-Cooled Reciprocating Compressors

Observed Abnormal Condition	Probable Cause or Contributing Factor
Cooling-Water Temperature Too High	Sludged cylinder jackets need cleaning Heat exchanger or cooling tower needs cleaning Jacket water pump not functioning Inadequate cooling-water supply
Compressor Knocks	Loose valve or unloader Broken unloader control spring Loose flywheel or sheave Excessive main or crankpin bearing clearance Loose piston rod nut Loose motor rotor on shaft Loose packing Piston hitting head; insufficient clearance
Compressor Vibrates	Improper grouting Incorrect speed; excessive discharge pressure Improperly supported piping Defective capacity control; unloader stuck Loose flywheel or sheave Loose motor rotor on shaft

Source: Plant Engineering, *October 12, 1978.*

(text continued from page 318)

The problem becomes compounded; the leakage gets progressively worse. The moving parts, that is, strips, channels or plates, may distort due to warpage and lack of clearance due to high temperatures. Carbon build-up occurs and the problem becomes even worse. Leakage may also be caused by faulty valve seat gaskets, eroded valve seats, or malfunction of the suction valve unloader.

Fortunately, there are inexpensive infrared devices that can speed up the task of monitoring valve temperatures and allow us to take corrective action when temperatures increase above normal. These temperature measuring devices are accurate and easy to use. One such device should be available at each plant. Valve cover temperature readings should be taken and recorded on a form as shown in Figure 5-1 at least once every operating shift. Contrary to popular belief, valves that start out as leakers

Valve Cover Temperatures

Compressor: Stage: 1st

Type Temperature Measuring Device:

#1-S #2-S HE #2-D #1-D

#1-S #2-S CE #2-D #1-D

Date	Time	Valve Location	Measured Temperature	Inspected By	Remarks

FIGURE 5-1. Compressor inspection report.

do not seat themselves. Therefore, the valve should perform perfectly the instant it is put into service.

If a valve performance problem is suspected, proceed as follows:

- With the compressor running loaded, check valve cover temperature of all suction valves in the cylinder in question.
- Check for wide temperature differences between valves. On a typical single-stage double-acting watercooled machine with 100 psi discharge, suction valve cover temperature may be approximately 115°–120°F. Variances of 5–8°F from cover to cover may be expected. With a leaking valve, cover temperature may be 135°F or more.

This procedure also applies to the discharge valves, which, of course, will be at a considerably higher temperature, 350°–375°F on a 100 psi single-stage air compressor. In any event, it is recommended practice to look for any valves exhibiting cover temperatures more than 15°F higher than usual.

INTERSTAGE PRESSURES

Multistage compressors are designed so that the amount of work performed per stage is exactly equal. In other words, a two-stage 100 BHP compressor will require a 50 BHP per stage.

The correct pressure ratio across each stage is determined by formulas employing absolute pressures:

$$2\text{ Stage}: \sqrt[2]{\frac{P_2}{P_1}} \qquad 3\text{ Stage}: \sqrt[3]{\frac{P_2}{P_1}} \qquad 4\text{ Stage}: \sqrt[4]{\frac{P_2}{P_1}}$$

Improper interstage pressures can cause problems with rod loads, capacity, vibration, etc.

On multistage compressors, the interstage pressure should be checked even before valve temperature readings are acquired. See page 25.

LOW INTERSTAGE PRESSURE

On multistage compressors, intercooler pressure deviations will point to the cylinder in which a problem is occurring. When a lower than normal interstage pressure is noted, a problem in the preceding cylinder, that is, first stage of a two-stage compressor exists. If the interstage pressure is low when the compressor is loaded and remains steady when unloaded, the fault is in the suction valves of the preceding cylinder (first-stage of a 2-stage compressor). Should the pressure drop to zero, the discharge valves of the preceding cylinder are at fault.

HIGH INTERSTAGE PRESSURE

A higher than normal intercooler pressure indicates problems in the succeeding cylinder, that is, second stage of a two-stage compressor. Should the pressure be higher than normal when the compressor is loaded, the suction valves of the succeeding cylinder are the problem. If the intercooler pressure is higher when unloaded, the discharge valves in the succeeding cylinder are at fault. If interstage pressure becomes too high, the safety relief valve will blow.

BELT DRIVES

On compressors that are belt driven, the V-belts will require some attention. Figure 5-2 is a guide for troubleshooting V-belts.

MOTOR CONTROLS

Troubleshooting of motor controls involves a point-to-point mechanical and electrical inspection of the motor control equipment. Table 5-4 is a guide for troubleshooting motor controls.

DIAGNOSTIC TESTS

Diagnostic testing is likely to pinpoint the source of a compressor malfunction.On small air compressors, proceed as follows:

TO TEST FOR UNLOADER BYPASSING

Remove suction filter. If pulsing is noted when hand is held over the air intake, the unloader is bypassing.

TO TEST IF FILTER IS CLOGGED

Remove filter for a short while and note if compressor output increases. If filter is clogged, clean with air and by washing with a caustic solution. Never use kerosene or gasoline for cleaning filters. Doing so is likely to cause an explosion.

INTERCOOLER PRESSURE

The pressure in the intercooler is a good indicator of compressor performance. At 100 psig discharge (atmospheric suction), the intercooler pressure is normally about 27 psig. If intercooler pressure is low, look for defective valves in the first stage. If intercooler pressure is high, look for defective valves in the second stage.

COMPRESSOR KNOCKS

Virtually all compressors will have an inherent gas noise, or pulsation. Intensity and sound patterns may be more or less pronounced, depending

BELT DRIVES

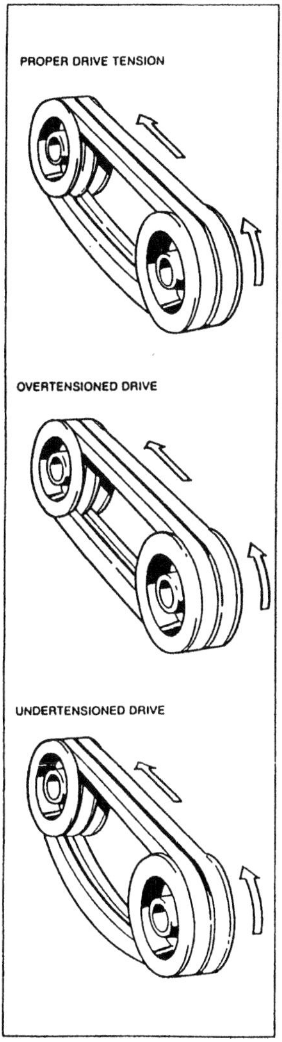

PROPER DRIVE TENSION

OVERTENSIONED DRIVE

UNDERTENSIONED DRIVE

Troubleshooting Guide for V-belt Drives

What Happened	Probable	What to Do
Short belt life	Spin burns from belt slipping on driver sheave under stalled load conditions or when starting	Tension belts
	Gouges or extreme cover wear caused by belts rubbing on drive guards or other objects	Eliminate obstruction or realign drive to provide clearance
	High ambient temperature	Use high-temperature belts; provide ventilation; shield belts
	Grease or oil on belts	Check for leaky bearings; clean belts and sheaves
	Worn sheaves	Replace sheaves
Belts turn over in grooves	Damaged cord section in belts; frayed or gouged belts	Replace belts
	Excessive vibration	Tension belts; replace belts if damaged
	Flat idler pulley misaligned	Realign idler
	Worn sheaves	Replace sheaves
	Sheave misalignment	Realign drive
Belt squeal	High starting load; belts not tensioned properly; excessive overload	Tension drive or redesign and replace drive
	Insufficient arc of contact	Increase center distance or use notched belts
Belt breakage	Foreign material in drive	Provide drive guard
	Belts damaged during installation	Follow installation instructions
	Shock or extreme overload	Eliminate overload cause or redesign drive
Belt stretch beyond takeup	Worn sheaves	Replace sheaves
	Underdesigned drive	Redesign and replace drive
	Takeup slipped	Reposition takeup
	Drive excessively tensioned	Properly tension drive
	Damaged cord section during installation	Replace belts and properly install
Excess vibration	Damaged belt cord section	Replace blets
	Loose belts	Tension drive
	Belts improperly tensioned	Tension drive with slack of each belt on the same side of the drive
Belts too long at installation	Insufficient takeup	Use shorter belts
	Drive improperly set up	Recheck driver and driven machine set up
	Wrong size belts	Use correct size belts
Belts too short at installation	Insufficient takeup	Use longer belts
	Drive improperly set up	Recheck driver and driven machine set up
	Wrong size belts	Use correct size belts

FIGURE 5-2. Guide for troubleshooting V-belts. (*Source:* Plant Engineering, *July 18, 1991*).

TABLE 5-4
TROUBLESHOOTING GUIDE FOR MOTOR CONTROLS

Problem	Possible Cause	Solution
Noisy magnet (humming or loud buzz)	Misalignment or mismating of magnet pole faces	Replace or realign magnet assembly
	Foreign matter on pole face (dirt, lint, rust)	Clean (do not file) pole faces; realign if necessary
	Low voltage applied to coil	Check system and coil voltage; observe voltage variations during startup
	Broken shading coil	Replace magnet assembly
Failure to pick up and seal in	Low voltage	Check system and coil voltage; watch for voltage variations during starting
	Wrong magnet coil or wrong connection	Check wiring, coil nomenclature, etc.
	Coil open or shorted	Check with an ohmmeter; when in doubt, replace
	Mechanical obstruction	Disconnect power and check for free movement of magnet and contact assembly
	Faulty contacts in the control circuit	Check switches in control circuit for proper operation
Failure to drop out or slow dropout	"Gummy" substance on pole faces or magnet slides	Clean with nonvolatile solvent, degreasing fluid
	Voltage to coil not removed	Shorted seal-in contact (exact cause found by checking coil circuit)
	Worn or rusted parts causing binding	Clean or replace worn parts
	Residual magnetism due to lack of air gap in magnet path	Replace worn magnet parts or accessories
	Mechanical interlock binding (reversing starters)	Check interlocks for free pivoting; new bushing or light lubrication may be required
	Welding contacts	Remove power at the source; check contacts for possible welding

on the length, size, and wall thickness of the suction pipe in relation to the rotational speed of the compressor crankshaft.

The air noise can be substantially reduced by one or several remedial measures:

1. Run the intake outdoors
2. Install a filter silencer
3. Use heavier piping on the suction line
4. Add or remove a short length of pipe on the suction line to avoid sympathetic vibration in tune with pulsation frequency of the compressor
5. Interrupt the suction line with a section of leather, canvas, or resilient material.

After becoming accustomed to the characteristic air noise of an air compressor, any unusual knock can be assumed to have its source in the mechanical running gear.

Valve Analysis

If a suction valve does not close and seal completely, the following occurs:

- During the compression stroke, some of the normally displaced gas will leak or blow by to the inlet side of the valve. The design volume of actual displaced gas will decrease. As the compressed hot gas passes over the valve in the wrong direction, the valve heats up. The inlet temperature of the gas on the suction side of the valve goes up, thus reducing the volumetric efficiency of the machine, that is, hotter than normal design suction temperature.
- The problem becomes compounded, the leakage gets progressively worse, the moving parts, that is, strips, plates may distort due to warpage from high temperatures. If enough heat is generated, carbon will build up on the valve and the problem will worsen.

Leakage may also be caused by eroded valve seat gasket surface or a jammed air unloading device.

Evaluating Reciprocating Compressor Condition Using Ultrasound and Vibration Patterns

As any gas (air, oxygen, nitrogen, etc.) passes through a leak orifice, it generates a turbulent flow with detectable high frequency components.

By scanning the test area with an ultrasonic detection device, a leak can be heard through the headset as a rushing sound or noted on the ballistic meter. The closer the instrument is to the leak, the louder the rushing sound and the higher the meter reading. Should ambient noise be a problem, a rubber focusing probe may be used to narrow the instrument's reception field and to shield it from conflicting ultrasounds.

Figure 5-3 depicts a device marketed under the name Ultraprobe. Produced by Elmsford (New York)-based UE Systems Company, this low-cost detector can spot a wide range of leakage-related defects and other abnormalities associated with fluid machinery.

Performance and condition analysis of compressor cylinders will provide substantial savings to the user. Analysis will:

- Reduce power consumption by 10% or more
- Increase compressor throughput
- Reduce maintenance costs

PRESSURE-VOLUME PATTERNS

Inspection of the pressure-volume patterns displayed on an instrument such as the Beta 250 Engine/Compressor Analyzer* will identify cylinder problems that cause loss of efficiency. These include:

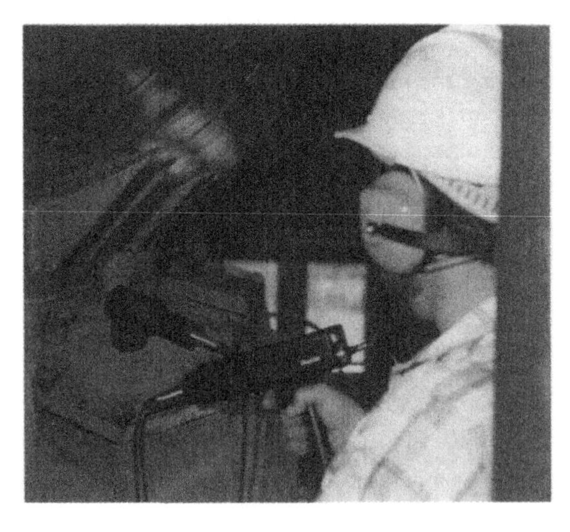

FIGURE 5-3. Ultrasonic detection in progress. (*Source: UE Systems Company, Elmsford, New York*).

*PMC/Beta, Natick, Massachusetts

- Leaking rings, suction valves, or discharge valves
- Suction or discharge valve springs too stiff
- Valve chatter

Figure 5-4 illustrates examples of faults shown on pV curves.

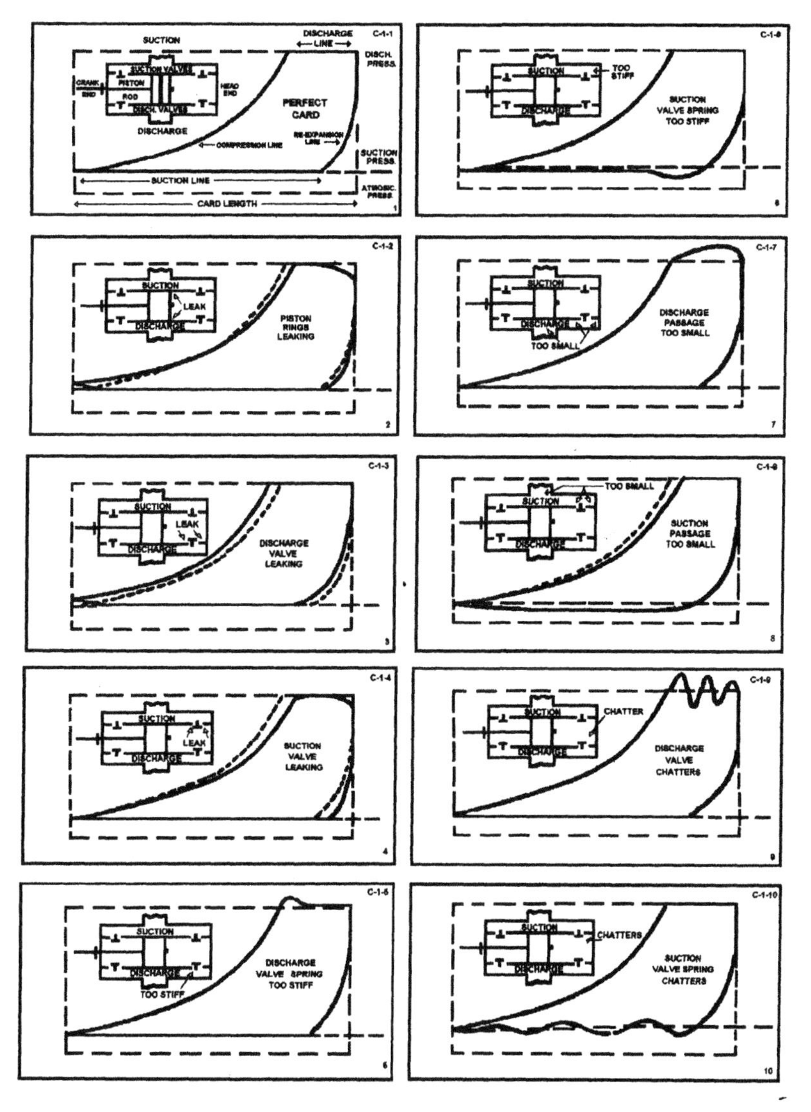

FIGURE 5-4. Typical compressor cylinder problems identified with pV displays.

The Beta 250 has three pressure channels, providing real time computation and display of cylinder HP and suction discharge HP losses. Excessive suction or discharge valve loss is a clear indication of poor valve condition.

VIBRATION VS CRANKANGLE PATTERNS

The use of vibration and ultrasonic patterns in combination with cylinder pV patterns will aid in diagnosing the condition of valves, rings, and packing. When valves open and close, they produce vibrations that are detected with an accelerometer. Figure 5-5 shows a compressor pressure vs crankangle trace. The vibration and ultrasonic patterns are also shown. These indicate valve timing and normal valve action.

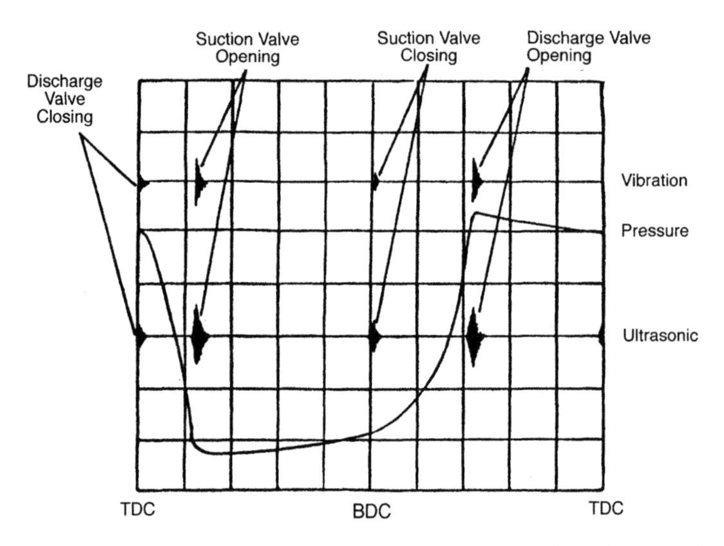

FIGURE 5-5. Pressure-time trace (pressure angle) with vibration and ultrasonic traces superimposed.

The accelerometer is attached to the valve cover with a magnetic clamp and is moved from cover to cover to detect faults. If the spring tension in the valves is too light, a late closure of the valve will result. If the spring is too heavy, multiple premature closures of the valve are likely to occur. These can be detected with the vibration trace on the C.R.T. screen as illustrated in Figure 5-6.

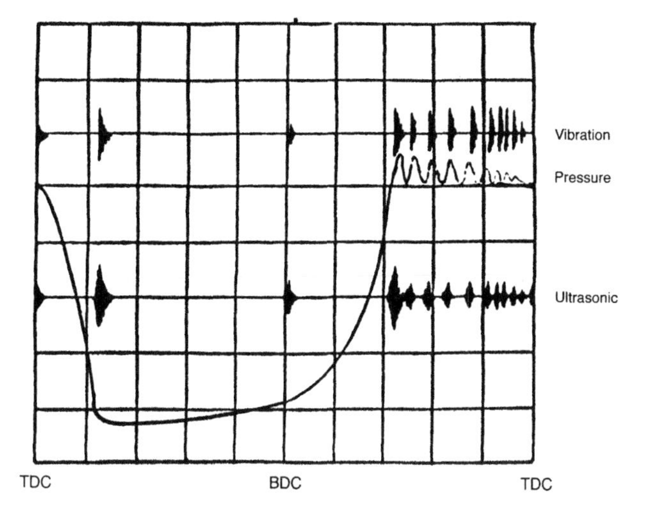

FIGURE 5-6. Springs in discharge valve too heavy. Repeated closures shown on vibration and ultrasonic trace.

If a valve is leaking, it will generate a vibration during the time the valve is supposed to be closed. Figures 5-7 and 5-8 show these events. With a little experience, other faults such as broken rings and rod problems can be detected as well.

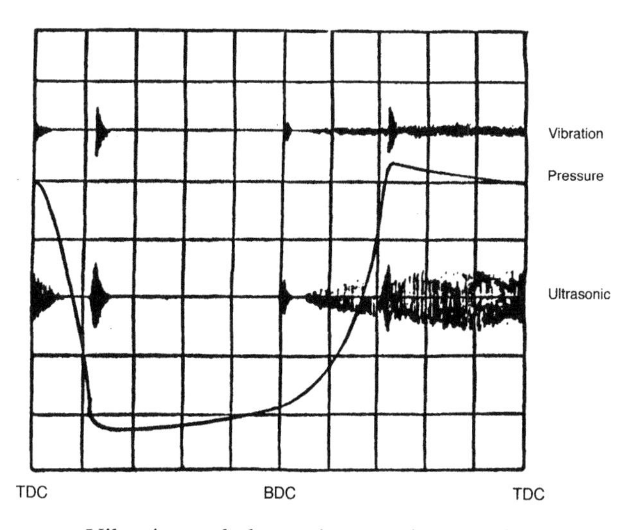

FIGURE 5-7. Vibration and ultrasonic trace show suction valve leaking.

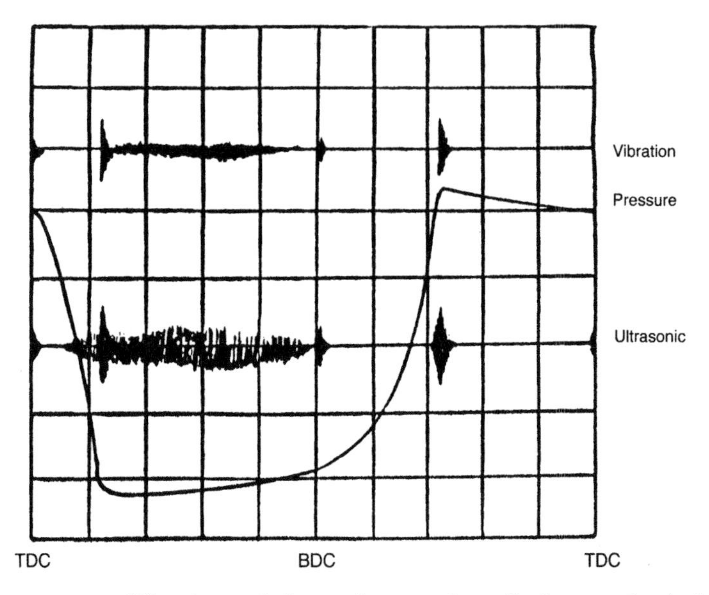

FIGURE 5-8. Vibration and ultrasonic trace show discharge valve leaking.

ULTRASONIC VS CRANKANGLE PATTERNS

Ultrasonic is inaudible vibration above the frequency range of the human ear. The frequency range of the ultrasonic pickup is 35,000 to 45,000 cycles per second. In this frequency range, leaking gases, certain types of abrasion, and clicking-type vibrations are easily detected. In a manner similar to the lower frequency vibration pickup (100–10,000 cps), the ultrasonic trace will show valve malfunctions, blow-by, scuffing, leaks, and inappropriate valve action.

One might question using both vibration and ultrasonic transducers. Neither device alone will detect all compressor malfunctions. Lower frequency knocks will only be seen by the accelerometer. The "hissing" of leaking gases is better seen using ultrasonics.

As with the vibration pickup, valve problems are diagnosed by moving the ultrasonic probe from valve cover to valve cover. Ring blow-by can best be seen next to a lubrication port at mid cylinder. Packing problems on double-acting cylinders are seen at the joint between the cylinder and distance piece, where the packing case is located (see Figures 5-5 through 5-8).

CASE HISTORIES FROM SERVICE TECHNICIAN REPORTS

Service technicians compiled examples of problems that could have been prevented by more-detailed or careful examination of early symptoms.

The following are some of the typical problems they described.

Symptom: Compressor would not unload, pumping until safety valve blew. Excessive blow-by existed at the piston rod packing.

Findings: The unloader piston was found to have rusted in its cylinder due to excessive condensation. Jacket water temperature at 50°F was increased to 100°F outlet. Unloader piping cleaned of dirt and moisture.

Symptom and Problem: Compressor forced shutdown due to extreme knocking. Found wiped crank pin bearing that had been recently installed.

Findings: Damaged parts were checked and crankpin was found to be .007" out of round.

Symptom: Two-stage compressor developed a knock. Preliminary inspection showed no apparent cause.

Findings: Cylinder heads of both stages were removed. Clearances of crosshead and rod were checked. Found piston rod nut of high pressure cylinder lacked cotter pin, allowing it to turn, loosening the piston and causing the knock.

Symptom and Problem: Horizontal non-lube compressor would not unload. Diaphragm unloader was "chattering," caused by diaphragms failing.

Findings: Diaphragm stem travel was insufficient to correctly unload.

Symptom: Single-stage, vertical-type compressor cylinder was worn excessively; piston was scored and machine was out of service.

Findings: Since all valve passages were found dry, it indicated lubricator was also dry, causing piston failure.

Symptom: Compressor piston rod was scuffed. Packing was worn and cylinders showed pitting from moisture.

Findings: Found piston rod packing cocked, impairing lubrication of the rod.

Symptom: Water leaking from the air side of the intercooler. Traps were bypassed to handle drain.

Findings: A number of tubes were corroded through from the water side due to improper water treatment.

Symptom: A recently overhauled, horizontal two-stage compressor developed a noisy suction valve.

Problem: The gasket under the suction valve had been left out making it loose on the seat.

Question: When intercooler pressure indicates a leaky or broken compressor valve, how do you tell which one is defective without removing all valves?

Answer: Leaking valves can be detected by cover temperatures that are somewhat greater than others, or by a slight difference in valve operating sound, or by both of these symptoms.

Question: How can the discharge temperature indicate the compressor mechanical condition?

Answer: A compressor, operating under the same air pressure and cooling water temperature, should register constant discharge air temperature. If the latter increases while the other conditions remain constant, leakage is definitely indicated, usually of the hot discharge back to an intermediate pressure. It may be either valve or piston ring leakage. A plant log recording interstage pressures and temperatures should be regular practice, because it definitely notifies the operator when to examine piston rings and valves. An increase in discharge temperature also puts him on guard concerning the mechanical condition of a single-stage compressor.

Question: What causes discharge air temperature to increase gradually and the cylinder to become somewhat hotter than formerly, when piston rings and valves are in first-class condition and jacket water temperature is normal?

Answer: After years of operation, if jacket water quality is not particularly good, the jacket cooling effects may have deteriorated considerably. The inside of jacket walls may be covered with heavy deposits of lime or magnesium, which interferes with normal heat transfer. The answer of course, is to thoroughly clean the jacket.

Service Condition Changes

A change in suction or discharge pressure may cause problems. Observing these pressures should be part of any troubleshooting effort.

A plant operator advised that he had purchased a 12 × 9 single-stage horizontal compressor and that he was using it for 100 psi pressure. He inquired about the working conditions for which this machine was suitable and, upon investigation, found that the compressor was designed and built for a maximum air pressure of 40 psi. The operator was, accordingly, cautioned that he was subjecting the machine to over twice the load for which it was designed and that continued use under these conditions might cause problems. That he had not already had trouble is undoubtedly due to the conservatism with which compressors are rated.

Like many other users, this person believed that the air cylinder on the machine appeared to be sufficiently heavy for 100 psi and that, therefore, there was no danger in operating the compressor at this pressure. That assumption brings us to the point: The maximum load allowable on a compressor is determined more frequently by the strength of the frame and the running gear rather than by the strength of the air or gas cylinder.

In the case cited above, it so happened that the 12 × 9 air cylinder is sufficiently heavy for 100 psi pressure. As a general rule, air cylinders are designed with quite a large margin of safety and are actually used over a wide range of pressures.

Compressor ratings are typically limited by the load on frame and running parts. The three points in the frame construction that are under heavy strain while transmitting the power from the pulley driving the machine to the air cylinder are the main bearings, the crank pin, and the crosshead pin. When these parts are under less than maximum load, they do their work without appreciable wear or damage risk, but when they are subjected to working strains greater than their ratings, the overload not only may cause considerable knocking, but excessive heating as well. Serious damage may be the logical outcome of operating with excessive load.

BASIC AIR COMPRESSOR SYSTEM EVALUATION

In this actual case, a plant had a system comprising four different reciprocating compressors feeding into a 100 psig receiver.

Reciprocating compressors at 100 psig

1–100 HP
1–60 HP
1–50 HP
1–40 HP
250 HP total, at approximately 5 CFM/HP or 1250 CFM

One day, a situation arose where the 100 HP compressor was shut down because it could not produce pressure. The 40 HP machine could not operate because of a highly audible noise.

A service person was dispatched to investigate the situation. The following are abbreviated entries from his notebook.

A. Checked 100 HP unit by opening discharge valve to pressurized systems (with compressor shut off). Crankcase pressurized (single-acting compressor) which indicated bad discharge valves and possibly worn or broken piston rings.
B. Checked 60 HP compressor by running and blocking off system and venting discharge to atmosphere, while maintaining 100 psig discharge pressure. Checked full-load amps, which were at 98% design. Therefore, this compressor at near full capacity of 300 CFM.
C. Checked 50 HP compressor the same as B. This unit also at full capacity or 250 CFM.
D. Checked 40 HP compressor by running to maximum design pressure (100 psig). Unit leaked severely at packing gland and could not be adjusted to correct. Unit also had loud rap or knock. Unit could not develop 100 psig, due to severe leaks. Required extensive repairs to cylinder and running gear.

Of the total 1250 CFM capacity, system would only produce 550 CFM.

Capacity Requirement Determined

At lunch time, the entire plant was shut down with no compressed air requirements. Both the 60 HP and 50 HP compressors were placed in operation to pressurize the system to 100 psig. But, only a maximum pressure of only 60 psig could be attained. A short walk through the plant revealed air leaks near every air user. The result was that over 100 HP was required to maintain the leaks in the plant. (If 100 psig was to have been maintained during this shutdown, precisely 110 HP was required to overcome the leaks).

The plant required at least 160 HP of compression to maintain 100 psig during plant operation. Without leaks, however, this plant required only 40 HP, or about 200 CFM.

The cost of this wasted energy is easy to calculate.

Using 110 HP, a utility charge of $.08/kw-hr, and assuming 8000 hrs of operation per year:

$$110 \text{ HP} \times .746 \frac{\text{KW}}{\text{HP}} \times \frac{\$.08}{\text{KW / HR}} \times 8000 \text{ Hrs} \div .9 \text{ motor efficiency}$$
$$= \$58,354 \text{ / year}$$

The importance of our recommendation to check your compressor system for leaks is intuitively evident.

For good measure, here's another short service report:

Underground air line leak was leaking through ground to surface. System pressure was at 100 psig. The service person exposed the leak to find a ¼-inch diameter hole in the line. A ¼″ diameter hole passes 104 cfm of air at 100 psig. This is equivalent to 21.5 Hp at 100 psig. The cost was calculated using the information in the above example:

$$\text{Cost} = \frac{21.5 \times .746 \times .08 \times 8000}{.9} = \$11,405.51/\text{year}$$

The bottom line is simply this: Air leaks can be prohibitively expensive. Conscientious maintenance makes economic sense!

CHAPTER 6

Preventive Maintenance for Reciprocating Compressors

It may be either impossible or utterly uneconomical to design and manufacture machinery for zero maintenance or perfect reliability with infinite life. Realistically, two principal maintenance philosophies are prevalent in industry:

- Do nothing until the equipment breaks down or an emergency occurs. The compressor is repaired as quickly or as cheaply as possible and is returned to service. Inevitably, the next emergency is just around the corner.
- Maintain the equipment in excellent condition, thus optimizing both equipment reliability and availability. Downtime events for preventive maintenance are planned, and the probability of an unexpected breakdown is minimized.

The first philosophy, sometimes called "breakdown maintenance," is rarely justifiable on economic and risk management grounds. Considering the safety risk alone should convince us of the potential danger of this approach.

The second philosophy will prove most profitable when used in conjunction with a conscientiously implemented program of predictive (monitoring and trending-based) maintenance. Although such a comprehensive program will require forethought and organization, its long-term profitability has been demonstrated beyond much doubt.

A well-structured compressor maintenance program will thus bring about several important benefits, including, of course, improved safety, reliability, efficiency, run-time, housekeeping and environmental/regulatory compliance. The final product cost will be materially decreased by this program. When increased production adds substantially to the profits of a plant, the minor expense of a well-structured maintenance program is insignificant. Dependability is a vital factor in any operation. The degree of dependability attained is in direct proportion to the effectiveness of the preventive maintenance program.

INTRODUCTION

Preventive maintenance encompasses periodic inspection and the implementation of remedial steps to avoid unanticipated breakdowns, production stoppages, or detrimental machine, component, and control functions. Preventive maintenance is the rapid detection and treatment of equipment abnormalities before they cause defects or losses. Without strong emphasis and an implemented preventive maintenance program, plant effectiveness and reliable operations are greatly diminished. Plants must minimize compressor failures.

In many process plants or organizations, the maintenance function does not receive proper attention. The naive perception is that maintenance does not add value to a product, and thus, the best maintenance is the least-cost maintenance. Armed with this false perception, traditional process and industrial plants have underemphasized preventive, corrective, routine maintenance, not properly developed maintenance departments, not properly trained maintenance personnel, and not optimized predictive maintenance. Excessive unforeseen equipment failures have been the result.

Maintenance is not an insurance policy or a security blanket. It is a requirement for success. Without effective preventive maintenance, equipment is certain to fail during operation.

COMPRESSOR MAINTENANCE

Five levels of maintenance exist. They are:

1. Reactive or breakdown maintenance. This type of maintenance includes the repair of equipment after it has failed, in other words, "run-to-failure." It is unplanned, undesirable, expensive, and, if the other types of maintenance are performed, usually unavoidable.

2. Routine maintenance. This maintenance includes lubrication and proactive repair.

   Lubrication should be done on a regular schedule.

   Proactive repair is an equipment repair based on a higher level of maintenance. This higher level determines that, if the repair does not take place, a breakdown will occur.
3. Corrective maintenance. This includes adjusting or calibrating of equipment. Corrective maintenance improves either the quality or the performance of the equipment. The need for corrective maintenance results from preventive or predictive maintenance observations.
4. Preventive maintenance. This includes scheduled periodic inspection. Preventive maintenance is a continuous process. Its objective is to minimize both future maintenance problems and the need for breakdown maintenance.
5. Predictive maintenance. This maintenance predicts potential problems by sensing operations of equipment. This type of maintenance monitors operations, diagnoses undesirable trends, and pinpoints potential problems. In its simplest form, an operator hearing a change in sound made by the equipment predicts a potential problem. This then leads to either corrective or routine maintenance.

Similarly, a predictive maintenance expert system can monitor machine vibrations. By gathering vibration data and comparing these data with normal operating conditions, an expert system predicts and pinpoints the cause of a potential problem.

Traditionally, industry has focused on breakdown maintenance, and unfortunately, many plants still do. However, in order to minimize breakdown, maintenance programs should focus on levels 2 through 5.

Emergency Repairs Should Be Minimized

Plant systems must be maintained at their maximum level of performance. To assist in achieving this goal, maintenance should include regular inspection, cleaning, adjustment, and repair of equipment and systems. On the other hand, performing unnecessary maintenance and repair should be avoided. Breakdowns occur because of improper equipment operation or failure to perform basic preventive functions. Overhauling equipment periodically when it is not required is a costly luxury.

At the same time, repairs performed on an emergency basis are three times more costly in labor and parts than repairs conducted on a preplanned schedule. More difficult to calculate, but high nevertheless, are costs which include the shutting down of the production process or time and labor lost in such an event.

Bad as these consequences of poorly planned maintenance are, much worse is the negative impact from frequent breakdowns on overall performance, including the subtle effect on worker morale, product quality, and unit costs.

EFFECTIVENESS OF PREVENTIVE MAINTENANCE

Preventive maintenance, when used correctly, has shown to produce maintenance savings in excess of 25%, but beyond this, its benefit quickly approaches a point of diminishing return. It has been estimated that one out of every three dollars spent on preventive maintenance is wasted. A major overhaul facility reports that "60% of the hydraulic pumps sent in for rebuild had nothing wrong with them." This is a prime example of the disadvantage of performing maintenance to a schedule as opposed to the individual machine's condition and needs.

However, when a preventive maintenance program is developed and managed correctly, it is the most effective type of maintenance plan available. The proof of success can be demonstrated in several ways:

- Improved plant availability
- Higher equipment reliability
- Better system performance or reduced operating and maintenance costs
- Improved safety

A plant staff's immediate maintenance concern is to respond to equipment and system functional failures as quickly and safely as possible. Over the longer term, its primary concern should be to systematically plan future maintenance activities in a manner that will demonstrate improvement along the lines indicated. To achieve this economically, corrective maintenance for unplanned failures must be balanced with the planned preventive maintenance program. Figure 6-1 shows that optimization of maintenance expenditures requires a sound balance between corrective and preventive maintenance of the compressor equipment.

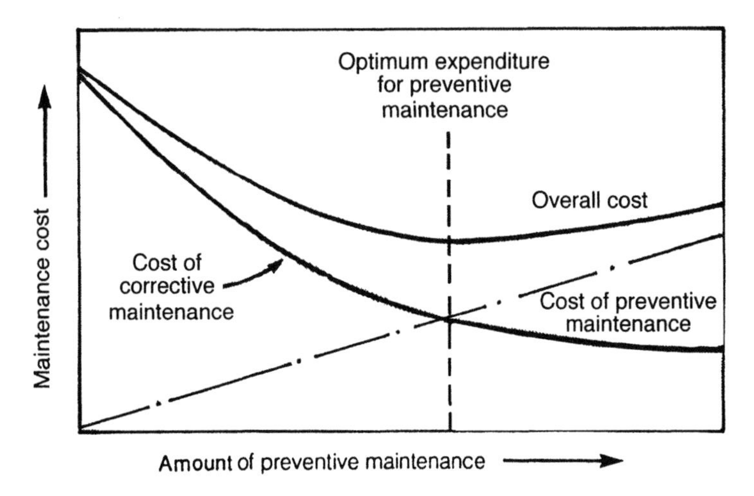

FIGURE 6-1. Factors contributing to the cost of maintenance.

The starting point for a successful long-term maintenance program is to obtain feedback regarding effectiveness of the existing maintenance program from personnel directly involved in maintenance-related tasks.

Such information can provide answers to several key questions:

1. What is effective and what is not?
2. Which time-directed (periodic) tasks and conditional overhauls are conducted too frequently to be economical?
3. What monitoring and diagnostic (predictive maintenance) techniques are successfully used in the plant?
4. What is the root cause of equipment failure?
5. Which equipment can run to failure without significantly affecting plant safety reliability?
6. Does any component require so much care and attention that it merits its modification or redesign to improve its intrinsic reliability?

It is just as important that changes not be considered in areas where existing procedures are working well, unless some compelling new information indicates a need for a change. In other words, it's best to focus on known problem areas.

To assure focus and continuity of information and activities relative to maintenance of plant systems, some facilities assign a knowledgeable staff person responsible for each plant system. All maintenance related information, including design and operational activities, flow through this system or equipment "expert," who refines the maintenance procedures for those systems under his jurisdiction.

Maintenance Improvement

Problems associated with machine uptime and quality output involve many functional areas. Many people, from plant manager to engineers and operators, make decisions and take actions that directly or indirectly affect machine performance. Production, engineering, purchasing, and maintenance personnel as well as outside vendors and stores use their own internal systems, processes, policies, procedures, and practices to manage their sections of the business enterprise. These organizational systems interact with one another, depend on one another, and constrain one another in a variety of ways. These constraints can have disastrous consequences on equipment reliability.

Program Objectives

An effective maintenance program should meet the following objectives:

- Unplanned maintenance downtime does not occur
- Condition of the equipment is always known
- Preventive maintenance is performed regularly and efficiently
- Maintenance needs are anticipated and planned for
- Maintenance department performs specialized maintenance tasks of the highest quality
- Craftsmen are skilled and participate actively in decision-making process
- Proper tooling and information are readily available
- Replacement parts requirements are fully anticipated and components are in stock
- Maintenance and production personnel work as partners to maintain equipment

Program Development

Preventive maintenance programs must reconcile the conditions that exist in the environment of the installation with the demands of the system for which the equipment was designed. Preventive maintenance programs provide schedules for inspection of the components at strategic opportunities. A complete program should be developed from data that include the following:

1. The process that the compressor serves
2. The conditions in the immediate vicinity of the machine
3. The atmosphere in the general geographic location
4. The user's operating schedule

The above requirements indicate that a preventive maintenance program for specific equipment should be tailored to the process involved.

COMPRESSOR PREVENTIVE MAINTENANCE PROGRAM

A compressor maintenance program should be developed for each major component or assembly. If followed, the program will assure total compressor reliability and eliminate unscheduled downtime of the machine. The program should be based on the following premises, which, if not fulfilled, will cause the program to be ineffective and not achieve the desired result of no unscheduled downtime.

1. Major components or assemblies are in "like-new condition." This does not mean that all new parts or components must be purchased and installed, but the existing parts or assemblies must be rebuilt and restored to "like-new," updated specifications. In some cases, new parts or assemblies will have to be obtained.
2. Training programs to update knowledge and skills on compressor maintenance and component reconditioning are developed and are ongoing for the maintenance personnel responsible for the compressor.
3. Procedures are developed for inspection and rebuild of all major components.
4. Proper and adequate records are maintained on the condition of components and on occurrences and locations of failures. These are necessary to allow a logical approach to solutions of problems and to determine what components will require reconditioning or replacement at the next scheduled shutdown.
5. Proper and adequate tooling is available to allow the maintenance personnel to properly service the compressor and perform the maintenance required.

6. Commitment of maintenance personnel to spend money on inventory of necessary spare parts and properly reconditioned parts. This is to prevent the re-use of questionable parts and reconditioning on a "crash" basis by personnel not qualified.

Planning for Preventive Maintenance

Preventive maintenance for reciprocating compressors focuses on:

- Restoring and upgrading compressor equipment.
- Involving operation and maintenance personnel and upgrading their knowledge and skills on compression equipment.
- Implementing effective maintenance planning and scheduling techniques.
- Identification of qualified vendors, suppliers, and contractors.

Preventive maintenance includes periodic inspection to detect conditions that might cause breakdowns, production stoppages, or detrimental loss of function. It follows that a key function of preventive maintenance is to eliminate, control, or reverse potentially costly breakdowns through the early detection of defects or malfunctions. This is where preventive maintenance and predictive maintenance overlap and complement each other. It is extremely important to realize that "best-of-class" companies use predictive techniques to determine when to plan for, or initiate, preventive measures.

Preventive maintenance includes the rapid detection and treatment of equipment abnormalities before they cause defects or losses. It has two very important aspects:

1. Periodic inspection
2. Planned restoration based on the results of the inspections

Daily routine maintenance to prevent deterioration is usually considered a part of the program.

Equipment Improvement

Breakdown and other losses cannot be eliminated unless basic equipment conditions are maintained and possible deterioration is checked for on a daily basis. Operators play a role in this effort by cleaning, lubricating, tightening bolts, and learning to conduct routine inspection.

Before an effective program can be developed, the compression equipment must be inspected to determine:

1. Existing condition
2. What is needed to bring it up to like-new condition or standard
3. New component parts required
4. What components require rebuilding and reconditioning

Consider the typical history of compressor equipment. When new and after installation, it went through start-up problems that were eventually corrected and the operation was then reliable and maintenance-free. This proceeded for several years but, if maintenance was not done or was marginally performed, the compressor became one of the sources of lost production due to unplanned, unscheduled shutdowns. To achieve the objective of no unscheduled shutdown of the compressor and to have an active and profitable preventive maintenance program, the compressor then must be brought back to like-new condition. The first step is a complete and detailed inspection. This should start with those components that cause the most problems and/or have the greatest wear rate, for example, valves, packing, piston/rider rings, piston rods, pistons and cylinders.

When a compressor is inspected, it is important to measure and record the actual dimensions of all components parts. Special inspection forms similar to those shown in Figures 6-2, 6-3, 6-4 are helpful and should be used in any inspection. These kinds of inspection forms assure a good documentation of the condition of the components. They make record keeping easier and form a permanent part of the compressor history.

Evaluating Inspection Data

Recording inspection data serves several purposes:

1. To establish the exact condition of all wearing parts.
2. To establish the wear rate of parts, which, if promptly replaced, will not deteriorate to such a degree that associated parts will be damaged and will also require replacement.
3. To determine which parts require reconditioning and which parts can be reconditioned to like-new condition.

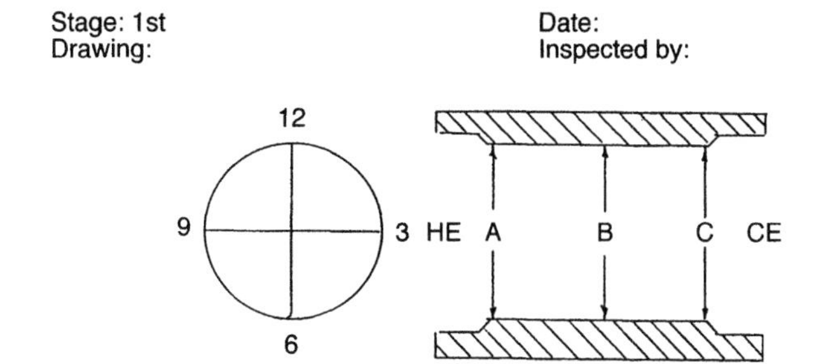

Cylinder Bore Diameters

Position	A	B	C	Drawing size
Vertical 6-12				15.250 + 2; -0
Horizontal 3-9				15.250 + 2; -0

Bore Condition

	Measured	Limit
Oversized		.035"
Out of Round		.012"
Taper		.006"

FIGURE 6-2. Cylinder inspection form.

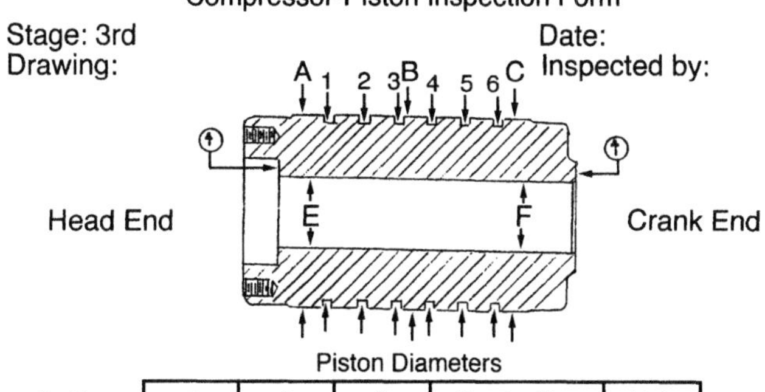

Position	A	B	C	Drawing Size	Limit
Vertical 6-12				5.492"/5.493"	5.482"
Horizontal 3-9					

	E	F		
Rod Bore			2.000"/2.001"	2.006"
Counterbore			3/4" x 3-1/116	- - -

Piston Ring Grooves

							Drawing Size	Limit
Depth							.2265"/.231"	- -
Width							.250"/.2505"	.255"

FIGURE 6-3. Piston inspection form.

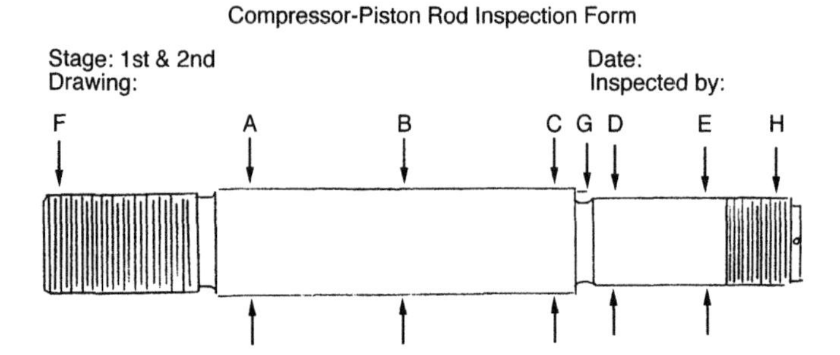

Piston Rod Diameters

Packing	A	B	C	Drawing Size	Limit
Vertical 6-12				2.999"/3.000"	U-2.995" O-2.998" T-2.998"
Horizontal 3-9					
Piston Fit	D		E	Drawing Size	Limit
Vertical 6-12				1.9975"/1.998"	1.996"
Horizontal 3-9					

Note: U=Undersize; O=Out Round; T= Taper

Piston Rod Runout=T.I.R.

F	A	B	C	G	D	E	H	Limit
								+-.001"

FIGURE 6-4. Piston rod inspection form.

4. To evaluate evidence found in order to detect causes of abnormally short life, for example, faulty lubrication conditions resulting in rapid bearing degradation, or excessive misalignment leading to component wear.
5. To evaluate data to determine inspection intervals, routines, and techniques.

Note: One of the objectives of preventive maintenance is to avoid performing unnecessary maintenance and unnecessary costs. Evaluating inspection data helps achieve this objective.

Maintenance Costs

Perhaps the most frequent problem maintenance departments face is rising cost and diminishing resources. While being constantly confronted with budget questions and the pressure to reduce costs, maintenance managers are still expected to raise the level of service in the plant. Maintenance often falls victim to budget cuts because management mistakenly believes maintenance can be deferred. It is thus important to wisely and effectively allocate and spend maintenance money. Inspection records and data can obviously be used to show what must be done to optimize the reliability of the compressor and to determine logical shutdown intervals.

GOOD MAINTENANCE IN A NUTSHELL

The following summarizes what good maintenance is all about.

- The design must be right.
- Correct materials must be used.
- Assembly must be without error.
- The compressor must be maintained properly.

SPARE PARTS

A very important part of compressor rebuilding, continuing reliability, and a factor in the preventive maintenance program is having an adequate inventory of parts for each compressor. Maintenance needs an effective parts supply. A good store is essential. You can't keep equipment running without parts. It is well known that the compressor manufacturer cannot be depended upon to furnish critical parts on a short-time basis. In order to properly overhaul and maintain the compressor, each user must have the required parts on hand.

Inspection data and records allow for the proper planning for the next overhaul and replacement of those worn components. If the proper parts required are not available at time of shutdown, compromise overhaul decisions are often made to use worn or damaged parts or non-standard and out-of-tolerance parts. The use of worn or non-standard parts invites premature failure and unscheduled shutdown of the compressor. It is false economy to use these parts if there is any question about their ability to endure the scheduled operational period. It is also false economy to reuse parts such as gaskets and O-rings which do not lend themselves to visual inspection methods. They should be replaced with new parts each time the assembly is disturbed.

CORRECT COMPRESSOR PARTS

The importance of obtaining and using correct compressor parts cannot be overemphasized. Compressor users traditionally buy materials from the lowest-price vendor. But this approach fails to consider true life cycle costs. Purchasing shoddy items only adds to total costs, causes more rejected parts, unscheduled shutdowns, and increases employee dissatisfaction because workers are forced to work with inferior parts.

Because of the compressor manufacturer's inability to supply parts on a timely basis and at reasonable cost, users try to find alternative sources of supply for critical parts. Unfortunately, there are many aftermarket suppliers of compressor parts whose products do not meet the safety and reliability requirements of modern process plants. Their parts are often not made with proper materials, properly engineered to perform the function required, or manufactured to proper standards (tolerances, squareness, parallelism, surface finish, etc.).

Many shops have basic machine tools but lack technical knowledge of compressors. Although they are often able to sell at a low price, their parts are actually higher priced when failure takes place. Usually a shop of this sort will take an old worn part and make a drawing by "reverse engineering" that part. This reverse engineering has two major drawbacks that a machine shop cannot overcome. It does not provide any information on optimum tolerances, and it does not tell whoever copies the part exactly how it was produced in the first place. This discrepancy can spell the difference between success and failure.

INSPECTION OF COMPRESSOR PARTS

Regardless of the source, all compressor spare or replacement parts should be inspected carefully. It has been found that 8% to 10% of all parts, including those from qualified suppliers, do not meet material or dimensional requirements or have defects such as cracks in coatings. This figure can be as high as 85% to 90% on least-cost or unqualified suppliers.

Therefore, an inspection program is mandatory for incoming compressor parts and materials. This inspection should be performed immediately upon receipt, and no material should be put into stores without inspection.

VENDOR SELECTION

Suppose we had to select a vendor to perform work on the compressor, or a supplier/repair shop for reconditioning compressor precision parts such as valves, packing, pistons, or piston rods, or to manufacture replacement components. In that case, we should consider factors other than cost. These factors include:

1. Background and technical qualification of the management or owners of the facility being considered. Do the owners have technical expertise and background or are they professional managers who have never repaired compressors and their components?
2. Will loss of one key person in the vendor's management threaten an established buyer-vendor relationship? Depth of knowledgeable management assures the continuity of current business practices and philosophy.
3. How long has the vendor been in this particular business?
4. Are the owners able and willing to furnish the names and addresses of clients for whom they have provided the services in question?
5. Is the vendor financially sound? Check with a business-rating organization such as Dunn and Bradstreet and the vendor's material suppliers.

FACILITIES

Answers to the following questions will help to evaluate the vendor's facility.

1. Does the vendor have an adequate and modern facility to perform the type of work being quoted? Adequacy and condition of equipment and facilities will partially determine the quality and efficiency of the work performed.
2. Does the vendor have in-house equipment to perform required operations to produce the component part? These operations include capabilities for dye penetrant inspection and similar NDT testing, performance testing, metal spraying, thread rolling, and cylindrical grinding.

   If the vendor does not have in-house capabilities and uses a subcontractor, qualification and capabilities of the subcontractor must be assessed. This is the responsibility of the vendor, since he is solely responsible for the completion and quality of the order.

3. Does the vendor regularly perform the type of work he is quoting or is this merely something he thinks he can do or which he would like to do? Has he supplied a list of clients for whom he has done the same sort of work?
4. Do the key shop personnel have technical qualifications and experience? Are they only machine operators or have they worked in the actual manufacture of the parts under consideration?

You will encounter ordinary machine shops that solicit compressor component manufacture and reconditioning. No matter how well-equipped the machine shop is, unless its employees have compressor experience, it should not be considered for anything other than plain machining work.

Procedures and Specifications

The following questions can be used to evaluate the vendor's procedures.

1. Does the vendor have a written quality control manual? Are its procedures and work processes adhered to by shop employees, or is the manual "just for show"?
2. Does the vendor have written specifications and procedures for reconditioning or manufacture of compressor components for which they are quoting? If the vendor has the specifications, have they been made a part of his quotation without being requested? Are the specifications issued to shop employees who are to perform the work?

   A qualified vendor will have procedures, specifications, and quality control manuals and will be eager to show them without being asked. If it is necessary to request these documents, beware.

   If the vendor does not have specifications and yet is selected to perform the work in spite of this inadequacy, the purchaser *must* supply the specifications for the work. Under no circumstances should a job be performed without specifications.
3. Does the vendor have copies of reports of work performed for other clients? The vendor is contractually obligated to provide a written

report on work completed, such as an overhaul or inspection report on reconditioned components. Copies of reports supplied to other clients should be requested. Unless a final inspection report is received, it is usually advisable to withhold payment.
4. Does the vendor supply mill test reports on the materials used in the manufacture of critical parts such as compressor piston rods? No critical part should be accepted without this report. This provision should be made part of the purchase order.

Vendor selection may require dealing with a sales force. Here, two questions may be of interest:

1. Is vendor's sales person an employee or is he an agent working on a percentage of the order received? An employee generally has a greater vested interest in the company than an agent.
2. What is the sales person's experience and technical knowledge? Has he actually worked in the shop in some capacity?

Sales people are by nature optimistic and feel that their company or shop can do anything. They are often reluctant to admit any shortcomings and don't like to miss out on what they consider an opportunity. Without proper knowledge and experience, a sales person may commit the vendor to something outside his area of competence.

Vendor Survey

It is absolutely mandatory to make a visit to the vendor's facility to review his capabilities. The previously mentioned items will serve as a checklist and help in the survey. In addition to management's attitude, workforce, equipment and facilities, it is important to observe the housekeeping in the shop and offices. Also, this is the time to review the vendor's documentation routines, work procedures, specification, drawings, shop equipment and testing procedures.

Personnel Training

Operational and maintenance personnel must be involved in an ongoing training program to achieve the greatest efficiency and reliability of the compressors. All too frequently, the only training that operators and maintenance personnel receive is by word of mouth and by on-the-job training. On-the-job training is excellent as long as what is taught is correct. Unfortunately, improper maintenance techniques and "traditions" are often passed on by the person in charge. Needless to say, this contributes to problems and loss of reliability. It is therefore necessary that all personnel be given training on proper and up-to-date techniques of operating and maintaining compression equipment. This training must include all personnel and supervisors as well as the hands-on maintenance people.

A training plan obviously is an important part of any preventive maintenance program. This detailed plan must describe the requirements for improving and/or maintaining the training required for long-term equipment reliability.

Effective training requires the development of technical training materials on the plant's specific compressor. Adding appropriate classroom and on-the-job training will allow employees to apply *relevant* knowledge and skills to the subject of compressor operation/maintenance.

MAINTENANCE CONTRACTORS

Many plants use maintenance contractors to perform defined maintenance on their compressors. As with every endeavor, there are advantages and drawbacks. A contract service enables the plant to perform planned maintenance instead of just reactive maintenance. This is often the primary reason for engaging a contract service.

There are other reasons for using a contract maintenance service, as well as reasons why a plant may not want to use one. Some plants argue that no one could understand their problems. Experience shows that the use of contractors may lead to a hands-off policy. This has sometimes led to an abdication of responsibility to the point where nobody is responsible for compressor maintenance. It is important to remember that hiring a maintenance contractor does not mean that the plant has no stake in compressor maintenance.

SELECTION OF A CONTRACTOR

Because the contract service business requires low capital investment, it occasionally attracts participants who have neither the knowledge nor the technology to properly service compressors. For this reason, it is imperative that the capabilities of every contractor be thoroughly examined. To work effectively with a contractor, the client must have confidence that the firm will be able to do the job properly and on time.

The points outlined for the selection of a vendor also apply in the selection of a compressor maintenance contractor. Considerations include:

1. What kinds of service does the company provide? Is it a full-service operation?
2. What will it cost, and how will the plant be billed? Price should never be the only determining factor in considering a contractor.
3. How long has the company been in business? What is its financial status? Does it have assets to fulfill its commitments? Does it carry sufficient insurance and liability coverage?
4. What is its turnover rate? A stable workforce is generally better trained, more safety conscious, and easier to work with.
5. Does the potential contractor insist on surveying the client's premises? A company willing to accept a job on specifications alone should be viewed with caution.
6. How dedicated to quality is the contractor? What kind of training does he provide?
7. How are his employees hired and how well-qualified are they? Are they permanent employees or are they "temporaries" hired for each job?
8. What are the backgrounds, skill levels and experience of the contractor's supervisors or lead persons? It is important and necessary that the lead person have a background in compressor maintenance.
9. What are the backgrounds of the owners or managers of the company?
10. Will the contractor supply a complete report and record of all work performed? Are samples available for review?

Initiating the Service

Selecting and hiring a maintenance contractor does not mean the plant no longer needs to have anything to do with the maintenance work being performed. Once the choice is made, steps must be taken to integrate the outside services into the plant's daily routine. The plant must provide guidance, outline what needs to be done, offer input, and make the decisions.

It is important to take time to evaluate at the outset how well the proposal blends the contractor's operations and management with those of the client. Because the service is being performed on the plant premises, the contractor must follow plant security and safety regulations. Supervisory interaction must occur daily.

Initially, expect apprehension among resident personnel that the contractor's presence may cause them to lose their jobs. This must be addressed at the outset, even before the contractor is hired. Careful evaluation and selection of the right contractor can help assure that this process will occur expeditiously. An experienced maintenance contractor will work with the plant to ensure that both parties benefit from the agreement.

SUMMARY

To summarize, a preventive maintenance program is a comprehensive program of reliability improvement and failure prevention carried out by all personnel, with the understanding and support of management.

A good preventive maintenance plan should:

1. Maintain or improve the reliability and efficiency of compression equipment.
2. Represent a planned maintenance program.
3. Include training to improve the maintenance skills of both operators and maintenance specialists.
4. Ensure the development and updating of proper and adequate records on the condition of the compression equipment.
5. Use proper and accurate tooling to maintain the compressors.
6. Ascertain availability of proper spare parts and properly reconditioned components for each type of compressor.

No single technique used alone can solve all the problems that result in high compressor maintenance. Undue emphasis is often placed on the predictive aspects of maintenance to the exclusion of preventive maintenance. Both have their place; they are complementary. The best available maintenance management practice is to apply predictive techniques in an effort to accurately define when to perform preventive maintenance.

Predictive Maintenance

The maintenance methodology known as predictive maintenance and condition-based maintenance is attracting attention as a highly reliable substitute for conventional periodic maintenance and overhaul.

Where preventive maintenance fails in efficiency, predictive maintenance offers improvement. This method involves the use of real-time or portable instruments such as vibration, infrared, and wear debris monitors to recognize the symptoms of impending machine failure. Predictive maintenance is an early warning technique to detect small amounts of damage before they lead to catastrophic failure.

In the design, operation, and maintenance of mechanical systems, quantitative wear data play varying roles. Unscheduled equipment shutdown can be very expensive due to lost production or service. Real-time monitoring of components has the added advantage of helping to identify replacement parts before they are actually needed.

Predictive maintenance won't stop things from wearing, but it will allow the freedom to schedule downtime to correct the problem. This is the smart way to do it, rather than permitting the problem to develop into a full-fledged failure.

Predictive Maintenance vs. Periodic Inspection Maintenance

Predictive maintenance evolved from several basic facts:

- Certain vital parts last longer and operate better if not frequently taken apart, but operation until complete destruction is not only foolish, but unsafe, and costly.
- 99% of all failures are preceded by signs, conditions, or indications of failure.
- Periodic inspection maintenance could disturb good parts.
- Assembly errors are possible whenever machines are taken apart.
- Running fit and finish are sometimes jeopardized during internal component inspection.
- Dirt intrusion is likely whenever a machine is opened.

Setting Up a Predictive Maintenance Program

Predictive maintenance is the continual monitoring of the condition and performance of operating equipment. A good predictive maintenance program needs three ingredients:

1. An organized program
2. Equipment (instrumentation/analyzers/monitors/detectors)
3. Responsibility

Each item is useless without the other two.

Program

Although monitoring of reciprocating compressors is not as simple and definitive as monitoring other rotating equipment, there are some things that can be and should be monitored. The first step in the program is to decide what is to be monitored. An effective predictive maintenance program should include the following:

1. Daily operating reports and logs. These are used to observe operating parameters, pressures, temperature, flows, etc. These are often overlooked and recorded as part of the operator's duties but not referred to until after a problem develops. Continual monitoring can show trends of developing problems.
2. Maintenance records and wear measurements. These are taken as part of the overall maintenance program covered previously. These records are the most important of all parts of the program. Inspection records will allow spotting of trends and prediction of possible component failure. These records serve as the basis of planning for shutdown and replacement of worn and failing components.
3. Infrared thermography of the valve covers temperatures (see Chapter 5). This very simple procedure is an important part of any maintenance program and used to predict valve problems that can be taken care of before a major failure occurs. It should be a part of every maintenance plan.
4. Lubricating oil monitoring. This procedure helps to detect the progressive deterioration of components such as bearings. It consists of

monthly sampling of the compressor lubricating oil and performing spectrographic analysis that provides an accurate quantitative breakout of individual chemical elements contained in the oil elements as oil additives and contaminants. A comparison of the amount of trace elements in successive oil samples can indicate wear patterns of all wetted parts in the equipment and warn of impending failure.

Full benefits of oil analysis can only be achieved by taking frequent samples and trending the data for each compressor. The basic data on each compressor allow the laboratory to build a unique database. Reports then include values from the current tests, the average for the particular compressor, and values from previous tests.

A spike in the content of one element indicates a sudden change in the conditions inside the compressor. A comparison with the plant and laboratory averages provides a means of judging the significance of the change.

Oil analysis can provide a wealth of information on which to base decisions. However, major payback is rarely possible without a consistent program of sampling in order that data can be trended. While oil sampling and analysis can provide an additional capability to existing preventive maintenance programs, it should not be depended upon to the exclusion of all other techniques. In other words, there are documented instances of bearing failures taking place in operating compressors that, for some reason, were not picked up by sampling the lubricating oil.

5. Vibration monitoring. This monitoring is particularly useful on those compressors that use anti-friction bearings that are the smaller sizes of reciprocating compressors. Because reciprocating compressors have relatively low rotative speeds, they produce low frequency vibrations and unfortunately require more than the traditional vibration velocity monitoring or frequency analysis. However, monitoring packages are available from experienced specialty firms (see pp. 244 to 245, also 262 to 268).
6. Acoustic emissions or ultrasonic detection of leaking gaskets, etc. (see pp. 329 to 334).
7. Oscilloscope analyzers. These devices can be used to observe what is happening internally in the compressor cylinder, and, by comparing the actual pressure, volume, time indicator card, to the theoretical indicator card, the analysis determines if components are malfunctioning.

An analysis of this sort was discussed in Chapter 5. It was noted that piston ring leaks can be detected by placing an ultrasonic microphone in the middle of the cylinder. Scuffing of piston rings or rider bands will show on the scope. A piston that is loose on its rod will show up at the end of the re-expansion and the end of the compression event.

It is obvious that reciprocating compressor maintenance programs should include predictive and preventive maintenance elements. Notice, again, that predictive maintenance alone is not enough. It would be foolhardy to completely depend on lubricating oil analysis and vibration monitoring to determine maintenance schedules for compressors. We must take effective practices from both types of programs and merge them into an overall preventive maintenance program.

INTEGRATED CONDITION MONITORING SYSTEMS

Figure 6-5 shows an integrated condition monitoring system. Comprised of modules that track several relevant parameters, the system can transfer data to a computer in two ways:

- Digitally, via direct connection to the process computer, or:
- By direct link initially connecting to data manager modules that make it feasible to view both static and/or dynamic data.

VALVE TEMPERATURE MONITORING

A modern 32-channel temperature monitor provides increased productivity and efficiency. The monitor detects changes in valve condition by using differential measurement of valve temperatures. This allows you to make shutdown decisions based on real machine conditions rather than preset maintenance schedules. Reduced machine maintenance cost is also a benefit of this system. Temperature trends can be used to schedule maintenance to minimize downtime and to provide information to help diagnose problems more quickly. Parts can be changed out as needed instead of "across the board" at scheduled intervals.

The 32-channel temperature monitor is designed for measuring valve suction and discharge temperatures on reciprocating compressors. A bad valve will typically be a few degrees warmer than the other valves in its

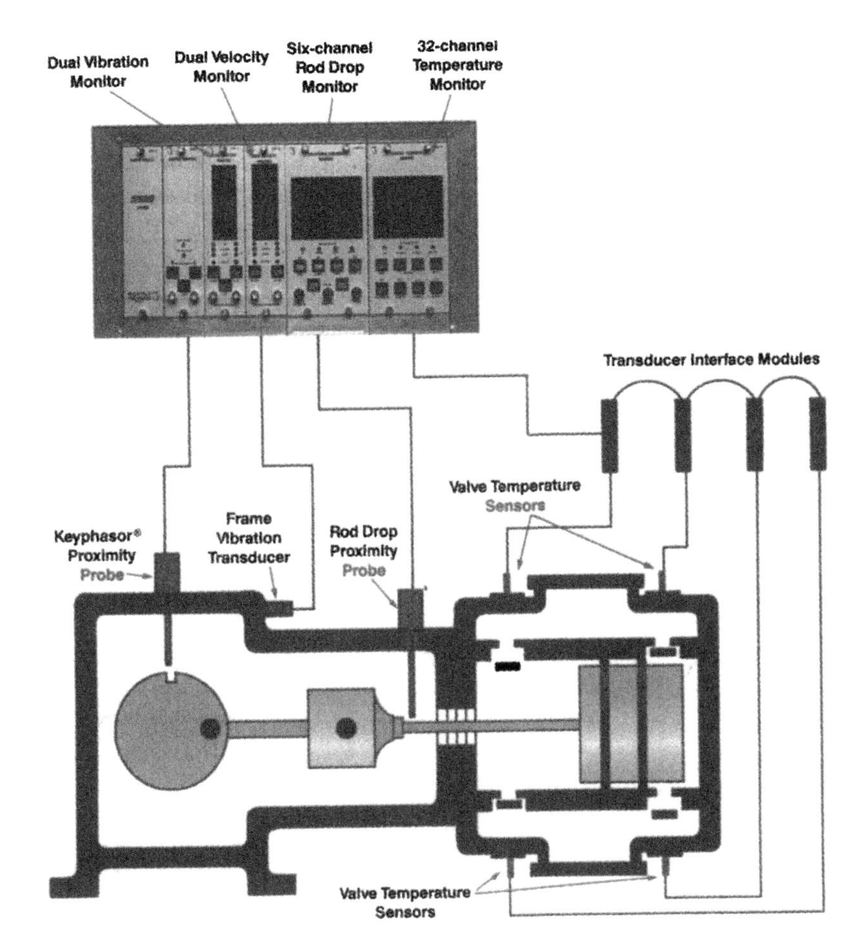

FIGURE 6-5. Modern monitoring system for reciprocating compressors. (*Source: Bently-Nevada Corporation, Minden, Nevada*).

group. Two levels of alarms can be programmed for any combination of over, under, or differential temperature for each channel. If the differential temperature alarm is selected, an alarm occurs when the channel is a specified number of degrees greater (or less) than the setpoint, compared to the average of the channels in the group. Actual temperatures and the average temperature of the jumper-programmed group are also available.

A single cable design from transducer interface modules to the monitor is used to minimize installation wiring costs and labor.

Frame Vibration Monitoring

The benefits of frame vibration monitoring are the prevention of catastrophic failure and the reduction of the amount of damage to a machine. This is accomplished by detection of small changes in the case motion. With modern 32-channel reciprocating compressor monitoring systems and associated seismic transducers, companies such as Bently-Nevada offer the user many advantages over the traditional, and often unreliable, "g-switch" type of detector.

Bently-Nevada, one of a number of capable manufacturers of machinery condition monitoring equipment, furnished an interesting case history.

A system comprising accelerometers and xy-proximity probes (see Figure 6-6) proved beneficial on a remotely operated, four-cylinder reciprocating compressor driven by a 2000 hp, 712 rpm electric motor. Since the machine operates at a fixed speed, the load is changed through unloaders that can effectively reduce throughput by holding valves open on one or more cylinders.

Proximity probes are mounted 90° apart in an xy-configuration at the motor bearings. A single accelerometer is mounted horizontally at the midspan of the compressor crankcase between two cylinder housings.

During normal operation, the compressor's acceleration amplitude was constant at approximately 3.5 to 4.0 g's. When the problem developed,

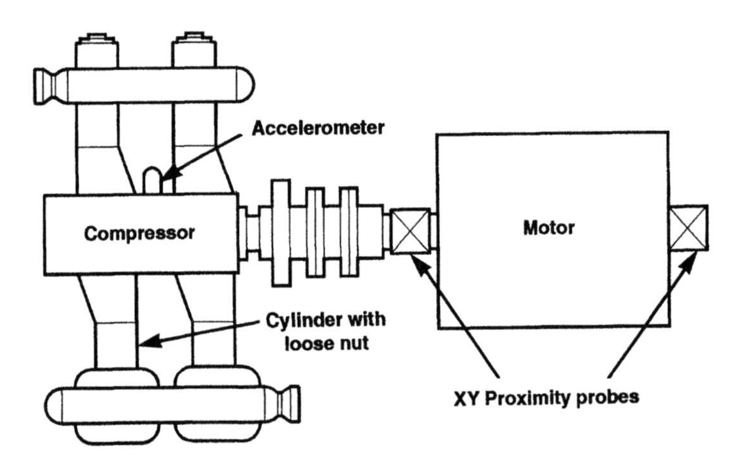

Figure 6-6. Machinery configuration diagram. (*Source: Bently-Nevada Corporation, Minden, Nevada*).

the acceleration level increased to over 6 g's, which caused alarms to actuate in the remotely-located control room.

The company's vibration specialists were called in to help analyze and determine the extent of the problem. They quickly determined that the acceleration level changed from approximately 5.5 to 7.8 g's, depending on the load. This increase was unusual. An oscilloscope was then connected to the buffered transducer output of the dual accelerometer monitor. By observing the timebase waveform, plant personnel had a clear picture of what was being observed by the accelerometer.

Figures 6-7 through 6-9 show acceleration amplitudes at 50%, 75%, and 100% load. Note the varying acceleration levels. Figures 6-7 and 6-8

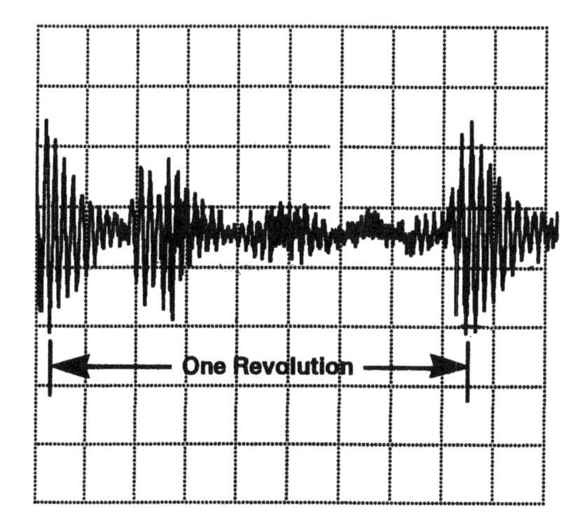

FIGURE 6-7. 50% of load. Acceleration level = 6.2 g. (*Source: Bently-Nevada Corporation, Minden, Nevada*).

indicate that at the end of the stroke of each piston, a hit occurred with an additional intermittent hit within the single cycle. At these loads, this intermittent hit was not consistent from cycle to cycle. At 100% load (Figure 6-9) when the acceleration increased to its maximum level of approximately 7.8 g's, the hitting became very dominant, along with the intermittent hitting that occurred halfway through the cycle. All of these patterns are characteristic of a problem in the compressor, probably between the piston and the crankshaft. The compressor was shut down, so an internal inspection could be done to determine the source of the suspected problem.

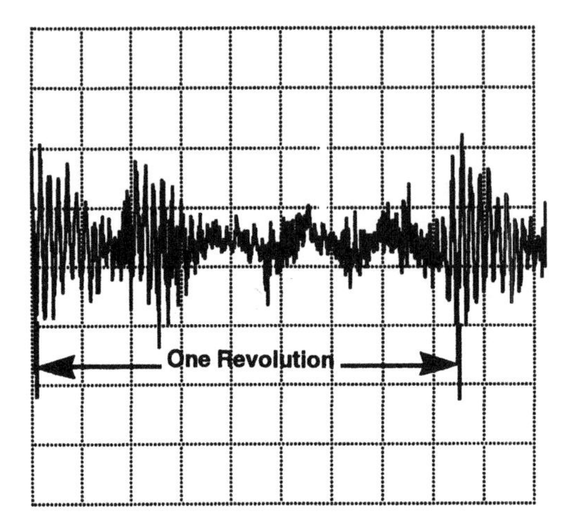

FIGURE 6-8. 75% of load. Acceleration level = 5.8 g. (*Source: Bently-Nevada Corporation, Minden, Nevada*).

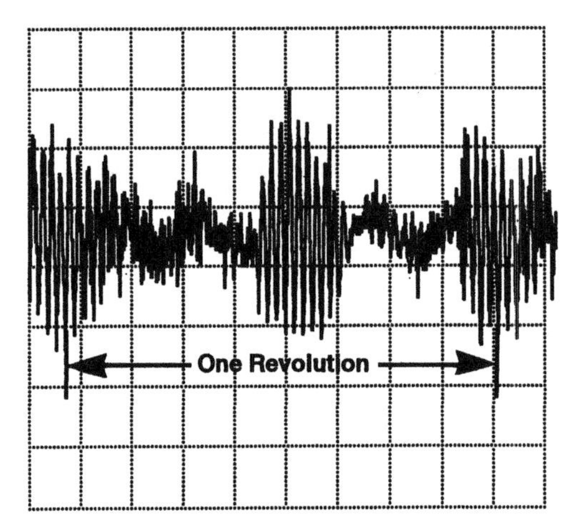

FIGURE 6-9. 100% of load. Acceleration level = 7.8 g. (*Source: Bently-Nevada Corporation, Minden, Nevada*).

The inspection revealed a loose locknut on one of the piston rods. This locknut locks the piston to the piston rod; the nut is secured to a specified torque. The nut had loosened, allowing the piston to become loose on the piston rod. This was the source of the hitting that was seen on the oscilloscope. Fortunately, no damage had occurred. The nut was retorqued, along with other mechanical checks, and the compressor was put back in service.

Figure 6-10 shows the timebase waveform after the machine was repaired at 100% load. A completely different pattern is displayed, with no evidence of hitting or looseness. The acceleration also returned to normal levels of 3.5 to 4.0 g's.

It is interesting to note how the accelerometer was able to detect the looseness. If you look at the action on the timebase waveforms, you will notice a tremendous amount of high frequency noise. However, whenever there was an abnormal impact, the effect is obvious.

Compare Figure 6-10 with Figures 6-7 through 6-9. The abnormal amplitude increases twice per shaft revolution. This indicates that there is a problem every time the piston changes direction.

Returning to our description of this particular, and obviously useful monitor, we find it capable of providing both "alert" and "danger" alarms. It can be wired to provide voting of two transducers for shutdown. The monitors also have "OK" circuits that reduce the possibility of false trips that can result from erroneous signals. A continuous indication of current vibration levels is provided, giving you an indication of even slight changes in machine condition. A computer interface can be provided by using a Dynamic Data Manager® or serial interface to

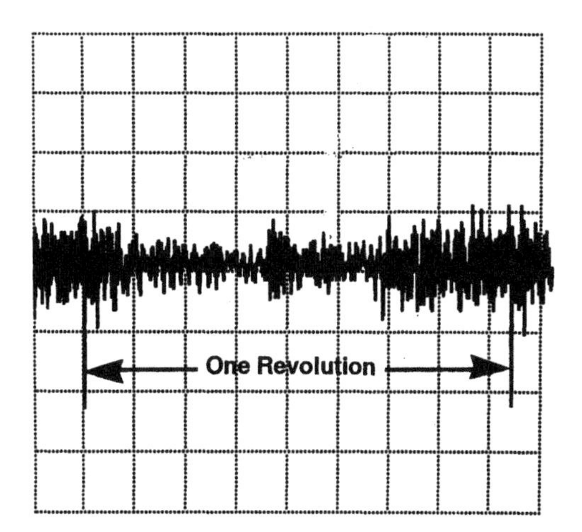

FIGURE 6-10. After the machine repair. 100% of load. Acceleration level = 3.5 g. (*Source: Bently-Nevada Corporation, Minden, Nevada*).

process computers for trending of values and machinery problem diagnosis. Portable analysis equipment can be connected to the buffered output terminals on the monitor front panel for machine analysis.

ROD POSITION MONITORING

This monitor enables the user to detect when the rider ring bands have worn away enough to require shutdown before the piston contacts the cylinder. A proximity probe is mounted vertically next to the cylinder to observe the piston rod position. The new monitor provides two modes of operation: average and instantaneous rod position. The instantaneous position measurement uses a Keyphasor® transducer to provide a once-per-turn crankshaft revolution reference signal for sampling the rod drop probe. This improves accuracy and minimizes problems caused by chrome plating or scratches on the surface of the rod.

Dynamic Data Manager and Keyphasor are registered trademarks of Bently-Nevada Corporation.

CHAPTER 7

Safety In Operation and Maintenance

BASIC SAFETY RULES

During the operation and maintenance of compressors, safety must never be compromised. All basic common sense rules must be strictly observed to prevent serious accidents.

1. Prior to operation or maintenance of the compressor, operating and maintenance personnel must be thoroughly familiar with plant procedures, as well as the compressor manufacturer's procedures.
2. Operators and maintenance personnel must familiarize themselves with the compressor controls and starting equipment.
3. Before starting, all protective guards should be in place and clear of moving parts.
4. The compressor should never be left unattended until all controls and safety devices are operating properly.
5. Practice all safety rules at all times.

Table 7-1 lists typical compressor safety devices.

LOCK-OUT/TAG-OUT PROGRAM

It is important that the compressor be locked out before any maintenance work is performed to prevent accidental starting.

TABLE 7-1
TYPICAL COMPRESSOR SAFETY DEVICES

Name	Function
Relief Valves	On compressor discharge side to relieve excessive pressure. NEVER use any shutoff valves between compressor and safety valve.
Overspeed	Trips out drive when compressor exceeds pre-determined safe speed.
Oil Failure Shutdown	For large compressors fitted with pressure lubrication, this device protects bearings by stopping the unit when oil pressure is insufficient for any reason.
Jacket-Water Valve	Shuts down compressor if water pressure fails. It is operated by either pressure or temperature.
Over-Pressure Shutdown	Stops compressor when discharge pressure goes above pre-set value.
Excessive-Temperature Shutdown	This gives protection against high discharge temperature by automatically stopping the unit.
Main Bearing Protection	Thermal shutdown devices stop compressor if bearing temperature goes too high.
Multistage Temperature Protection	Recording thermometers for each stage give continuous reading of each stage outlet temperature.

LOCK-OUT OR TAG-OUT PROCEDURES

Procedures for lock-out/tag-out situations should include these actions:

Step 1. Familiarize employees with the energy source and hazard.
Step 2. Have authorized employee de-energize and secure equipment.
Step 3. Notify affected employees.
Step 4. Turn off equipment.
Step 5. Shut off main energy source to the affected machine or equipment, including, but not limited to, the electrical disconnect switch (air system).
Step 6. Have authorized employee assess and control the possibility of stored energy being present.
Step 7. Secure, by locking out or tagging out as required, all incoming energy sources, ensuring that all control measures are in effect.
Step 8. Dissipate residual energy in the machine.

Step 9. Check previous steps; try to operate the machine to ensure it will not work.

Step 10. Proceed with servicing.

The following steps are required to restart locked-out or tagged-out equipment.

Step 1. Remove all nonessential tools, materials, or parts from the immediate area or the machine.

Step 2. Ensure that all guards and safety equipment have been reinstalled.

Step 3. Ensure that employees are completely clear of machinery or equipment.

Step 4. Remove all lock-out or tag-out devices.

Step 5. Turn on main power, electrical disconnect switches, air system, etc.

Step 6. Announce aloud that machinery is going back on. Ensure that employees in area understand.

Step 7. Turn equipment on.

Compressors can be dangerous if not designed, installed, operated, and maintained properly. Consequently, a periodic review of common problems that could lead to downtime or safety incidents is helpful to operating and maintenance personnel.

DEPRESSURIZE

Depressurization is required following shutdown. Purging may be required where gases are involved. Isolation is best assured where piping can be disconnected from the system. Block valves and slip blanks can be used where safety procedures are in place. Be absolutely certain that there is zero pressure in the compressor cylinder before valve removal is begun. One hundred psi on a 6″ inlet valve applies a force of over 2800 pounds.

Suppose the valve was stuck on the valve seat even with the center bolt and cover removed. What would happen if one attempted to dislodge the valve by hand and it suddenly blew out of the port? Never trust a closed gate valve that reads charge line and never believe a pressure gauge that reads zero. A safety valve is located between the compressor cylinders and the gate valve. Remove it so that any trapped gas leaking by a gate block valve can vent to atmosphere.

Piston Safety Precautions

Pistons are vented in order to relieve any gas pressure that may have built up inside the piston. Gas can be forced into the piston rod clearance. Venting is accomplished through 1/16" vent holes at the bottom of the piston ring grooves.

Warning!! All hollow pistons may become pressurized during operation. To be sure that all the pressure is relieved before piston and rod are disassembled, carefully loosen piston nut and wait. Do not remove completely!!

Block Flywheel

The flywheel should be locked in place when the engine or compressor is opened to prevent injury to personnel who must put their hands and arms inside the compressor cylinder. The guard should be on the flywheel while the machine is in service to prevent personnel or equipment from contacting the wheel.

Drain Water

Water should be drained from the compressor cylinder jacket when servicing the cylinder end. If the water is drained from the engine radiator during a compressor shutdown because of repair work, fill the radiator as soon as repairs are completed. If it isn't, the deposits inside the radiator will dry. Then, when the engine is put back into service, deposits will circulate through the system. A filter ahead of the water pump will catch those materials. Outwardly the water pump will appear to be doing its job, but its filter could be nearly plugged off, resulting in a low circulation rate and a hot engine or compressor.

Prevent Gas Leakage

A means of preventing gas from leaking into the compressor during overhaul is necessary. Any gas trapped inside the compressor may leak into the room when the equipment is opened up.

When opening up the compressor part of the machine, pancake/skillet blinds should be placed between flanges or a blind flange installed on existing piping. Those positive shutoffs are easier said than done at most installations. A double block and bleed valve combination is satisfactory if the bleed line (for example, a vent line to atmosphere) consists of more than small-diameter tubing. Remember, very high pressures are held back by the block valve. If there is leakage around a gate or plug in a valve, a small vent line may not be able to fully bleed down the line. This can result in gas leaking through the second block valve into the compressor room.

Replacement Bolts and Torque

Another problem occurs if the replacement bolt and nuts are of a different hardness and steel composition. Compressor manufacturers specify nut and bolt properties. They also specify torque values aimed at evenly distributing the load on all nut-bolt connections. Impact wrenches are able to exert enough tension on the bolts to cause the cast iron casing to crack!

Disassembly/Assembly

Disassembly should not be undertaken with tools such as crescent wrenches, pipe wrenches, cheater bars, or worn/incorrect size tools. Many compressors require special tooling for disassembly and assembly.

During assembly, all clearances should be double checked and recorded. All bolts and nuts must be rechecked for tightness. Internal areas of the cylinders and running gear must be free of any fasteners, tools, rags, or other foreign material.

Oil systems must be flushed and verified to be clear. Cleaning of oil sumps or crankcases should be done with lint-free rags and vacuumed clean. With clean oil installed and the compressor prepared for start up, manually turn over the unit to check free rotation. Manually (or remotely) run cylinder lubricator to furnish ample oil to cylinder and packings. Large compressors require frame oil circulation. Electrically bump unit to check rotation. Start up to less than full speed and shut down. Check conditions. Start up and run in accordance with manufacturer's start-up procedures. Again verify that all conditions are as anticipated.

Inspection and Cleaning

Thorough cleaning and inspection often reveal problems that would not normally have been expected. During the inspection and repair procedures, assure that any measuring instruments (feeler gauges, micrometers, vernier calipers, etc. are calibrated and in good condition. One wrong reading can result in severe problems.

CAUTION!! Never use kerosene or gasoline in a compressor cylinder for cleaning purposes. This highly dangerous practice can easily lead to explosions.

Housekeeping

Spilled oil is a common problem around many compressors. It can be found as you walk (or slip) in the doorway, on the floor around the compressor-engine block and on the working platform at the compressor. Some oil stems from overflow at lubricators and drainage from hoses attached to portable lube-oil barrel pumps.

Oil and grease should be wiped up immediately to prevent operator hazards, especially on smooth compressor floors.

Lubricators

Check the lubricator to be certain all lubricator cylinders are filled and are pumping oil. Check also for cracked lubricators. Most lube-oil systems have rupture discs that relieve when the oil pressure becomes high due to flow stoppage. The release of system pressure causes a no-flow condition and a shutdown.

Vibration

Observe forces that influence compressor operation. Forces in the machine go in many directions. These can cause damaging vibration plus excessive deflection of parts of the equipment. For that reason, grout is added between the compressor and its foundation.

Oil, vibration, and heavy stress loading frequently cause grout to fail or break up. Epoxy grouts are popular, having replaced cement grouts both at the initial setting of the machinery and when there is no recourse but to regrout. Be certain grout is in good condition.

Foundation bolts should be in place with nuts tight.

Alarm and Shutdown

Monitors can shut down a compressor if physical operating conditions change. Vibration switches are one type of monitor. Some others are high lube-oil temperature, low lube-oil level, low oil pressure, high compressor discharge temperature, low first-stage suction pressure, low or high discharge pressure, overspeed, high engine temperature and high liquid level in the gas scrubber ahead of the compressor first stage.

Compressors should have alarm and/or shutdown devices. Shutdown devices are preferred; they protect the compressor. Note, however, that shutdowns can cause operating emergencies.

At manned installations, when possible, an alarm should precede the compressor shutdown. Operators may be able to keep the machine running by taking specific remedial action whenever alarm annunciations occur.

Cooling Problems

Watch for cooling problems. Compressor cylinders are usually water or antifreeze cooled. Water is used for the intercooler between stages. The water is supplied from cooling towers, fin fan coolers, or vertical radiators at one end of the compressor-engine block. Poor cooling water condition and quality can result in poor heat transfer, plugging of small ports, corrosion, and increased cylinder operating temperatures. Cooling towers, because they are not a closed system, will accumulate bugs, leaves, etc., as well as pieces of the tower material that sometimes circulate through the equipment and can create heat transfer problems

Safe Maintenance Procedures Restated

Prior to doing any inspection, servicing or disassembly, take the following precautions:

1. Shut the machine down.
2. Cut off the electrical supply from the motor by opening a manual disconnect switch in the power line to the motor, and tag the disconnect switch so it will not be closed accidentally.

3. Lock the belt wheel or block crankshaft to prevent rotation.
4. When servicing the cylinder end, drain the water from cylinder jackets.
5. Never install a shut-off valve between the compressor and air receiver or aftercooler unless a safety valve is put in the piping between the valve and compressor.
6. Do not attempt to service any parts without first relieving the entire system of air or gas pressure.

Maintenance Hints

1. Never pull the piston rod through the oil scraper rings or piston rod packing rings.
2. Never use a chisel or other sharp instrument to open any joint between piston rod packing cups.
3. Never use kerosene, gasoline, or any contaminating safety solvent as a cleaning agent.
4. Never start the compressor after cylinder overhaul without first barring the unit through one complete revolution to be certain everything is free.
5. Never fail to shut down to investigate a new or unusual noise or knock.
6. Never insert hands into a cylinder or running gear until the driver is inoperative and the crossheads have been blocked to prevent rotation.
7. Never use a pipe wrench on any part of a piston rod.

Valve Installation

Compressor valve installation is critical to safety because many compressors will accept suction or discharge valves in all valve ports. Valve installation must ascertain that a suction valve is *not* installed in a discharge port, or a discharge valve is *not* installed in a suction port.

Noninterchangeable Valves

Generally, suction and discharge valves are the same and can be used in either port by merely turning them over. If care is not taken, a suction valve could be put in a discharge port in the wrong flow direction thereby causing pressures in the cylinder to build up to a dangerous level.

To prevent this possibility, noninterchangeable valves are available. Valve polarization and similar methods have been developed to prevent accidental interchanges.

Full-Pin Polarization

In full-pin polarization, the seat and guards are fitted on the circumference with locating pins. The pins on the guard are located 180° apart while those on the seat are located 90° apart. The valve ports are slotted for insertion of the pins. The discharge port contains these slots at 90°, and the suction port contains these slots at 180°. Only suction valves can thus be placed in the suction valve port and discharge valves placed in the discharge valve port.

Semi-Pin Polarization

Semi-polarized valves are similar to the polarized valves, except they have only one locating pin, located on the guard circumference. The suction valve ports are slotted to accept this pin. Since the discharge valve ports are *not* slotted, semi-polarized suction valves will fit only in the suction ports.

Diametrical Polarization

Diametrical polarization prevents accidental assembly of discharge service valves in suction ports, or the reverse. Different concentric diameters, machined on guard and seat, interfere with valve port cavities when a valve is placed in the wrong port. Valves properly seated in suction ports will operate as suction valves, and valves seated in discharge ports will operate as discharge valves (see Figure 7-1).

Valve Cover Bolting

Compressor valve covers should be checked to make sure they are fully bolted rather than short bolted, meaning they have only some of the threads of the nut in contact with the stud bolt. Full-thread contact is necessary to reduce the risk of the compressor valve cover being blown off.

There is a remote possibility that a longer stud could have been substituted, which gives the appearance of short bolting. But this is unlikely.

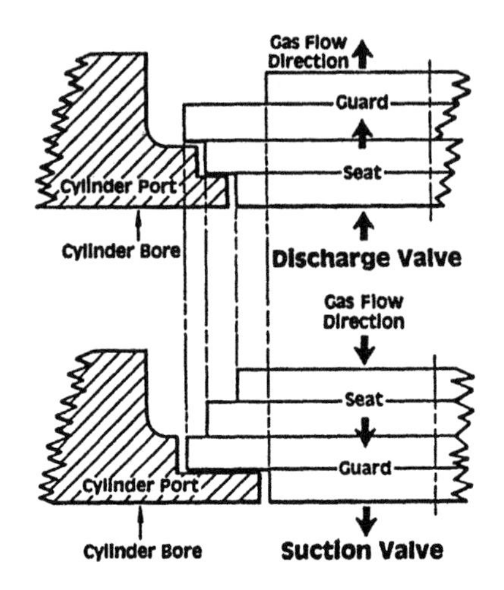

DIAMETRICAL POLARIZATION

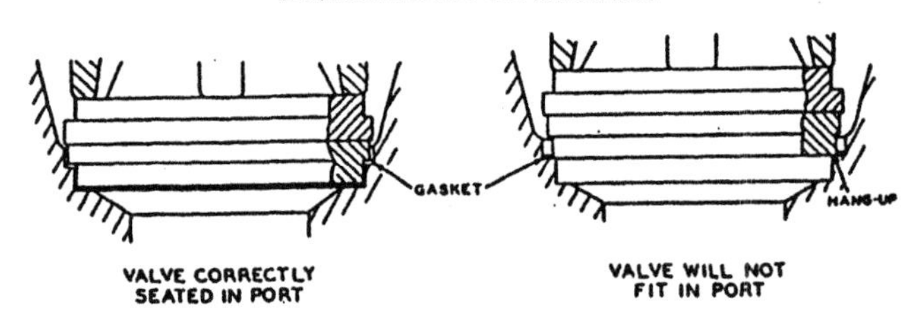

FIGURE 7-1. Noninterchangeable valves.

Look at the other nuts to see what size they are. Stud bolts should be replaced when their threads become worn, corroded, or split because the holding ability of the nut-bolt connection is reduced. Be certain that the new stud is long enough to engage all of the nut threads.

FIRES AND EXPLOSIONS

AFTERCOOLER AND COMPRESSOR FIRES

Compressor fires are always difficult to analyze when the compressor is in proper operating condition and cylinder jackets and coolers are clean, particularly on the water side.

What We Know

The few fires we know of and have investigated were caused by the following:

1. Dirty, scaled-up cylinder water jackets.
2. Oil collecting in low points in discharge piping.
3. Overheating due to compressor valve failure.
4. Overheating due to cooling water failure.

However, it is not possible to attribute a definite cause to all fire events.

How Fires Start

The exact cause of a fire or explosion in a compressor discharge system is seldom, if ever, known. However, excessive deposits and high discharge temperatures are practically always involved.

It was formerly thought that oils of very high flash point should be used in compressors to minimize the danger of fires and explosions, but this view is no longer accepted. The flash point of a correct compressor oil is in the neighborhood of 400°F, but a flash point test is performed at atmospheric pressure and the negligible quantity of vapor formed must be ignited with a flame in order to burn.

The presence of liquid films on discharge valve surfaces, which are the hottest parts of compressors, indicates that only negligible quantities of oil are normally vaporized. Temperatures much higher than the flash point would have to be reached before large amounts of oil vapor would be formed, and still higher temperatures would be necessary for autoignition of the oil to occur. (Autoignition temperatures of oil are about 750°F or more at atmospheric pressure and less at higher pressures.)

These higher temperatures could develop only from a combination of factors. For example, a combination of recompression because of leakage due to a broken valve and high cylinder pressure due to severe restriction of the discharge line by deposits could cause an extremely high temperature. However, it is believed that fires or explosions may occur even though the high temperature necessary for autoignition of the lubricating oil is not reached.

Analyses have shown that the bulk of compressor deposits usually consists of a variety of foreign contaminants, many of which are combustible. For example, coal, flour, or paper dust and other finely divided solids are often present. These contaminants are held together by oil and oil oxidation products.

Oxygen from the air is also present. At temperatures below autoignition of compressor lubricating oil, but nevertheless abnormally high, it is believed that oxidation of combustible contaminants can proceed within a layer or mass of deposit and that the attendant generation of heat can cause a portion of the deposit to glow. A portion of glowing deposit, however small, is often sufficient to start a serious compressor fire or explosion.

Under certain conditions, a considerable mass of deposit may reach glowing temperature in the manner just described, and the heat generated within this may so weaken the walls of a discharge pipe or some other part of a discharge system that an explosive rupture may occur.

In some cases, fires or explosions occur at points remote from the compressor discharge area, for example, in an intercooler or receiver. It is probable that these fires or explosions originate near the compressor discharge area in the manner previously described. A fire, once started, may follow along the discharge pipe, or a fragment of glowing deposit may be carried by the gas stream to a point where a combustible mixture exists. In this connection, it should be noted that aftercoolers, because of their cooling effect, have considerable ability to prevent the starting and spreading of fires.

Fires or explosions have sometimes occurred shortly after compressors were cleaned with kerosene or other light cleaners. Combustible products such as these should never be used to clean compressors. Instead, a strong soap solution or other noncombustible cleaner should be used.

Measures that should be taken to prevent fires or explosions are identical to those that should be taken to prevent abnormally high operating temperatures. In addition, on air machines, adequate filters should be provided to assure clean suction air. It is possible that compressed air may contain traces of inflammable gases, which may collect in a small area and become a fire hazard. Fumes or vapors coming off an oil may sometimes cause the same risk.

Consider Safety Devices

For extra safety, a high air discharge temperature alarm and shutdown switch may be necessary. These would be in addition to the low water pressure shutdown switch.

What To Watch

The following checkpoints should be watched:

1. Oil specifications
2. Air discharge piping drains
3. Water system clogging
4. Carbon formation in piping
5. Compressor valve condition

An example of the seriousness of these fires and explosions can be seen with the case of a soot-blowing compressor. This was a 200 HP, three-stage machine with a 500 psi discharge. It experienced an explosion in the discharge line between the compressor discharge and the aftercooler. The explosion ruptured a 4″ pipeline.

In another case, a 2000 HP air compressor in an automobile plant had an explosion in the air discharge of the high pressure cylinders. The explosion and ensuing detonation waves blew out a portion of the building, several sections of the parking lot and, unfortunately, killed two workers.

Fire and Explosion Protection—Air Compressors

Fires or explosions involving air compressors can be classified into two general categories:

1. Those in which compressor lubricating oil is in contact with air stream.
2. Those involving closed-loop operation.

Lubricating Oil in Contact with Air Stream

The majority of fires or explosions in air compressor systems have involved reciprocating machines. The fuel for air compressor fires is the cylinder lubricating oil itself, or the carbonaceous products formed by oxidation of the lubricating oil. The formation of carbonaceous deposits in air compressor systems depends on the amount and type of the lubricating oil used and the temperature of the metal surface on which the oil is deposited.

These effects appear to be interrelated, for example, an operating temperature that is satisfactory with the correct amount of oil may cause carbon deposition if excess oil is used.

The mechanism by which the fuel in air compressor fires is ignited is not definitely known. However, a factor common to all theories of ignition is excessive temperature, which may involve either the gas itself or a localized condition resulting from mechanical friction. High temperature is also important because it promotes deposition of carbon in the compressor system.

Excessive temperatures are generally caused by valve or cooling water failure or by operation at unusually high compression ratios. High operating speeds combined with an ineffective jacket also promote high cylinder temperatures.

To minimize the risk of fires in reciprocating air compressors, the following precautions are suggested:

1. Limit operating temperature to 350°F.
2. Provide temperature recorders or high temperature alarms in discharge.
3. Use the minimum amount of lubricating oil that will lubricate cylinders satisfactorily. Use the least viscous oil that will satisfy operating conditions. Strongly consider diester-based synthetic lubricant (not feasible with certain types of plastic pipe).
4. Train operators to detect faulty valves and have repairs made promptly.
5. Take inlet air from a cool, clean location. Provide adequate air filters (preferably of a dry type) and service at regular intervals.
6. Provide adequate intercoolers to maintain interstage suction temperatures at lowest practical level. Keep intercoolers and cylinder jackets free of deposits.
7. Inspect discharge piping, reservoirs, cylinders, etc., regularly. Remove all carbonaceous deposits and oil accumulations.
8. Check valves on a regular basis.

SUMMARY

Safety should always be paramount in operation and maintenance. The following are basic rules.

1. Before attempting any maintenance, be certain the compressor cannot be started accidentally. Tag the switch and then pull the fuses or disconnects.
2. Blow down the receiver and all intercoolers and aftercoolers. Isolate the compressor. Tag the shutoff valve.
3. Be certain there is no pressure in the machine before opening. Loosen, but do not remove, all nuts on a discharge valve cover. Pry it loose so that any trapped air may escape; then remove all nuts and the cover.
4. Place a wooden block in the cylinder between the piston and each head to prevent its movement. Do this through a valve hole. Sometimes it may be more convenient to block the crosshead in the frame.
5. Always use a safety solvent for cleaning compressor parts and dry them thoroughly before replacing.
6. Pop safety valves manually at least once a week.
7. If a safety valve blows during operation, stop the unit immediately and determine the cause. Safety valves on the receiver will normally blow only if the capacity control is not functioning correctly. An intercooler safety valve will blow when there is unusual leakage in the higher-pressure stage. Any blowing safety valve indicates trouble somewhere.
8. Read the instruction book.
9. Review safety precautions as shown in Figure 7-2.

AIR PIPING

More and more compressed air piping systems are being installed using plastic pipe. Plant engineers recognize the advantages of specifying plastic pipe. It is easier to fabricate, install, and maintain than metal. But, plastic pipe must be carefully selected when intended for air pressure system.

Most plastic pipe is limited to pressures below 150 psi and temperatures below 200°F to 300°F. The pipe used for plant compressed air distribution is ABS (acylonitrile butadiene styrene copolymer). It can handle air pressure to 185 psi at ambient temperature. Because it cannot rust, it does not contaminate wet compressed air with corrosion products.

However, some type of lubricating oil used in compressors can attack ABS pipe. ABS is especially sensitive to many synthetic oils; therefore,

SAFETY PRECAUTIONS
READ CAREFULLY BEFORE INSTALLING THE COMPRESSOR

Where lubricating oil is present in the compressor discharge, an aftercooler should be installed in the final compressor discharge line. It should be mounted as close as possible to the compressor.

When installing a new compressor, it is essential to review the total plant air system. The use of plastic or carbonic bowls on line filters without metal guards can be hazardous.

A pressure relief valve must be installed in the discharge piping between the compressor and any possible restriction, such as a block valve, check valve, aftercooler, or air dryer. Failure to install a pressure relief valve could result in overpressure, pipe rupture, damage to the compressor, and personal injury. Refer to instruction book for specific information.

On belt driven compressors, a belt guard, conforming to OSHA, state, and local codes, shall be installed by the user.

Those responsible for installation of this equipment must provide suitable grounds, maintenance clearance, and lightning arrestors for all electrical components as stipulated in OSHA 1910.308 through 1910.329.

When a receiver is installed, it is recommended that occupational safety and health standards as covered in the Federal Register Volume 36, Number 106, Part II, Paragraph 1910.169 be adhered to in the installation and maintenance of this receiver.

All electrical installation must be in accordance with recognized electrical codes. Before working on the electrical system, be sure to cut off the electrical supply from the system by use of a manual disconnect switch. Do not rely on the starter to cut off the electrical supply.

Before starting the compressor, its maintenance instructions must be thoroughly read and understood.

Do not remove the covers from the compressor while the unit is in operation. Severe injury from moving parts can result.

Compressed air and electricity can be dangerous. Before doing any mechanical work on the compressor:

FIGURE 7-2. Safety precautions.

A. Shut the machine down.
B. Remove power from the motor by opening a manual disconnect switch in the power line to the motor. Tag and lock the disconnect switch so no one will close it accidentally. Do not rely on the motor starter to cut off the electrical supply.
C. Do not attempt to service any compressor parts without first relieving the entire system of air pressure.
D. Ensure that the belt wheel has been locked to prevent rotation and subsequent personal injury.

Periodically, all safety devices must be checked for proper operation. Use only safety solvent for cleaning the compressor and auxiliary equipment.

FAILURE TO HEED ANY OF THESE WARNINGS MAY RESULT IN AN ACCIDENT CAUSING PERSONAL INJURY OR PROPERTY DAMAGE.

FIGURE 7-2. Continued.

it is important that the pipe manufacturer be aware of the type of compressor oil used. There is no problem if the air is produced in oil-free compressors.

When plastic ABS pipe is used for compressed air systems, it is extremely important that the compressor is properly maintained. Cylinder cooling is important, and when cylinder jackets become fouled, increasing discharge temperatures will result. Valves must be correctly maintained, and the compressor not allowed to operate with leaking discharge valves. Elevated discharge air temperatures would otherwise result.

The aftercooler must also be properly maintained so that the air discharge temperature to the system does not increase. Close control of air discharge temperatures is necessary as the strength of the plastic pipe reduces dramatically with an increase in temperatures (see Figure 7-3).

PVC Pipe

Not all plastic pipe is suitable for compressor air systems. PVC pipe should not be used. Here is an example of what can happen:

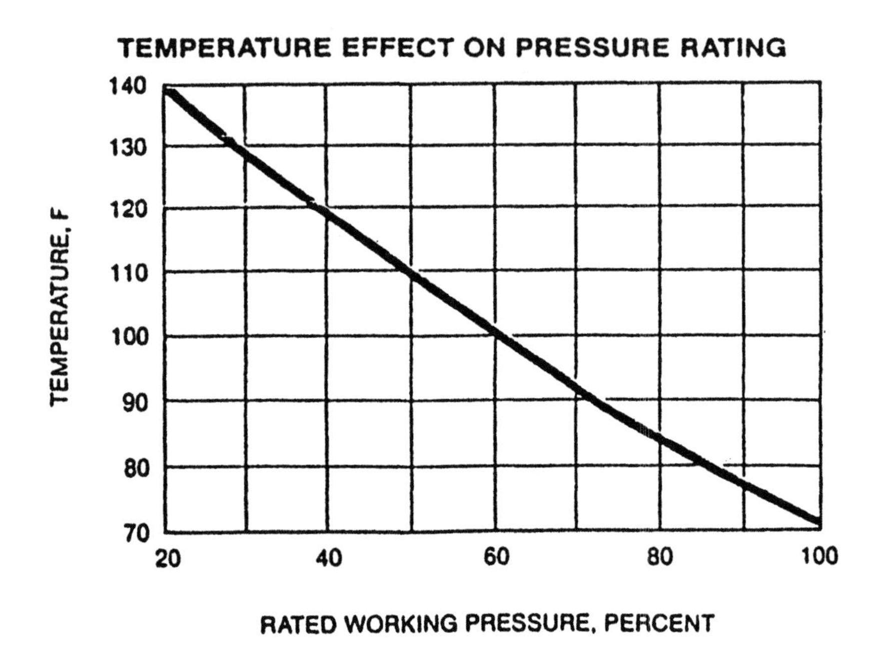

FIGURE 7-3. The effect of temperature on pressure rating of plastic piping.

"Plastic polyvinyl chloride (PVC) cannot be used in compressed air piping systems without the risk of explosion," warns the Washington State Department of Labor & Industries (L&I) in a Hazard Alert bulletin issued several years ago.

The bulletin states that when PVC piping explodes, plastic shrapnel is thrown in all directions. Paul Merrill, senior safety inspector in L&I's Spokane office, is quoted: "We're seeing more incidents of explosive failure, and we're citing more employers for using PVC air system piping. It's probably just a matter of time before someone gets seriously injured in one of these explosions unless everyone pays more attention to the manufacturer's warnings."

In Washington, according to the bulletin, a section of PVC pipe being used for compressed air exploded 27 ft above a warehouse floor. A fragment of the pipe flew 60 ft and embedded itself in a roll of paper. Fortunately, nobody was in the area at the time. In another incident, a PVC pipe explosion in a new plant in Selah, WA, broke an employee's nose and cut his face. In a Yakima, WA, incident, PVC piping buried 3 ft underground exploded, opening up a crater 4 ft by 3 ft.

Only one type of plastic pipe has been approved for use with compressed air. That pipe, says the bulletin, is acrylonitrile-butadiene-styrene (ABS) and the pipe is marked as approved for compressed air supply. By law, employers in Washington must protect their workers by avoiding the use of unapproved PVC pipe in compressed air systems. Existing systems using PVC piping must be completely enclosed, buried, or adequately guarded according to specifications approved by a professional consulting engineer.

Bibliography

Bently-Nevada Corporation. *Technical Bulletin on Model 3300/81 Six-Channel Rod Drop Monitor,* Minden, Nevada, 1995.

Bloch, Heinz P., "Consider a Low-Maintenance Compressor," *Chemical Engineering,* July 18, 1988.

Bloch, Heinz P. and Fred K. Geitner. *Practical Machinery Management, Volume 3, 2nd Edition: Machinery Component Maintenance and Repair.* Houston: Gulf Publishing Company, 1990.

Caldwell, James H. *Preventive Maintenance for Reciprocating Gas Engines and Compressors.* Cooper Bessemer Bulletin No. 129, 10M-7-69.

Indikon Company with contributors James McNabb and Leon Wilde. *Web Deflection Measurement,* Somerville, Massachusetts.

PMC/Beta. *Application Note B 816,* Natick, Massachusetts.

APPENDIX*

Reciprocating Compressor Calculations

CYLINDER SIZING

Cooling jackets, lubricant, packing and ring material limit cylinder temperatures. 275°F (135°C) is set as ideal maximum with 375°F (190°C) as absolute limit of operating temperatures. Using Equation A-1 with the above limits, the design temperature ratio per cylinder can be set.

$$T_{2ad} = T_1 \left(\frac{P_2}{P_1} \right)^{\frac{k-1}{k}} \tag{A-1}$$

Figure A-1 is a plot of Equation A-1.

T_2 actual may be obtained from these values where isentropic or adiabatic efficiencies are known from:

$$T_2 = \frac{\Delta T_{ad}}{\varepsilon_{ad}} + T_1 \tag{A-2}$$

The design compression and tension load on the piston rod must not be exceeded and, therefore, rod load must be checked on each cylinder application. Rod load is defined as follows:

* Source: Fred K. Geitner, P. E., Sarnia, Ontario, Canada

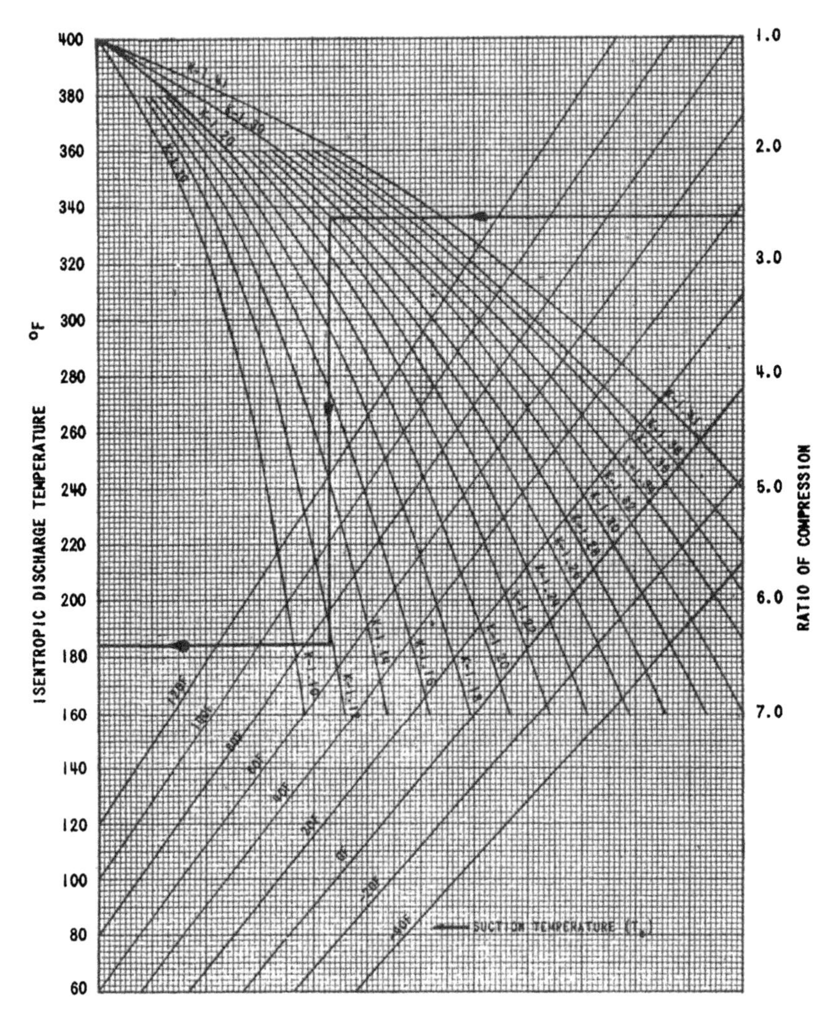

FIGURE A-1. Isentropic discharge temperature as a function of suction temperature, k-value, and compression ratio.

$$RL = P_2 \times A_{HE} - P_1 \times A_{CE} \quad \text{(A-3)}$$

where RL = rod load in compression in pounds.
A_{HE} = cylinder area at head end in any one cylinder.
A_{CE} = cylinder area at crank end, usually $A_{CE} = A_{HE}$ − rod area.
P_2 = discharge pressure
P_1 = suction pressure

The limiting rod loads are set by the manufacturer. Some manufacturers require lower values on the rod in tension than in compression. For this, the above limits are checked by reversing A_{HE} and A_{CE}. However, for most applications, the compression check will govern.

Compressors below 500 psig, such as air compressors, are usually sized by temperature. Rod load usually becomes the limiting factor on applications above these pressures.

VALVES

Compressor valves are the most critical part of a compressor. Generally, they require the most maintenance of any part. They are sensitive both to liquids and solids in the gas stream, causing plate and spring breakage. When the valve lifts, it can strike the guard and rebound to the seat several times in one stroke. This is called valve flutter, and leads to breakage of valve plates. Light molecular weight gases, such as hydrogen, usually cause this problem. It is controlled in part by restricting the lift of the valve plate, thus controlling valve velocity. An earlier edition of API 618 specified valve velocity as:

$$V = \frac{D \times 144}{A} \qquad \text{(A-4a)}$$

where V = average velocity in feet/minute.
D = cylinder displacement in cubic feet/minute.
A = total inlet valve area per cylinder, calculated by valve lift times valve opening periphery, times the number of suction valves per cylinder, in square inches.

At one time, compressor manufacturers objected to the above, because it gave a valve velocity for double-acting cylinders one half the value compared to equivalent single-acting cylinders. Therefore, manufacturers' data on double-acting cylinders often indicated a valve velocity double the API valve velocity. Care must be taken to find out the basis upon which valve velocity is given.

The fourth edition of API 618 (June 1995) defines valve velocity as:

$$V = 288 \cdot \frac{D}{A} \tag{A-4b}$$

For heavier mol weight gases (M = 20), API valve velocities are selected about 3,580 fpm or 18.2m/sec. and lighter mol weight gases, (M = 7), 7,000 fpm (35.6 m/sec.).

Manufacturers often have interchangeable suction and discharge valves. This can lead to putting valves in the wrong port, which can result in massive valve breakage or broken rod or cylinder. It would be well to specify that valves must not be interchangeable. However, a protruding non-reversal feature can be lost or broken off; correct valve placement should thus always be checked.

CALCULATIONS

General

Solving compressor problems by the use of the ideal gas laws has been reduced to a relatively simple sequence of applying a few basic formulae and obtaining values from some basic curves.

In order to help in the understanding of these terms and equations, the first chapter of this text covers some basic thermodynamics. Also, following the sequence of events as they occur in a compressor cylinder may help develop an understanding of the terms encountered in most reciprocating compressor problems (see Figure A-2).

Piston Displacement

Piston displacement is the actual volume displaced by the piston as it travels the length of its stroke from Position 1, bottom dead center, to Position 3, top dead center. Piston displacement is normally expressed as the volume displaced per minute or cubic feet per minute. In the case of the double-acting cylinder, the displacement of the crank end of the cylinder is also included. The crank end displacement is, of course, less than the head end displacement by the amount that the piston rod displaces. For a single-acting cylinder:

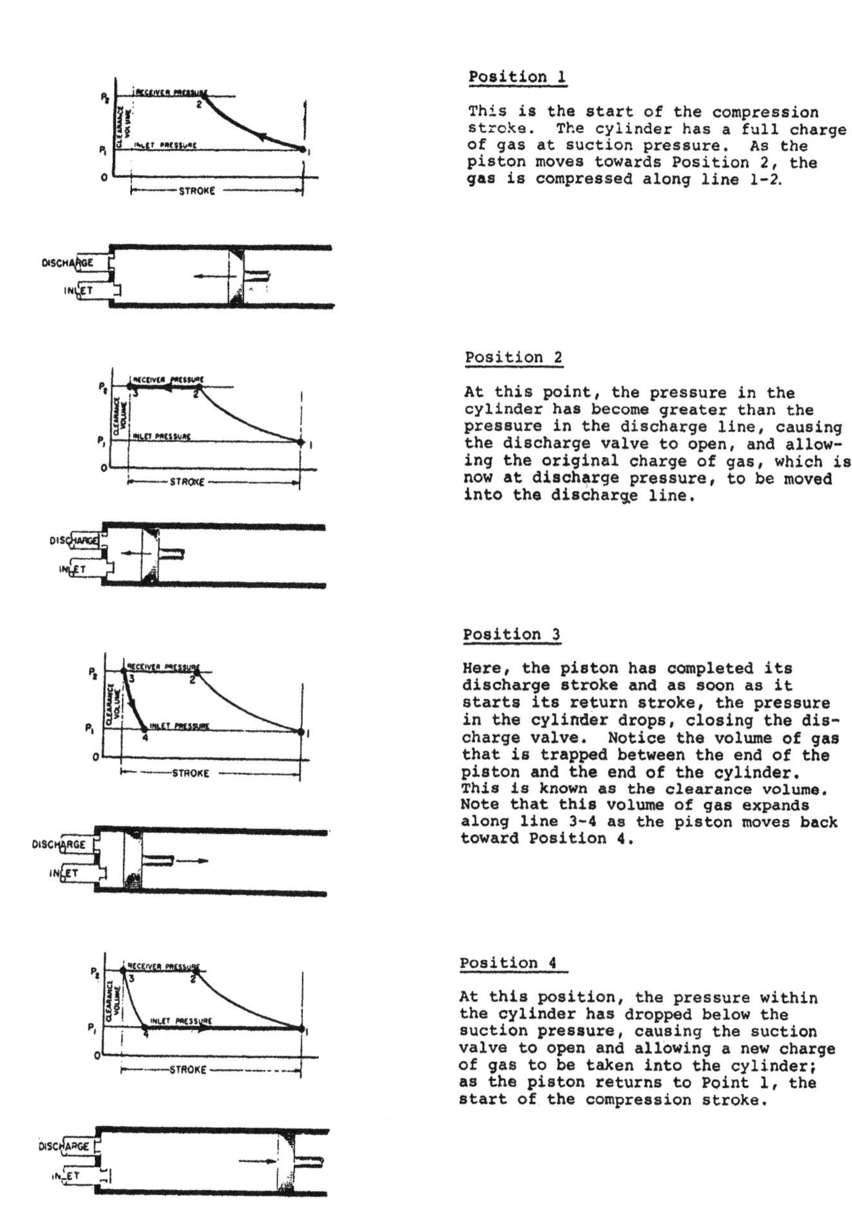

Position 1

This is the start of the compression stroke. The cylinder has a full charge of gas at suction pressure. As the piston moves towards Position 2, the gas is compressed along line 1-2.

Position 2

At this point, the pressure in the cylinder has become greater than the pressure in the discharge line, causing the discharge valve to open, and allowing the original charge of gas, which is now at discharge pressure, to be moved into the discharge line.

Position 3

Here, the piston has completed its discharge stroke and as soon as it starts its return stroke, the pressure in the cylinder drops, closing the discharge valve. Notice the volume of gas that is trapped between the end of the piston and the end of the cylinder. This is known as the clearance volume. Note that this volume of gas expands along line 3-4 as the piston moves back toward Position 4.

Position 4

At this position, the pressure within the cylinder has dropped below the suction pressure, causing the suction valve to open and allowing a new charge of gas to be taken into the cylinder; as the piston returns to Point 1, the start of the compression stroke.

FIGURE A-2. Ideal pV-diagrams as they relate to piston position.

$$PD = \frac{A_{HE} \times S \times RPM}{1728} \tag{A-5}$$

where:
A_{HE} = head end area of piston, square inches.
S = stroke, inches.
PD = piston displacement in CFM.

For double-acting cylinders:

$$PD_{DA} = \frac{A_{HE} \times S \times RPM}{1728} + \frac{A_{CE} \times S \times RPM}{1728} \tag{A-6}$$

or:

$$PD_{DA} = \frac{S \times RPM \times 2}{1728} \times \left(A_{HE} - \frac{A_R}{2} \right) \tag{A-7}$$

where A_R = rod area, square inches.

Compression Ratio

P_2/P_1 is the ratio of the discharge pressure to the suction pressure with both pressures expressed as absolute pressures. In the pV diagram (Figure A-2, Position 3), the line 2-3 is at discharge pressure, and line 4-1 is at suction pressure.

It should be stressed that the compression ratio is not controlled by the compressor, but by the system conditions at compressor suction and discharge. During compression, the cylinder pressure will rise until it is high enough to discharge the gas through the valves and into the system. During expansion, the cylinder pressure will fall until the system can force gas into the cylinder and maintain the pressure. Naturally, the compression ratios cannot be infinite. As the compression ratio is increased at constant suction pressure, the piston rod loading will increase and the piston rod will fail. Further, the cylinder cooling provided may not be adequate for the discharge temperature associated with higher compression ratios. Nevertheless, the machine will always attempt to meet the system compression ratios.

On multistage compressors, the actual compression ratios on each stage are self-determining as all the gas from the low pressure cylinders must pass through the higher pressure cylinders. The compression ratio on each stage will not change in proportion to the system pressure ratio, but it will change appreciably if the efficiency of any of the stages decreases because of piston ring blow-by or valve failure.

Percent Clearance

Percent clearance is used to calculate the volumetric efficiency of a cylinder. It is the ratio of clearance volume to piston displacement, expressed as a percentage:

$$\%Cl = \frac{\text{Clearance Volume (in.}^3)}{\text{Piston Displacement (in.}^3)} \times 100\% \tag{A-8}$$

Clearance volume is the volume left in the cylinder end at position 3, Figure A-2, including valves and valve ports.

Percent clearance varies between each end of the cylinder, so each is considered separately in calculations, although it can be averaged. Percent clearance is vital information for reciprocating compressor calculations. Therefore, it should be specified as name plate information required on all purchases of compressors. For estimating purposes, if clearance is unknown, a reasonable assumption is 15% clearance.

Volumetric Efficiency

Volumetric efficiency is the ratio of actual gas drawn into the cylinder piston displacement, as a percent. Referring to Figure A-2, Position 3, volumetric efficiency represents the distance 4 to 1, divided by the stroke. Theoretically, this is expressed by the equation

$$\varepsilon_v = 100 - Cl\% \left[\left(\frac{P_2}{P_1} \right)^{\frac{1}{k}} - 1 \right] \tag{A-9}$$

In actual machines, other factors affect the volumetric efficiency. These factors are leakage across the valves, across the piston rings, and through the packing in addition to the heating effect of the incoming gas from the residual heat in the cylinder and the varying compressibility of the gas. A better formula has been developed as shown in Equation A-10:

$$\varepsilon_V = 100 - L - Cl\% \cdot \left[\frac{Z_1}{Z_2} \cdot \left(\frac{P_2}{P_1} \right)^{\frac{1}{k}} - 1 \right] \qquad \text{(A-10)}$$

where L is taken from Figure A-3.

Basic Equations

Reciprocating compressor calculations are performed in stages. Therefore, the number of stages must be determined first. This is based on lim-

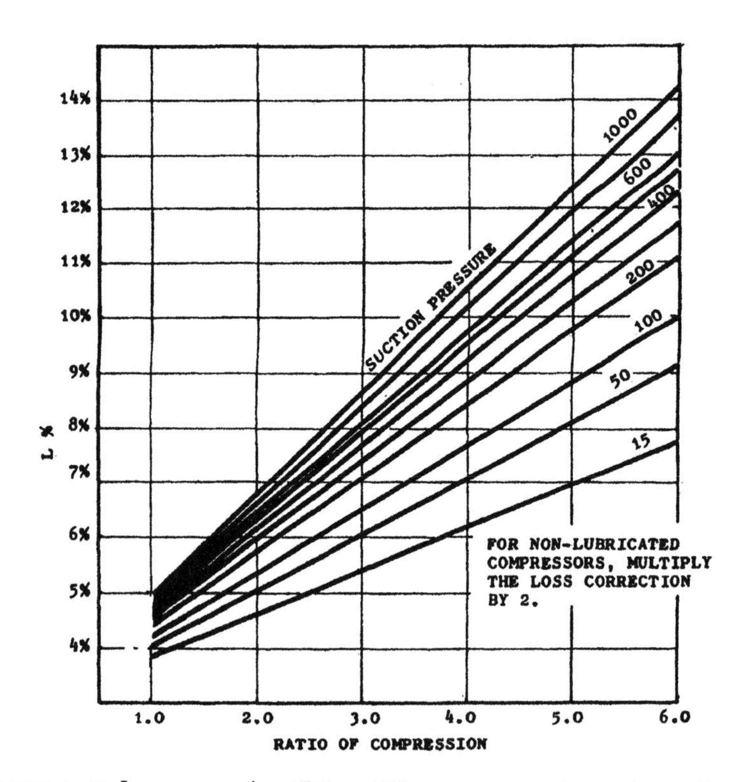

FIGURE A-3. Loss correction "L" at different compression ratios and suction pressures.

iting temperature and rod load, discussed earlier. In addition, horsepower savings can be made by having the same ratio for each stage. This ratio may be calculated by the following equation:

$$\text{Stage ratio} = [\text{Overall ratio}]^{1/n} \tag{A-11}$$

where n = number of stages

Generally speaking, in process work, the ratio per stage seldom exceeds 3.5 or 3.6. The basic equation for reciprocating compressor horsepower is:

$$HP_{ad} = \frac{ACFM \times 144 \times P_1}{33000} \times \frac{k}{k-1} \times \left[\left(\frac{P_2}{P_1} \right)^{\frac{k-1}{k}} - 1 \right] \tag{A-12}$$

This equation has been modified and drawn as a curve, using capacity in million cubic feet per day at 14.6 psia and suction temperature. This curve is given in Figure A-4.

This horsepower is theoretical and must be modified by compressibility, valve efficiency, and mechanical efficiency.

Valve efficiency allows for the pressure drop across the valves, which results in higher discharge and lower suction pressure within the cylinder than is supplied to the compressor. This may be found from Figures A-5 and A-6.

Mechanical efficiency is based on the power loss in the crankcase, and efficiency is usually expected to be 95%. Then, actual horsepower may be calculated as follows:

$$HP_{Act.} = \frac{MMSCFD}{Z_{Std.}} \times \frac{T_1}{T_{Std.}} \times \frac{BHP}{MMCFD} \times \frac{Z_1 + Z_2}{2} \times \frac{1}{\text{Valve Eff.}} \times \frac{1}{\text{Mech. Eff.}} \tag{A-13}$$

where, MMSCFD is million standard cubic feet per day at 14.4 psia and 60°F.

The above equation should be used for finding horsepower. For existing machines, displacement is known and volumetric efficiency can be calculated:

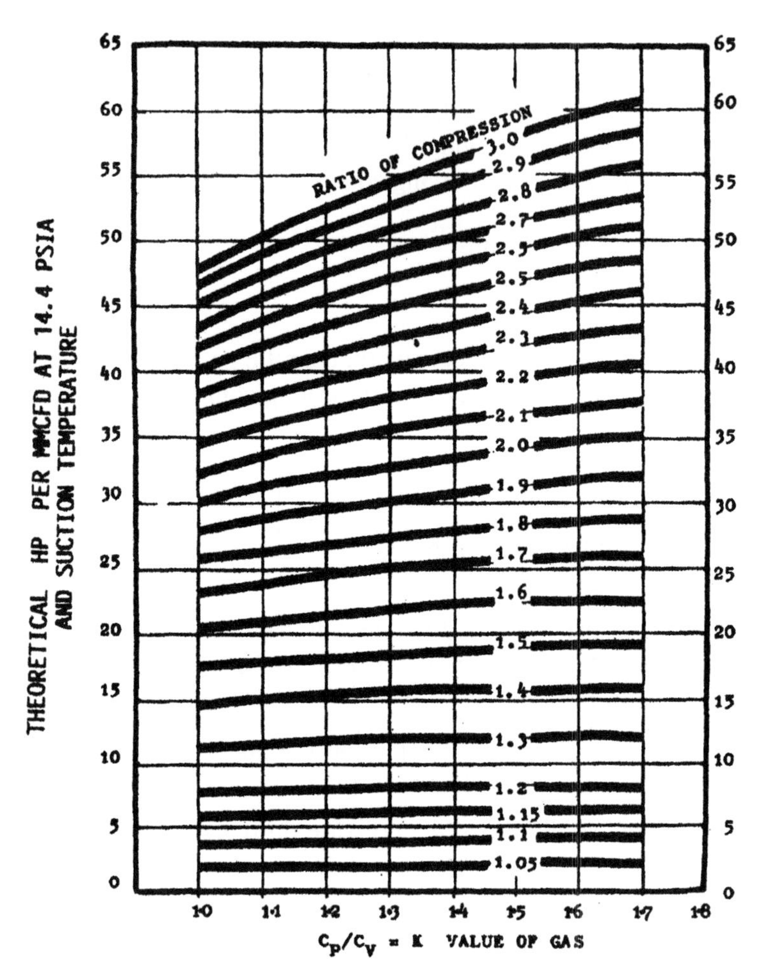

FIGURE A-4. Relationship of horsepower per MMCFD (million cubic feet per day), compression ratio, and k-value.

$$\text{MMSCFD} = \text{displacement (CFM)} \times \varepsilon_v \times \frac{T_1}{T_{\text{Std.}}} \times \frac{Z_{\text{Std.}}}{Z_1} \times \frac{P_1}{14.4} \times \frac{1440}{10^6}$$

Then

$$\text{HP}_{\text{Act.}} = \text{displacement (CFM)} \times \frac{\text{BHP}}{\text{MMCFD}} \times \frac{\varepsilon_v \times P_1}{10^4} \times \frac{Z_1 + Z_2}{2Z_1} \times \frac{1}{\text{Valve Eff.}} \times \frac{1}{\text{Mech. Eff.}} \qquad \text{(A-14)}$$

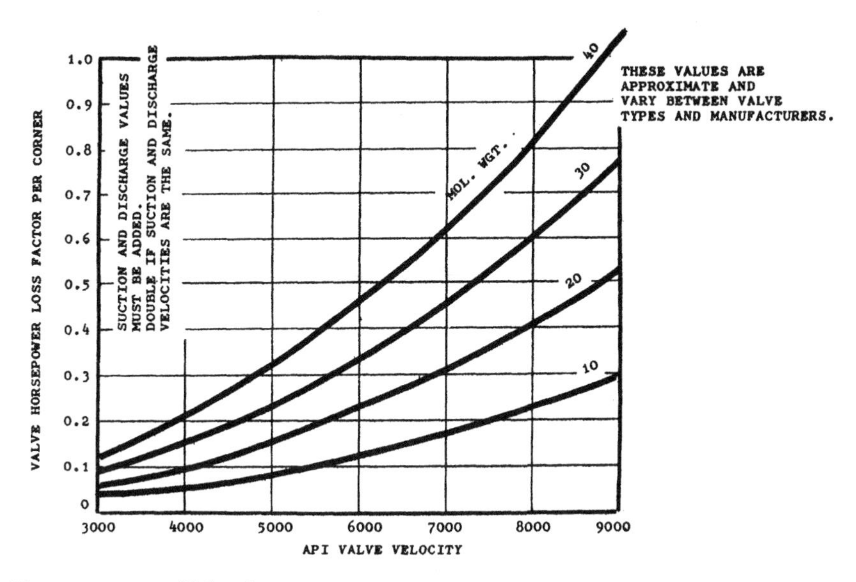

FIGURE A-5. Valve losses as a function of gas molecular weight and velocity.

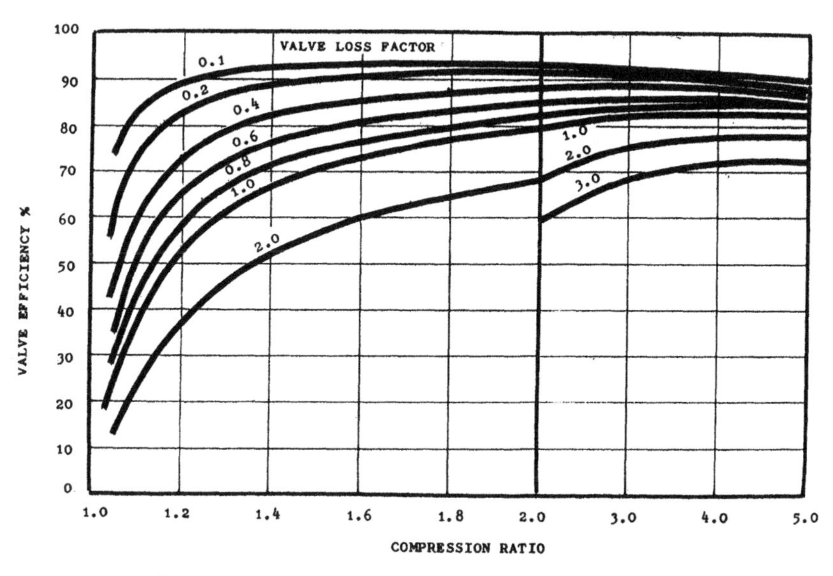

FIGURE A-6. Valve efficiency as it relates to valve loss factor and compression ratio on typical air compressors.

Machines may be sized on the basis of Table A-1. Different manufacturers may vary slightly from these sizes.

TABLE A-1
TYPICAL RECIPROCATING COMPRESSOR DESIGN DATA FOR US-BUILT MACHINES

Stroke	HP Per Stroke	Maximum HP Per Machine	Rod Size (In.)	Rod Load (Pounds)	Speed (RPM)
9	250	500	2-¼	20,000	600
12	300	1000	2-½	30,000	400
14	500	1500	3	45,000	327
16	750	3500	3-¾	75,000	300
18	1000	6000	4	1000,000	277

Table A-2 shows the relative influence or importance of several of the parameters discussed earlier.

TABLE A-2
HOW DIFFERENT DESIGN PARAMETERS RELATE TO COMPRESSION RATIOS

What is Important?		
Compression Ratio		More Important Factor
Very High	(10–30)	Clearance
High	(8–10)	Clearance principally, valving somewhat
Moderate	(5 Max.)	Balanced, all equal
Low	(2 or less)	Valving

SAMPLE CALCULATION

A New Machine

Design data:
Process gas molecular weight 17.76
k-value 1.26

Suction temperature	60°F
Suction pressure	150 psia
Discharge pressure	1000 psia
Flow rate	9 MMSCFD
Compression ratio	6.67 overall, or $\sqrt{6.67} = 2.58$ per stage

From Figure A-1, we obtain a discharge temperature of 308°F.

$$\text{Specific gravity of gas} = 17.76/28.97 = 0.613$$
$$Z_s = 0.98$$
$$Z_d = 0.97$$
$$Z_{avg.} = 0.975$$

$$\frac{k-1}{k} = \frac{0.26}{1.26} = 0.206$$

$$\text{Capacity} = 9\ \text{MMSCFD} \times \frac{10^6}{1440} \times \frac{P_{Std}}{P_s} \times \frac{T_s}{T_{Std.}} \times \frac{1}{Z_s}$$
$$= 9 \times \frac{10^6}{1440} \times \frac{14.4}{150} \times \frac{520}{520} \times \frac{1}{0.98}$$
$$= 612\ \text{ACFM}$$

Compression efficiency from Figure A-7, $\varepsilon_{ad} = 80\%$*
Compressibility factor Z_s is obtained from Figure A-8.
General HP formula:

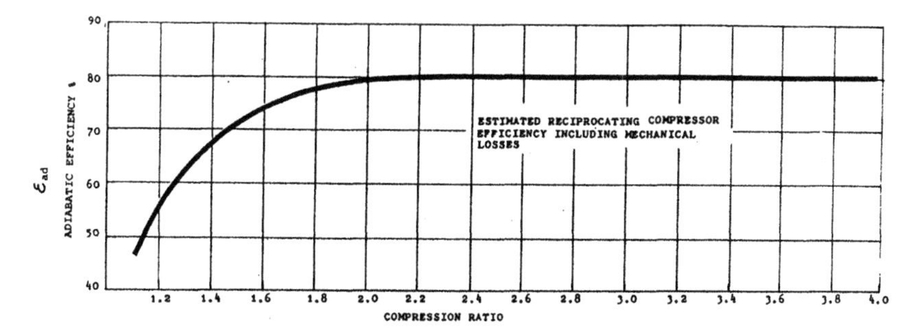

FIGURE A-7. Estimated efficiency as a function of compression ratio—reciprocating compressors.

* For comparison with centrifugal compressors, refer to Figure A-9.

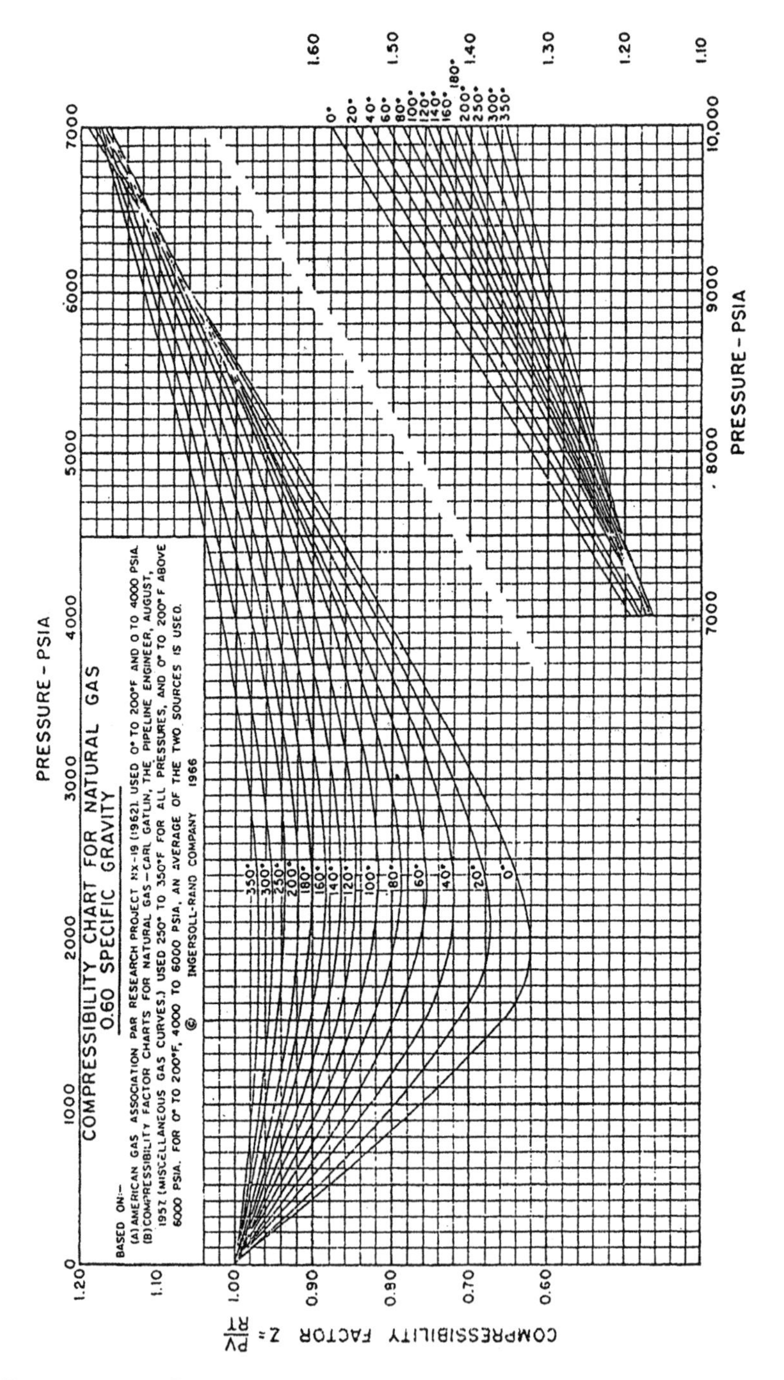

FIGURE A-8. Compressibility chart for natural gas 0.60 specific gravity.

$$
\begin{aligned}
HP &= \frac{ACFM \times 144 \times P_1}{33000 \times \dfrac{k-1}{k}} \cdot \left[\left(\frac{P_2}{P_1}\right)^{\frac{k-1}{k}} - 1\right] \cdot \frac{1}{\varepsilon_{ad}} \\
&= \frac{612 \times 144 \times 150}{33000 \times 0.206} \cdot [2.58^{0.206} - 1] \cdot \frac{1}{0.80} \\
&= 1160
\end{aligned}
\tag{A-13}
$$

Referring to Table A-1, we find that 1160 HP is too large for one throw, and a compression ratio of 6.67 is larger than the normal ratio of 3.0. Therefore, the machine is best configured as a two-stage machine. We decide on a two-cylinder, two-stage machine with a 14-inch stroke, 327 rpm, 45,000 lb rod load.

Using a blank Compressor Worksheet (Figure A-10A), we fill in the following data and ultimately obtain Figure A-10B.

$$
\begin{aligned}
\text{Ratio per stage} &= (6.67)^{0.5} \\
&= 2.58
\end{aligned}
$$

Interstage pressure = 150 × 2.58 = 386.0 psia

Interstage temperatures (from Figure A-1) = 172°F

Capacity of second stage:

$$
\text{Capacity} = 9 \times \frac{10^6}{1440} \times \frac{14.4}{386} \times \frac{632}{520} \times \frac{1}{0.98} = 289\ \text{ACFM}
$$

We are assuming a clearance of 15%.

Approximate volumetric efficiency:

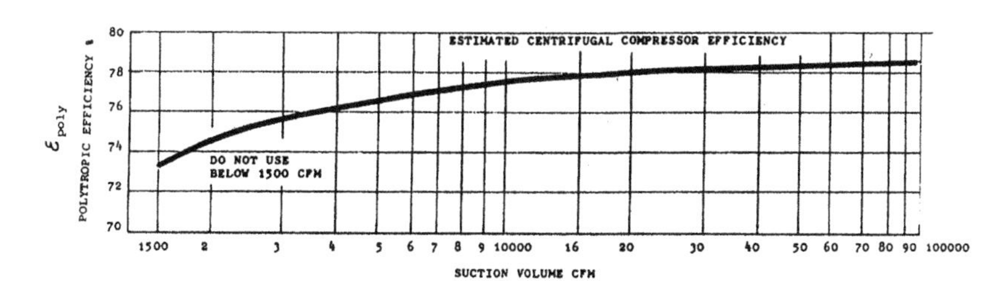

FIGURE A-9. Estimated efficiency as a function of compression ratio—centrifugal compressors.

	CYLINDER							
Displacement H.E. Piston Area/144 x Stroke/12 x RPM								
Displacement C.E. Piston Area/144 x Stroke/12 x RPM								
Ratio								
P_1 psia								
P_2 psia								
T_1 °F								
T_2 °F								
Z_1 Appendix								
Z_2 Appendix								
L								
$\mathcal{E}_v$								
$\mathcal{E}_v$								
ACFM H.E. Displacement H.E. x $\mathcal{E}_v$ H.E.								
ACFM C.E. Displacement C.E. x $\mathcal{E}_v$ C.E.								
ACFM Total								
SCFM ACFM x $520/T_1$ x $P_1/14.4$ x Z_{STD}/Z_1								
BHP/MMCFD								
API Valve velocity H.E.=Displacement H.E. x 144/Valve Area								
Valve Efficiency H.E. -								
BHP H.E. -								
API Valve Velocity C.E.= Displacement C.E. x 144/Valve Area								
Valve Efficiency C.E. -								
BHP C.E. -								
BHP Cylinder								
BHP Total								
Area H.E.								
Area H.E. x P_2								
Area C.E.								
Area C.E. x P_1								
Rod Load								

EQUATION 15 $\mathcal{E}_v = 100\% - L - \%\ \text{Clearance} \left[\frac{Z_1}{Z_2} \times \left(\frac{P_2}{P_1}\right)^{1/k} - 1\right]$

EQUATION 17 H.P Actual $=$ Displacement (cfm) x. $\frac{\text{B.H.P.}}{\text{MMCFD}} \times \frac{\text{Vol.Eff.} \times P_1}{10^4} \times \left[\frac{Z_1 + Z_2}{2Z_1}\right] \times \frac{1}{\text{Valve Eff.}} \times \frac{1}{0.95}$

FIGURE A-10A. Reciprocating compressor worksheet.

	CYLINDER							
	13	9-3/4	13	9-3/4	13	9-3/4		
Displacement H.E. Piston Area/144 x Stroke/12 x RPM	368	207	368	207	368	207		
Displacement C.E. Piston Area/144 x Stroke/12 x RPM	337	177	337	177	337	177		
Ratio	2.58	2.58	2.10	3.17	2.29	2.92		
P_1 psia	150	386	150	315	150	342		
P_2 psia	386	1000	315	1000	344	1000		
T_1 °F	60	174	60	150	60	160		
T_2 °F	174	307	150	307	160	307		
Z_1 Appendix	.98	.98	.98	.98	.98	.98		
Z_2 Appendix	.98	.97	.98	.97	.98	.97		
L	6.3	7.0	5.7	7.7	5.9	7.4		
$\mathcal{E}_v$ H.E.	72.2	75.7	79.6	70.1	77.0	72.8		
$\mathcal{E}_v$ C.E.	73.7	76.9	80.6	71.6	78.1	74.2		
ACFM H.E. Displacement H.E. x $\mathcal{E}_v$ H.E.	265	157	293	145	283	151		
ACFM C.E. Displacement C.E. x $\mathcal{E}_v$ C.E.	248	190	272	127	263	132		
ACFM Total	513	347	565	272	546	283		
SCFM ACFM x $520/T_1$ x $P_1/14.4$ x Z_{STD}/Z_1	5460	7820	6000	5170	5820	5820		
BHP/MMCFD					392	522		
API Valve velocity H.E.=Displacement H.E. x 144/Valve Area					6510	5720		
Valve Efficiency H.E. -					88	90		
BHP H.E. -					197	309		
API Valve Velocity C.E.= Displacement C.E. x 144/Valve Area					5780	4810		
Valve Efficiency C.E. -					89	90		
BHP C.E. -					181	249		
BHP Cylinder					378	558		
BHP Total						936		
Area H.E.					133	74.6		
Area H.E. x P_2					45800	74600		
Area C.E.					122	63.6		
Area C.E. x P_1					18300	21700		
Rod Load					27500	52900		

EQUATION 15 $\mathcal{E}_v = 100\% - L - \%\ \text{Clearance} \left[\frac{Z_1}{Z_2} \times \left(\frac{P_2}{P_1}\right)^{1/k} - 1\right]$

EQUATION 17 H.P Actual $=$ Displacement (cfm) x. $\frac{\text{B.H.P.}}{\text{MMCFD}} \times \frac{\text{Vol.Eff.} \times P_1}{10^4} \times \left[\frac{Z_1 + Z_2}{2Z_1}\right] \times \frac{1}{\text{Valve Eff.}} \times \frac{1}{0.95}$

FIGURE A-10B. Reciprocating compressor worksheet (sample problem).

$$*\varepsilon_V = 100 - L - \%Cl \cdot \left[\left(\frac{P_2}{P_1} \right)^{\frac{1}{k}} - 1 \right] \qquad \text{(A-16)}$$

L may be obtained from Figure A-3.

$$\text{First stage } \varepsilon_v = 100 - 6.3 - 15 \cdot \left[(2.58)^{\frac{1}{1.26}} - 1 \right] = 76.9\%$$

$$\text{Second stage } \varepsilon_v = 100 - 7.0 - 15 \cdot \left[(2.58)^{\frac{1}{1.26}} - 1 \right] = 76.2\%$$

$$\text{Displacement} = 2 \times \frac{\text{cylinder area}}{144} \times \frac{\text{stroke}}{12} \times \text{rpm} \times \varepsilon_v$$

$$\text{Cylinder area} = \frac{\text{displacement} \times 144 \times 12}{\text{stroke} \times \text{rpm} \times 2 \times \varepsilon_v}$$

$$\text{Area first-stage cylinder} = \frac{612 \times 144 \times 12}{14 \times 327 \times 2 \times 0.769} = 150 \text{ in}^2$$

First-stage diameter = 14 in.

$$\text{Area second stage} = \frac{288 \times 144 \times 12}{14 \times 327 \times 2 \times 0.769} = 71.4 \text{ in}^2$$

Second-stage diameter = 9-⅝″

The machine selected would be a 14 × 9-⅝ × 14 two-stage compressor driven by a 1200 HP motor. Similarly, a four-cylinder machine could be used, which would be 9-¾ × 9-¾ × 6-¾ × 6-¾ × 14. This would be also a two-stage compressor at 400 rpm with 45,000 lb rod load. Rod loads would have to be checked by using Equation A-3.

*Comparison with the more exact Equation A-10 will show that compressibility is neglected in Equation A-16.

CYLINDER SIZING:

14″ cylinder

$$\begin{aligned} R.L. &= P_2 \times A_{HE} - P_1 \times A_{CE} \\ &= (386 \times 154) - (150 \times 147) = 37{,}400 \text{ lbs.} \end{aligned}$$

9-⅝″ cylinder

$$R.\ L. = (1000 \times 72.8) - (386 \times 65.8) = 47{,}400 \text{ lbs.}$$

This exceeds the allowable rod load of the two-cylinder machine! Therefore, the four-cylinder machine must be checked for road load:

9-¾″ cylinder

$$R.L. = 386 \times 74.7 - 150 \times 67.6 = 18{,}700 \text{ lbs.}$$

6-¾″ cylinder

$$R.L. = 1000 \times 35.8 - 386 \times 28.7 = 24{,}700 \text{ lbs.}$$

This is within the limit of the machine. Therefore, the compressor for the application will be a 9-¾ × 9-¾ × 6-¾ × 6-¾ × 14, two-stage with a 1,200 HP, 400 rpm driver. Note that the dew point of the gas should be checked at interstage condition to make certain that no fraction has gone into the two-phase region. Interstage pressure must be altered by cylinder sizing if this occurs.

Existing Machine for a New Service

You have an existing compressor available. It is a 13 × 9-¾ × 16 compressor with a 1200 HP driver at 300 rpm. What will be the performance of this machine for the service in the preceding example?

Existing machine	13″ cylinder	9-¾″ cylinder
Bore, inches	13	9-¾
Stroke, inches	16	16
Cylinder design pressure, psi	900	1300
Head end clearance, %	18.4	14.8
Crank end clearance, %	17.1	13.8
Suction valve area, per end, in^2	8.4	5.3
Discharge valve area, per end, in^2	8.4	5.3
Rod size, inches	3-¾	3-¾
Maximum rod load, lbs.	75,000	75,000
Speed, rpm	300	300

Refer to Figure A-10B, Compressor Work Sheet.

Result

Horsepower and rod load are acceptable for this machine.

Capacity = $5820 \times 1440/10^6 = 8.4$ MMSCFD, which is less than required. Therefore, process capacity requirements could be decreased to accept this capacity. Alternatively, clearance could be reduced to increase ε_v to bring capacity closer to process requirements.

Index

Printed and bound by CPI Group (UK) Ltd, Croydon, CR0 4YY

08/05/2025

01864838-0006